STATISTICAL ANALYSIS IN THE BEHAVIORAL SCIENCES

Second Edition

JAMES C. RAYMONDO

TENNESSEE TECHNOLOGICAL UNIVERSITY

Kendall Hunt
publishing company

Cover image © 2015 Shutterstock, Inc.

www.kendallhunt.com
Send all inquiries to:
4050 Westmark Drive
Dubuque, IA 52004-1840

ISBN 978-1-4652-6967-6

Printed in the United States of America

This book is dedicated to my wife
Sherry A. Raymondo

Contents in Brief

PREFACE xv

PART I Some Basics 1
Chapter 1 Basic Issues in Statistics 3
Chapter 2 Sampling 21

PART II Descriptive Statistics 41
Chapter 3 Data Reduction: Frequency Distributions, and the Graphic Display of Data 43
Chapter 4 Measures of Central Tendency 71
Chapter 5 Measures of Variation 87

PART III The Bridge to Inferential Statistics 111
Chapter 6 The Normal Distribution 113
Chapter 7 Probability 143
Chapter 8 Hypothesis Testing 163

PART IV Inferential Statistics 185
Chapter 9 Correlation 187
Chapter 10 Linear Regression 221
Chapter 11 Hypothesis Tests for Means 247

Chapter 12 Analysis of Variance 283
Chapter 13 Nonparametric Statistics 303

APPENDICES
Appendix A: Statistical Tables 339
Appendix B: Answers to Selected Problems 369
Appendix C: The General Social Survey Data Set 377
Appendix D: How to Use SPSS 389

Glossary 403
Index 411

Detailed Table of Contents

PREFACE		xv
PART I	**Some Basics**	**1**
Chapter 1	**Basic Issues in Statistics**	**3**
	Introduction	3
	What Is Statistics?	3
	The Role of Statistics in the Research Process	4
	The Research Process	4
	Basic Terms in Statistical Analysis	7
	Measurement	11
	Scales of Data or the Levels of Measurement	12
	Why Does the Level of Data Matter?	14
	Common Symbols and Mathematics Used in Statistics	14
	Commonly Used Statistical Symbols	15
	Introduction to Computer Applications	17
	Computer Applications	18
	Summary of Key Points	18
	Questions and Problems for Review	19
Chapter 2	**Sampling**	**21**
	Introduction	21
	What Is Sampling, and Why Do It?	22

Sampling Strategies 23
Some Basic Sampling Concepts 23
Nonprobability Sampling Methods 24
Probability Sampling Methods 25
The Sampling Distribution 29
How Much Change Is in Your Pocket? 29
What Is the Sampling Distribution? 32
How Many Sample Means Can Be Drawn from a Given Population? 34
Calculate the Odds of Winning the Lottery! 36
Computer Applications 37
Summary of Key Points 37
Questions and Problems for Review 39

PART II Descriptive Statistics 41

Chapter 3 Data Reduction: Frequency Distributions, and the Graphic Display of Data 43
Introduction 43
The Construction of Frequency Distributions 44
What are the Advantages of a Simple Frequency Distribution? 45
Frequency Distributions Where $i > 1$ 46
Guidelines for the Construction of a Frequency Distribution 46
An Example of Constructing a Frequency Distribution 47
Frequency Distributions May Be Created with Any Type of Data 48
Midpoints and Limits in a Frequency Distribution 50
Proportions, Percentiles, Deciles, and Quartiles 52
The Relationship between Percentiles, Deciles, and Quartiles 54
Finding the Score Corresponding to a Given Percentile 54
Steps in Finding a Percentile 56
Finding the Percentile Corresponding to a Given Score 57
The Graphic Display of Data 60
The Bar Graph, the Histogram, and the Frequency Polygon 60
Some Common Graphic Patterns Seen in Data 64
Computer Applications 66
Summary of Key Points 67
Questions and Problems for Review 68

Chapter 4 Measures of Central Tendency 71
Introduction 71
Mode 72

	Median	74
	Mean	79
	The Mode, Median, and Mean Compared	83
	The Grand Mean	83
	Summary of Key Points	85
	Questions and Problems for Review	85
Chapter 5	**Measures of Variation**	**87**
	Introduction	87
	The Range, Interquartile Range, and Semi-interquartile Range	88
	The Variance and the Standard Deviation	92
	Computing Variance and Standard Deviation for Frequency Distributions	101
	Variance as Prediction Error (or Cabo San Lucas Here I Come!)	103
	Computer Applications	107
	Summary of Key Points	107
	Questions and Problems for Review	108
PART III	**The Bridge to Inferential Statistics**	**111**
Chapter 6	**The Normal Distribution**	**113**
	Introduction	113
	The Normal Distribution	115
	The *Z* Table, Areas Under the Normal Curve	116
	A Normal Distribution	122
	Predicting the Distribution of Scores in A Normal Distribution	123
	Types of Normal Distributions	128
	With So Many Possible Samples How do We Know We Have a "Good" One?	130
	Areas Under the Sampling Distribution	132
	Point Estimation and Interval Estimation	134
	Formula for Computing a Confidence Interval	135
	A Computational Example of a Confidence Interval	135
	Computing the 95% Confidence Interval	136
	The Sampling Distribution Is the Foundation for Two Important Statistical Concepts	136
	Sometimes the Sampling Distribution is Not Normal: The t Distribution	137
	Computer Applications	139
	Summary of Key Points	139
	Questions and Problems for Review	140

Chapter 7	**Probability**	**143**
	Introduction	143
	Origins of Probability Theory	144
	Probability	145
	The Link Between Probability, Hypothesis: Testing, and Statistical Inference	160
	Summary of Key Points	160
	Questions and Problems for Review	161
Chapter 8	**Hypothesis Testing**	**163**
	Introduction	163
	The Sampling Distribution	164
	Hypotheses and Types of Relationships	169
	A One-Tail Test of the Hypothesis or a Two-Tail Test of the Hypothesis	170
	Devising a Research Strategy	171
	Hypothesis Testing Is Conducted Indirectly: The Research Hypothesis and the Null Hypothesis	173
	A Summary of the Steps in Testing a Statistical Hypothesis	174
	Some Key Points to Keep in Mind	175
	Two Examples of the *Z* test	175
	A Second Example (with an Important Twist)	177
	Types of Error in Hypothesis Testing: Type I and Type II Error	178
	Why Do We Need Statistical Tests, and What Does a Finding of Statistical Significance Really Mean?	180
	The Need For Statistical Tests	180
	What Does a Finding of Statistical Significance Mean?	181
	What Does Statistical Significance NOT Mean?	181
	Summary of Key Points	182
	Questions and Problems for Review	182
PART IV	**Inferential Statistics**	**185**
Chapter 9	**Correlation**	**187**
	Introduction	187
	A Brief Review of Levels of Measurement	188
	Choosing the Proper Correlation Coefficient	188
	Bivariate Data Plots–Graphing Two Variables to Reveal the Relationship between Two Variables	189
	The Pearson Correlation Coefficient	193
	Logic of the Pearson *r*	194
	Computing the Pearson *r*	195
	Testing the Pearson *r* for Statistical Significance	196

Brief Review of the Steps in Testing a Pearson r for Statistical Significance 198
Computation and Interpretation of r^2 199
Some General Guidelines for the Interpretation of Correlation Coefficients 200
A Pearson *r* Example Using Education and Income 201
The Point-Biserial Correlation Coefficient 203
The Logic of the Point-Biserial Correlation Coefficient 203
The Formula for the Point-Biserial Correlation Coefficient 204
Computing the Point-Biserial Correlation Coefficient 205
Testing the Point-Biserial Correlation for Statistical Significance 206
Using the Point-Biserial Correlation Coefficient to Measure the Relationship between Sex and Physical Dexterity 207
The Spearman Rank Order Correlation Coefficient 208
The Logic of the Spearman Rank Order Correlation Coefficient 208
The Formula for the Spearman Rank Order Correlation Coefficient 208
Computing the Spearman Rank Order Correlation Coefficient 209
Testing the Spearman Rank Order Correlation Coefficient for Statistical Significance 210
An Example of the Spearman r_s Involving Tied Ranks 210
An Example of the Spearman Correlation Coefficient r_s with Rank on Physical Attractiveness and Rank on Perceived Intelligence 211
Using a Correlation Coefficient to Control for the Effects of a Third Variable 213
The Partial Correlation Coefficient 213
Formula for the Partial Correlation Coefficient 214
Computer Applications 215
Summary of Key Points 216
Questions and Problems for Review 217

Chapter 10 Linear Regression 221
Introduction 221
Fitting a Straight Line to Describe a Linear Relationship 223
The Regression Equation 225
Formula for "b", the Regression Coefficient 225
Predicting Values of *Y* 228
Assessing the Quality of the Regression Model 229
The Standard Error of the Estimate 231
A Conceptual Formula for the Standard Error of the Estimate 232
Explained and Unexplained Variance 232
A Second Example: Using Education to Predict Income 233
Standardized Regression Analysis and Outliers 236
Outliers 237

An Alternative Method of Calculating "*b*" 237
Assumptions for Linear Regression 238
Two Important Alternatives to Linear Regression 239
A Conceptual Example of Multiple Regression 240
Predicted Annual Sales 242
Computer Applications 242
Summary of Key Points 242
Questions and Problems for Review 244

Chapter 11 Hypothesis Tests for Means 247
Introduction 247
A Brief Review of the Logic of Hypothesis Testing 248
A Quick Review of the Steps in Testing a Statistical Hypothesis 249
The *Z* Test for Comparing a Sample Mean $\bar{X}$ to a Known Population Mean μ 250
A Finding of Statistical Significance Involves More than the Mean 253
The t Distribution 256
The t-Test for Comparing a Sample Mean $\bar{X}$ to a Known Population Mean μ 258
Statistical Tests for Two Independent Sample Means 260
The *Z* test for Two Independent Samples 266
The *Z* test for Significant Differences between Two Proportions 267
Computing Proportions and the Standard Error of the Difference 268
Formula for the *Z* test for a Significant Difference between Two Proportions 269
The t-Test for Two Related Samples 272
Computer Applications 278
Summary of Key Points 279
Questions and Problems for Review 280

Chapter 12 Analysis of Variance 283
Introduction 283
One-Way Analysis of Variance 284
Basic Terms and Assumptions for Analysis of Variance 286
Computing the Sum of Squares 287
A Computational Example of Analysis of Variance 288
Computing the *F* Statistic 290
The ANOVA Summary Table 290
Obtaining the *F* Critical Values 290
Post Hoc Tests for Significant Differences 291
Tukey's HSD Multiple Comparison Test 291
Variance Explained by the Independent Variable: Eta Squared, and Omega Squared 293

Computational Formula for Eta Squared 293
Computational Formula for Omega Squared 293
An Example of ANOVA with Unequal Sample Sizes 294
Fisher's Protected t-Test 296
Comparing the Placebo Group and the High-Dose Group 297
Comparing the Placebo Group and the Low-Dose Group 297
Comparing the Low-Dose Group and the High-Dose Group 298
Computing Eta Squared and Omega Squared to Estimate Variance Explained 298
Variations on a Theme in Analysis of Variance 299
Computer Applications 299
Summary of Key Points 300
Questions and Problems for Review 300

Chapter 13 Nonparametric Statistics 303
Introduction 303
The Construction and Presentation of Data in a Contingency Table 304
Difficult to Interpret Tables 308
Degrees of Freedom in a Contingency Table 309
The Chi Square Test for Independence 311
Assumptions for the Chi Square Test 311
A Computational Example of Chi Square 312
The Easier Way to Compute Expected Results 313
The Formula for The Chi Square Test for Independence 314
Critical Values from the Chi Square Distribution 315
Some Additional Issues Regarding Chi Square 317
Yate's Correction for Continuity for 2 × 2 Tables 317
Measures of Association for Contingency Tables 318
The Coefficient of Contingency (c) 318
The Phi Coefficient 319
Guttman's Coefficient of Predictability, Lambda 320
Goodman's and Kruskal's Gamma 322
A Second Example of Chi Square and Measures of Association 323
Computing the Measures of Association, How Strong Is the Relationship? 325
Computing Gamma 326
Nonparametric Tests of Significance 327
The Mann–Whitney U Test for Two Independent Samples 328
An Application of the Mann–Whitney U Test 329
The Wilcoxon T Test for Two Related Samples 330
An Application of the Wilcoxon T Test 331

Computer Applications 332
Summary of Key Points 332
Questions and Problems for Review 334

APPENDICES

Appendix A: Statistical Tables 339
Appendix B: Answers to Selected Problems 369
Appendix C: The National Opinion Research Center General Social Survey 377
Appendix D: How to Use SPSS 389

Glossary 403
Index 411

Preface

I want to discuss several things in the Preface of this book including: the target market; the organization of the book; how you can learn the most from reading it; and finally, I want to acknowledge the help I received from a number of people while I was writing it. Let's begin with the purpose of the book, and the target audience.

The Target Market of This Text

The market for this text includes the disciplines of the behavioral sciences. My own course in statistics that I have taught for over 25 years at the college level has enrolled at one time or another sociology majors, criminal justice majors, psychology majors, business administration majors, and students from social work. My goal in writing this text is to provide you with a textbook that includes all of the basic elements expected in a statistics text for the behavioral sciences, but to also go beyond that and give you a far better understanding of what statistics is, what the statistical procedures included in the text really mean, and just as importantly, what they do not mean.

The Organization of the Text

The book is organized into four major sections. Part I, Some Basics, includes the first two chapters. In Chapter 1, I try to provide you with a sense of what statistical analysis is all about, introduce you to the language and terminology of statistical analysis, and provide you with a short mathematics review that will enable you to perform all of the statistical procedures included in the text. One misconception that students often have about statistics is that statistics is a "math course." Statistics is not a math course; statistics is better thought of as a process that uses mathematics to help us analyze information, and I think you will be pleasantly surprised by the mathematical presentation in the text. Chapter 2 provides you with a fairly comprehensive overview of sampling. This is one of the distinguishing features of this text compared with many of the others. Almost everything we do in statistical analysis is based on the distinction between a population and a sample, so I think you need to be aware of those differences and exposed to the idea of the sampling distribution early on in the text.

Part II, Descriptive Statistics, consists of Chapters 3, 4, and 5. In this part of the text, you will begin to see how statisticians reduce a set of data to a form that is more easily interpreted. Chapter 3 includes a discussion of frequency distributions, and several major types of graphing techniques. I also include a brief discussion of bivariate analysis where you learn how to look for evidence of a relationship between two variables. Chapter 4 includes measures of central tendency such as the median, and the mean (commonly called the average). Part II ends with Chapter 5 on measures of variation. Chapter 5 includes several statistical techniques, which indicate how much difference or "variation" there is, in a set of scores.

Part III The Bridge to Inferential Statistics consists of Chapters 6, 7, and 8. The chapters in Part III include statistical concepts that are important in their own right, but they also serve to link us to the final section of the text, dealing with inferential statistics. In descriptive statistics, we are trying to "describe" a population or a sample. In inferential statistics, we are wanting to take information from a sample and make generalization (inferences) about the population from which the sample was drawn.

Part IV, Inferential Statistics, is the final section in the text, and consists of Chapters 9, 10, 11, 12, and 13. Chapter 9 presents several types of correlation coefficients, which are useful in telling us if there is an association between two variables. Chapter 10 deals with linear regression which is an important technique used to predict the level of one variable based on the level of another variable. Chapter 11 deals with statistical tests for means, and includes several statistical tests used to tell us if two means are statistically significantly different. Chapter 12 presents a discussion of one-way analysis of variance, which allows us to examine a group of means for the presence of statistically significant differences. Finally, Chapter 13 presents a number of nonparametric statistical techniques. Nonparametric statistics are inferential statistics that allow us to make inferences about a population based on the examination of a sample, but do not require us to meet some of the rigorous assumptions required for the parametric statistical procedures covered in Chapters 9 through 12.

Computer Applications

Almost all of the statistical procedures presented in this text can be performed by software dedicated to statistical applications, and many of the procedures can be performed by statistical applications found in popular spreadsheet software. My goal is to provide you with a text that conveys a basic understanding of statistical analysis, with real-world examples and exposure to current technology. The instructor's edition of this text includes access to a dataset consisting of an excerpt from the 2000 to 2012 General Social Survey; an ongoing project by the National Opinion Research Center of the University of Chicago, based on a national probability on the non-institutionalized population of the United States. Appendix C provides you with a brief description of the General Social Survey, and a list and description of the variables included in the dataset. Most chapters of the text have a section at the end called "Computer Applications," with suggested computer exercises, and many of them allow you to work with the General Social Survey data. The exercises are keyed to use with one of the leading statistical analysis packages on the market, Statistical Package for the Social Sciences (SPSS), but you may also complete the exercises by using any of several alternative software packages. The "Computer Applications" exercises are followed by a "How to do it" section providing directions and tips for those of you with access to SPSS. In addition, Appendix D (How to Use SPSS) serves as an informational resource providing you with a brief overview of the major features and operation of SPSS.

How to Get the Most From This Text

First, I would suggest that you read it. A textbook is not like a novel where you can just scan or maybe even skip the boring parts. Unfortunately, in most textbooks, the boring parts often contain some of the more fundamental material, and if you do not understand it, or do not read it at all, you will not be able to fully understand the material that comes later. You might even need to read some sections more than once.

Second, do not get discouraged if something is not clear to you. Statistical analysis is an integrated set of ideas, and you will not fully understand the entire process until you have been exposed to all parts of it. The problem is that the parts of the process have to be presented one piece at a time, so you might legitimately have some confusion, or at the very least not understand why something is being presented, until you have been exposed to a later concept or idea. Have confidence that you will succeed, and everything will fall into place.

Third, ask questions and get help when you need it. Most professors appreciate students' questions because they are a sign that you are working and thinking about the material, and not a sign of limited ability. At the beginning of every semester, I tell my students to feel free to come see me if they need some help with the material, and every semester, I have one or two students who come in after failing the first exam saying, "I was pretty sure that I was confused and did not understand the material." Get help if you need it, and get it before the exam, not afterward.

Finally, I have tried to write a text that helps you understand what statistical analysis is all about, not just how to do it. Pay particular attention to the purpose of the statistical techniques and what the results mean. My goal is for you to get more from the text than being able to plug a set of numbers into a formula and generate a correct answer.

Acknowledgements

No textbook, not even one such as this carrying the name of only one person as the author, is the result of an individual effort. A number of people have helped me during the writing and production of this book, and I want to take a moment to publicly thank them for their help. I am very grateful to the professionals at Kendall Hunt, Inc., and I particularly want to thank Mr. Paul Carty, Director of Publishing Partnerships; Ms. Angela Willenbring, Senior Development Coordinator; Ms. Lynne Rogers, Production Editorial Supervisor; Mr. Sudheer Purushothaman and Mr. Nandhakumar Krishnan Project Editors; and, the entire editorial team for their assistance, support, and work that improved the text at every step along the production process. Finally I would like to thank Ms. Sherry A. Raymondo, my wife and statistically significant other, for her unending support and encouragement during the writing of this book, and for her valuable comments along the way.

Statistical Analysis in the Social Sciences

Part I Some Basics

Part I consists of Chapters 1 and 2. Chapter 1 provides an overview of what statistical analysis is all about, and introduces you to the language and terminology of statistical analysis. The introductory material can be quite important because many of the terms and concepts described in the first chapter are used throughout the remainder of the text. The first chapter also includes a short mathematics review that will enable you to perform all of the statistical procedures included in the text.

Chapter 2 presents a fairly comprehensive overview of sampling. The material on sampling is presented earlier and in more detail than it is in most other texts. One reason for this is that almost everything we do in statistical analysis is based on the distinction between a population and a sample, so you need to be aware of those differences. I also use Chapter 2 to introduce you to the idea of the sampling distribution, which will be a very important concept when we turn our attention to inferential statistics later in the text.

CHAPTER 1

Basic Issues in Statistics

Key Concepts

Population
Concept
Variable
Hypothesis
Census
Sample
Parameter
Statistic
Descriptive Statistics
Inferential Statistics
Discrete Scale
Continuous Scale
Ratio Level Data
Interval Level Data
Ordinal Level Data
Nominal Level Data

Introduction

In this chapter, you will be introduced to statistical analysis. You will see the role that statistical analysis plays in the larger context of the research process as it is practiced in the behavioral sciences, and be given an overview of the basic terms and concepts commonly used in statistical analysis. The chapter concludes with a brief mathematics review that will demonstrate the basic mathematical operations commonly used in statistical analysis, and the special symbols we use to represent them.

What Is Statistics?

Statistics is really two things; it is both an end product and a process. You have dealt with the product aspect of statistics many times. If you are a sports fan, you may keep track of the performance of your favorite player or team by looking at a statistical summary of some type. If you are interested in business, you may track the performance of a particular company or the financial markets in general by following some of the many statistical reports that are generated on a regular basis. If you have an interest in ecology, you might track the reports of ozone levels in the upper atmosphere or the percentage increase in deforestation of the South American rain forest. There are all sorts of numbers that we refer to as "statistics." In one sense then, statistics refers to numerical information about some individual, group, organization, or other entity.

The Role of Statistics in the Research Process

Statistics is more than just a collection of numbers, it is also a process. It is a way of analyzing information, and it is the process of statistical analysis that plays a vital role when we conduct research in the behavioral sciences. The importance of the process of statistical analysis can be seen by taking a brief look at the research process, and the role that statistics plays in it.

The Research Process

The research process can be summarized easily enough. It begins with an idea which is then refined into a specific researchable question. The target group for the research is identified, and a portion of that group is selected for observation. The resulting data are analyzed, and a report is generated. Of course, a research project is much more involved than that. Let's look at what we are attempting to accomplish at each stage in the research process in a little more detail.

(1) Identify a general area of interest—The research process begins with a question that needs to be answered or a problem that needs to be solved. The first step is to identify the general area of interest, or in other words, what are we trying to find out? For those of us in the behavioral sciences, what we are usually trying to find out invariably involves an explanation of human behavior of some sort. Common examples of areas of interest for research in the social or behavioral sciences include crime, juvenile delinquency, economic success, authoritarianism, abnormal behavior, normal behavior, and prejudice, just to name a few.

(2) Identify the population of interest—The population of interest is the entire group that is the subject of your research. The majority of the research conducted in the social sciences involves a population of interest consisting of human beings. We are usually interested in finding out why people behave in a certain way, but the population of interest can just as easily be organizations, objects, or cultural artifacts. For example, we might be interested in conducting research on the characteristics of colleges and universities in the United States. In that case, our population of interest would be defined as all of the colleges and universities in the United States.

(3) Identify the key concepts—The third step in the research process involves specifying the research problem. Once you have decided on the general area of interest, you need to specify which aspects of the problem that you will actually examine. These general aspects are also called **concepts**. Concepts are usually things at an abstract level that we can conceive in our mind and name. Perhaps you have used the term "conceptualize." Essentially what you are trying to do is get a "mental handle" on something so that you can express the idea to others more clearly. In other cases, a concept may be more concrete, or easily seen and directly experienced. Financial success is an example of an abstract concept, and we might be interested in designing a research study to discover the factors that cause some people to be financially successful and others unsuccessful. Fertility is also an example of a concept, but one that is more concrete in nature (a woman has either had a child or she has not). We might be interested in designing a research study to discover why some couples have large numbers of children and other couples have few or none. A **proposition** is a statement of cause and effect between two concepts.

Of course, not all research deals with individuals, and not all concepts are defined at the individual level. For example, we might be interested in conducting research on colleges in the United States with particular emphasis on the differences between "large institutions" and "small institutions." In this case, the size of the institution would be one of the key concepts in the study. Communities have characteristics we might want to study, and some of the key concepts might include the rate of crime, the age distribution, the geographic location within the United States (north, south, east, or west), and so on. Each of these aspects can be thought of as a concept. On an intellectual level, you know what I mean when I use the term "financial success," "fertility," "size of the institution," or, "age distribution," but for the purposes of research we need to specify the meanings more precisely in a form that lends itself to observation and measurement. This is accomplished in the next step.

(4) Define variables to serve as indicators of concepts—Variables are precisely defined and measurable indicators of concepts. While concepts are at the general level and often abstract, variables are at the concrete level and can actually be observed and measured. You probably know what I meant when I used the term "financial success," but how would I actually measure financial success for the purposes of research? One good variable to measure the concept of "financial success" would be annual income. What about a variable to serve as an indicator of a "large institution?" First we would have to specify what "large" or "small" refers to. Do we mean "large" with respect to the number of students; the number of faculty; the size of the campus; or some other aspect of the college? If we are interested in size in terms of the number of students, we could define a "large college" as one with 10,000 or more students, and a "small college" as one with fewer than 10,000 students. We would then have an exact indicator for the "size" of a college or university suitable for conducting research. What about "age distribution" of a community? In a general sense, you probably know what I mean, but how would we specifically measure "age distribution?" It might be something as simple as the percentage of the population under age 30 years, or the percentage over age 65 years. In other cases, we might want to compute the average age of the community.

(5) State hypotheses—Since research is intended to answer a question or solve a problem, we are usually interested in determining a cause and effect relationship. We begin to establish the basis for a cause and effect relationship by examining pairs of variables. A **hypothesis** is a statement of relationship between two variables. For example, we might suspect that one's income is caused to some degree by one's level of education. A hypothesis relating the variables income and education could be stated as follows. The higher the level of education the higher the level of income. What types of variables might be effected by the size of a university? One logical hypothesis might be as the size of the university increases, the amount of time students spend talking to professors outside of class decreases.

Keep in mind that demonstrating two variables is related to each other is only the first step in demonstrating a true cause and effect relationship. If we want to demonstrate that one factor causes another factor, we begin by showing that the two factors are related or associated with one another. This is sometimes expressed by saying that we must show that the two variables covary. That is, when one variable changes we should see a corresponding change in the other variable. However, covariation alone is not sufficient to demonstrate cause and effect. We must also show that the time order between the two variables is logical. A presumed cause should occur before the presumed effect. We also need to account for the action of other variables that might have an impact on the relationship under examination. Finally, we should be able to build a logical case as to why one variable should be the cause of another. Our logical case is often based on reviewing the literature of previous research, and on some type of theoretical rational. Statistical analysis can help us with the first step by allowing us to demonstrate that two variables are associated with one another.

(6) Select a sample for observation—Earlier in the research process we had to identify the population of interest for our research. We are seldom in a position to examine each element of a population when we do research since many populations are quite large. When entire populations are examined we are said to be conducting a census. The census of population conducted in the United States every 10 years is an example of an attempt to count and describe every member of the U.S. population. Of course, this is a very costly and time consuming task. As an alternative to conducting a census we will usually select a small subgroup of the population called a **sample**. If selected properly, the sample will serve as a suitable representation of the population, and we will be able to generalize the results from our observation of the sample to the total population. In other words, we will be able to say that what is true for the sample is true for the population from which it was selected. A sample may be selected in a variety of ways, and many of the common sampling techniques are discussed in Chapter 2.

(7) Select a suitable research method—Once the sample has been selected, we are ready to begin collecting information. We must select a method of conducting the research, or way to collect data from the sample. There are many methods of research to choose from, but usually there will be only one or two methods that are well suited to the particular research you are conducting. Psychologists often use an experimental design to gather data. Sociologists often use a survey design that employs the use of a questionnaire or interview schedule to gather data. A research method appropriate to the type of research should be selected, and then used to collect data from each member of the sample.

(8) Statistical analysis and interpretation—It is at this point that statistical analysis begins to play a central role in the research process. Once the data have been collected, we use a variety of statistical techniques to help organize and interpret it all. Most of the rest of this book deals with the variety of statistical methods that are available to the researcher. Even if you never expect to be conducting research yourself you will still benefit from a knowledge of statistical analysis, because it will enable you to know how those who did research arrived at their conclusions. After all, you are exposed to the results of other's research all the time.

(9) Writing and dissemination of results—The final step in the research process involves the writing and dissemination of the results of your research. It is in this final step that you are sharing your information with others and contributing to the body of knowledge in your field. It is often quite difficult to convey quantitative information clearly in a written form, but it is important. As you progress in your career you will often find that it is not always the best idea that is accepted by others. Sometimes, it is the idea that is expressed best.

As you can see, statistical analysis plays an important role in the research process. Statistical analysis is the major way for us to make sense of the data that we have collected. The results of statistical analysis are often very persuasive, especially when the results are communicated clearly. However, keep in mind that statistical analysis is seldom an end in itself; usually it is part of a larger process where other factors will also play a part. You should also be aware that statistical results can be misrepresented even by individuals with the best of intentions. Other times data are misrepresented in a deliberate fashion to support a particular point of view. While behavioral scientists do not have the equivalent of the physician's Hippocratic Oath, we are expected to follow the cannons of ethical behavior. Deliberately misrepresenting data or the results of a statistical analysis to support some hidden agenda is a violation of commonly accepted ethical behavior. Unfortunately, the distinction between ethical and unethical behavior is not always immediately clear. Consider the hypothetical data below consisting of crimes per 100,000 population on an annual basis.

Year:	2008	2009	2010	2011	2012	2013	2014
Crime Rate:	125	135	145	155	165	175	185

Suppose that you are employed by the city administration, and part of your job is to keep track of crime statistics and to generate a report reflecting the level of crime. Clearly, the rate of crime has increased from 2008 to 2014. Over the 7 year period, there has been a 48% increase in the crime rate, and your report reflects that fact. Now suppose that your supervisor, the mayor, comes to you and points out that this is an election year and that a key plank in the mayor's reelection campaign is success in crime reduction. The mayor also points out that the draft of your crime report does not seem consistent with the campaign theme. You begin by taking the ethical position that the data indicate an increase in crime, and that there is nothing you can do about that. The mayor responds by saying, "I'm not asking you to lie, but can you find a way to present the data in the report that makes my record on crime look a little better?"

As you continue to examine the data you notice something. Yes, there has been an increase in crime over the 7 year period, but the *rate of increase* has been declining slightly from year to year. As a matter of fact, from 2008 to 2014 the rate of increase in the crime rate has dropped by 28.75%. This does not mean that crime has been reduced. Quite the contrary, the crime rate has increased by a steady 10 points per year. However, when the increase in crime is expressed as a percentage of the previous year's rate we see smaller and smaller increases; not because crime has been reduced, but because the basis for the percentage is getting larger year by year. A 10 point increase from 2008 to 2009 (10/125) represents an 8% change. A similar 10 point increase from 2012 to 2013 (10/175) represents only a 5.7% change because the basis for the percentage in 2013 is 175. Since 5.7% is 28.75% less than 8%, you can technically say that the percent increase in crime is declining.

Year:	2008	2009	2010	2011	2012	2013	2014
Crime Rate:	125	135	145	155	165	175	185
Percent Change:	—	8.0%	7.4%	6.9%	6.5%	6.1%	5.7%

You take your new analysis to the mayor, and explain the results. The response is "That looks more like it. I tell you what, just take out all those numbers from year to year and have the report say 'Percent increase in crime from 2008 to 2014 drops by 28.75% under Mayor's administration.' That's more in line with what I am looking for in the report." Technically, the statement that the percent increase in crime from 2008 to 2014 has dropped by 28.75% is true, but is it an accurate description of the crime situation in the city? Would it be ethical for you to present the data in that form just to support the political goals of the mayor? These are questions that obviously go beyond statistical analysis, but they are worth considering. Once you lose your reputation for honesty and integrity, it is unlikely that you will be able to get it back.

Basic Terms in Statistical Analysis

Statistics is like many other academic areas in that it has its own specialized vocabulary. There are a number of basic terms describing fundamental concepts in statistics, which you will encounter throughout this book. Several of the terms will be introduced here to provide you with a foundation, but others will be incorporated throughout the text as we move on to more advanced topics.

The scope of topics suitable for research in the academic disciplines represented by the behavioral sciences is so vast that it can legitimately include almost anything. But observation and measurement are central to the task of behavioral science research no matter what topic is chosen. Those things that we observe directly and measure, and which can take on a range of values are called variables. Examples are endless, and could include almost anything. Variables that we might measure among a group of college students would include age, major, grade point average (GPA), classification, ethnic origin, gender, income, occupation of parent, IQ, and many others.

Remember that **variables** may be thought of as measurable indicators of **concepts**. The value observed or measured for a particular variable for a given member of the sample is referred to as an **attribute**. The following example may clarify the distinction among a concept, variable, and attribute.

Let's suppose you are interested in conducting research on factors leading to academic success among college students. In this case, academic success is a concept. Again, on the immediate intellectual level you probably know what I mean by academic success, but for the purposes of research we need to find a way to quantify it. There are any number of variables that might serve as suitable indicators of academic success: GPA, class rank, receipt of a degree, and any number of others. Suppose we choose GPA as our indicator or variable of academic success. Each individual that we observe will have a particular value for his or her GPA. The individual's particular GPA is an attribute. If we are interested in intelligence, and choose a particular IQ test as our indicator of intelligence, then the individual's specific score on the IQ test is the attribute of that individual.

Suppose we have collected data on the following variables: GPA, gender, academic major, and age. The data we collect would consist of the attributes of each variable for each of the members of our sample. For a sample of $n = 10$, a typical data file might look like the following:

We have measured the same variables for each of our 10 individuals (or cases in the sample). Case number 01 has the following attributes: a GPA of 3.25, a female gender, a major of accounting, and an age of 21 years. In an actual research project, we might use numbers to code the variables of gender and major. We might assign the number 1 to the attribute "female," and the number 2 to the attribute "male." Similarly, we might have a coding scheme that assigns a unique number to each major. Computer applications are often easier when we code data with numbers rather than using what is often termed alpha numeric data such as words. Keep in mind that the fact that we might assign a number to represent gender

Concept	Variable	Attribute
Academic Success	Grade Point Average	Some value for each individual within a range of 0.00–4.00
Intelligence	IQ test	Some specific score for each individual within the range of the test
Economic success	Annual income	Some level of income reported by each individual
Ethnicity	Country of origin	The identity of one's country of origin

Figure 1.1 Relationship among concept, variable, and attribute

Case	GPA	Gender	Major	Age
01	3.25	Female	Accounting	21
02	2.23	Female	Economics	18
03	2.78	Male	Art History	22
04	3.58	Male	Mathematics	21
05	2.20	Female	Anthropology	19
06	3.75	Male	Psychology	22
07	2.95	Male	Biology	18
08	3.80	Female	Sociology	20
09	3.21	Female	Engineering	23
10	2.85	Male	Marketing	20

Figure 1.2 Hypothetical data file

or academic major does not elevate gender or academic major from the nominal level of measurement (this point is expanded on later in the chapter). We have simply changed the way we represent the data in a computer data file.

Since research is intended to answer a question or solve some type of problem, we are often interested in determining a cause and effect relationship. To do this, we make a distinction between two major types of variables: **independent variables**, and **dependent variables**. An independent variable is so named because the particular value that it takes on is independent of outside influences. A dependent variable is one whose value is determined or dependent on the value of some other factor. In the context of a cause and effect relationship, you can think of the independent variable as the cause, and the dependent variable as the effect.

In some cases, it is very clear which variable is the independent one and which variable is the dependent one. In other cases, the relationship between the two variables is not so clear. For example, we might be interested in seeing if the children of divorced couples are more likely to have a marriage end in divorce than children who grow up in a home where the parents did not divorce. In this case, it is very clear which variable is the independent one and which variable is the dependent one, because the time order of the variables is obvious. The marital status of the child's parents occurs first, and is potentially the cause of the marital status of the child as an adult. (Keep in mind that even if we found such a result, it is not by itself proof of a cause and effect relationship.) Similarly, the greater the weight of the object that you drop on your foot, the greater the amount of pain you will feel. Again, the time order is obvious. It would make no sense to suggest that the amount of pain you experience is causing the weight of some object to change.

What about the relationship between the two variables fertility and education? In general, the two variables move in opposite directions. Individuals with high fertility (many children) tend to have lower levels of education, but which is the cause and which is the effect? You might argue that having children at an early age results in an individual leaving school which would make fertility the independent variable, and education the dependent variable. That would certainly make sense. But could you not also argue that individuals with a high level of education decide to postpone having children to take advantage of the benefits of higher education that exist in the labor market? In that case, the level of education is determining the amount of fertility. Sometimes the direction of the relationship, or identity of the independent and dependent variables is not easy to determine.

Independent and dependent variables are often linked together in a statement called a hypothesis. A hypothesis is defined as a statement of relationship between two variables. Hypotheses may be stated in many different ways, and we have already seen several examples. The higher the level of fertility the lower the level of education; or stated in an alternative form, the higher the level of education the lower the level of fertility. You might be thinking that both of those statements cannot be hypotheses since we are not sure which variable is the independent one and which variable is the dependent one. In fact, they both are hypotheses. A hypothesis does not have to be correct; it just has to be testable. Finding out which hypotheses are correct, and which are false is the whole point of doing research. Once we can establish the basic relationship, we can develop additional research projects to specify or examine the relationships more fully.

Research is often conducted on a sample drawn from a population. As you recall, a population is the entire group of interest in a research project, and a sample is a subgroup drawn from the population, and which we actually observe or measure when conducting the research. Each of the variables that we include in our research will apply equally to the population and the sample. For example, we might want to do research on the undergraduate students at your college or university. The group of all undergraduate students comprises the population, or universe as it is sometimes called. The size of the population is usually symbolized by the uppercase letter N. To conduct the research, we would select a subgroup of the population called the sample. The sample size is usually symbolized by the lowercase letter "n."

One of the variables we might be interested in is the average age of the undergraduate students. The population of undergraduate students will have a single value that represents their average age. We also might be interested in the average number of credit hours earned per student as of the end of last semester. In each case, there will be a single value that *exactly* represents the average age of all students, and the average number of credit hours earned per student. These characteristics of the population are called population **parameters**. You can think of the population parameter as being the true value of some measured characteristic of the population.

We seldom know what the actual population parameters will equal, since most research is conducted on a sample. And in fact, a large number of unique samples may be taken from a given population. The variables we have selected may be measured among the sample just as they could be in the total population. One given sample will have an average age, and an average number of credit hours earned per student. These characteristics of the sample are called sample **statistics**. Just as we thought of the population parameter as the true value of some characteristic, we can think of the sample statistic as an estimate of the true population value. For example, by asking a few of the students who sit near you in class what their age is would allow you to estimate the average age of the entire class. It is unlikely that your estimate would be exactly right, but it might be pretty close to the true average age of the entire class. This is especially true if there is no pattern by age as to the way students sit in the class. Your estimate would be less accurate if older students tended to sit in one area of the room, and younger students tended to sit in another area of the room.

In most types of research, we will select and measure only one sample from a particular population. However, a large number of unique samples can be drawn from the same population, and each sample may have slightly different characteristics. For example, the true average age for the population of undergraduate students of your college or university might be 22.7 years. We might select one sample of $n = 250$ undergraduate students and observe an average age of 22.3 years. We could select a second sample of $n = 250$ undergraduate students and observe an average age of 22.8 years. The total number of unique samples that we can draw from a given population can be quite large.

Think of yourself and the other students in your statistics class as a population. Some classes are larger than others, but let's assume there are a total of $N = 40$ students in the population represented by your class. Now suppose we wanted to draw a sample of $n = 5$ students from the population of $N = 40$. How many unique samples of $n = 5$ do you think we could draw? Make a note of how many samples you think we could draw and write it down on the first page of your class notes. In Chapter 2, you will see how to calculate the number of samples of a particular size that it is possible to draw from a population of a particular size, and you might be interested to see how close your guess is to the actual number.

You have seen that one way we use the term statistic is to refer to the measured characteristics of a sample. However, the term statistic is used in two other ways. Statistical analysis is often split into two major categories: **descriptive statistics**, and **inferential statistics**. The term descriptive statistics refers to a variety of techniques that are intended to describe the general characteristics of a sample or population. Examples include the percentage of observations in one group or another, or the mean (the statistical term for the average), the mode, or the median value of some distribution of numbers.

Inferential statistics involves the distinction we make between the population and sample. In some cases, our research efforts are only intended to be descriptive in nature, and what we discover about a particular sample is sufficient to satisfy our needs. However, we are more often interested in conducting research on a particular sample because our observations will provide valuable insight into the characteristics of the population from which the sample is drawn. In this case, we are using our knowledge of the sample to make inferences about the population. So inferential statistics represent another set of statistical techniques that allow us to make generalizations about a population based on our limited observation of the sample.

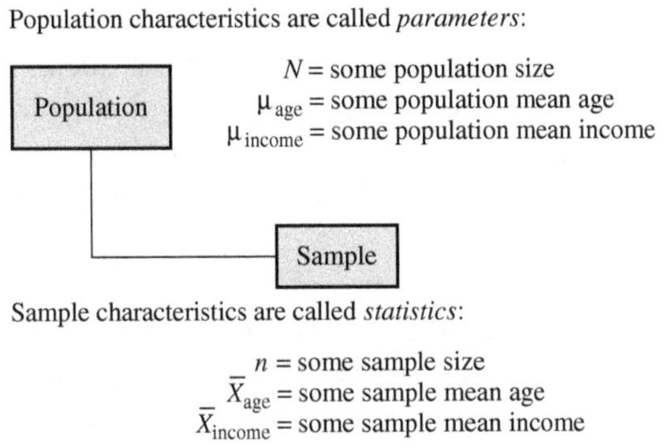

Population characteristics are called *parameters*:

Population

N = some population size
μ_{age} = some population mean age
μ_{income} = some population mean income

Sample

Sample characteristics are called *statistics*:

n = some sample size
$\bar{X}_{age}$ = some sample mean age
$\bar{X}_{income}$ = some sample mean income

Figure 1.3 The relationship between a population and a sample

By careful observation of the sample we may be able to make very close approximations of what the entire population is like.

Measurement

Think about the various types of characteristics of college students that we could measure in the course of a research project. We could easily name hundreds, if not thousands, of variables to measure. Some possibilities include age, gender, classification, income, ethnic origin, social security number, and GPA, just to name a few.

These variables are similar in that they each represent characteristics of typical college students, and they are easily measured. However, the way in which they are typically measured is not always the same. You might notice that we can arrange the variables into two groups:

Group 1	Group 2
Gender	Age
Classification	Income
Ethnic origin	GPA

The variables in Group 1 have one thing in common. They are all measured by means of categories. For example, gender is measured by the two categories: male or female. Classification is measured by the four categories: freshman, sophomore, junior, or senior. Ethnic origin is measured by a large number of categories including French, English, German, Italian, Spanish, and many others. In each of these cases, a given student can logically be placed in one and only one category, and more importantly, it does not make any sense at all to think of someone being placed between categories. A student is either a sophomore or not. It does not make any sense to think of a student as a sophomore and a half. Likewise, a student is either female or male. It does not make any sense to consider someone female and a half.

Now consider the variables in Group 2. They also have something in common. Age, income, and GPA can be measured on a scale with many more categories. For example, we could measure age in years using a scale marked off in single years of age. Certainly it makes sense to think of someone as being 20 years of age, but it also makes perfect sense to think of them as 20 1/2, or 20 3/4 years of age. We might measure income on a scale marked off in thousands of dollars. Someone might earn $55,000 per year, but it would make sense to measure someone at $33,742 per year, or $78,762.50 per year.

You may have noticed that I neglected to place one variable in either of the two groups. Which group would social security number more logically belong? Since your social security number is a series of digits from 0 to 9, you might think that it belongs in Group 2 with the other variables that are measured by numbers. Actually, social security number is a variable that logically belongs in Group 1. Even though a social security number is a series of nine digits, it is nothing more than a categorical variable similar to your name, gender, or ethnic origin.

A more formal way to describe the two groups of variables is to say that the variables in Group 1 are examples of **discrete measurement** typically measured on a **discrete scale**. A discrete scale is one in which the variable is measured by a series of separate categories. Some variables may have more categories than others, but it does not make sense to be between any two categories on the scale. The variables in Group 2 are examples of **continuous measurement** typically measured on a **continuous scale**. A continuous scale will have some number of measurement points, but it makes perfect sense for an observation to fall between two points on the scale. In most cases, we are only limited in the number of points we can place on a continuous scale by our technical ability to achieve a particular level of precision in our measurement. For example, we can construct a continuous scale to measure time which is marked off in years, or months, or days, or hours, or minutes, or seconds, or milliseconds, and so on. At some point, we will lack the technical ability to measure time in any shorter intervals, but theoretically we could continue to smaller and smaller units. And in each case, it would make perfect sense to observe some length of time that fell between two adjacent points on the scale. Even when we measure the passage of time in nanoseconds (billionths of a second) it would still make sense to observe an interval of 1.324 nanoseconds.

Scales of Data or the Levels of Measurement

We have seen that variables may be measured on either a discrete or a continuous scale. Discrete and continuous scales may each be further classified into two types. The resulting four categories constitute the four **scales of data**, or levels of measurement as they are commonly called. It is important to be aware of the type of data you have collected. Certain statistical procedures may be legitimately applied to any type of data, but other statistical procedures will require data of a certain level. Examples will follow, but first examine the four scales of data. Ratio and interval level data are each examples of continuously measured variables. Ordinal and nominal level data are each examples of discretely measured variables.

Ratio level data are measured on the most rigorous scale possible. A variable is considered at the ratio level if it meets the following two criteria. It must be measured on a continuous scale with equal appearing intervals; and, the scale must have a meaningful zero point. A scale with equal appearing intervals is one in which the amount of change in the measured variable represented by the movement from one unit on the scale to an adjacent unit on the scale is the same anyplace along the scale. For example, suppose we are measuring annual income with a scale marked off in units of $1,000. To be a ratio level scale, the difference in annual income represented by a move from 1,000 to 2,000 must be equal to the

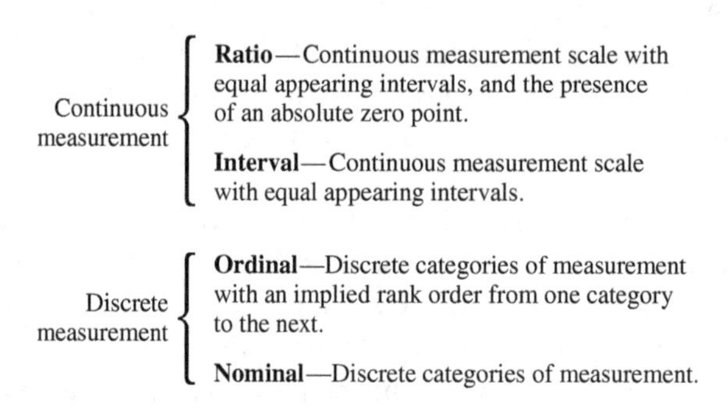

Figure 1.4 Scales of data identified

difference in annual income represented by a move elsewhere on the scale such as from 8,000 to 9,000. In both cases, the change represents a net difference of $1,000.

The second criterion is the existence of a meaningful zero point. A meaningful zero point is sometimes called an absolute zero point, and it is the point on the scale that represents a complete absence of what is being measured. On the income scale a meaningful zero point would be the point that indicates absolutely no income. The variable income is one that can be measured on a ratio scale as just described. However, as you will see later, income does not have to be measured in that way.

One of the advantages of measurement at the ratio level is that it allows us to make true ratio statements. An annual income of $60,000 is twice as much as an annual income of $30,000. A raise in annual income of $5,000 is half as much as a raise of $10,000. Sixty years is three times as long as 20 years.

Interval level data are almost as rigorously measured as ratio level data, except there is no requirement of an absolute zero point. We must still have a measuring scale with equal appearing intervals where the net change represented by the movement from one unit on the scale to an adjacent unit on the scale is the same anyplace along the scale. Examples of true interval level scales are not very common. One of the best examples is that of a temperature scale marked off in degrees Fahrenheit, or in degrees Celsius. What we are actually measuring with a temperature scale is the amount of heat in the surrounding environment, or in some object. Suppose our scale is marked off in single degrees Fahrenheit. The net change in heat represented by a move from 20°F to 21° F is exactly the same amount of change in heat represented from move from 81°F to 82° F. In other words, a degree change in heat is the same anyplace along the scale. What prevents a Fahrenheit scale from representing measurement at the ratio level is that the zero point on the scale is simply an arbitrary point, and does not represent an absolute zero point. When it is 0°F it does not mean that there is no heat in the surrounding air. In fact there is heat. Compared to a temperature of 100°F there may not be very much, but compared to a temperature of -40°F there is quite a lot of heat. Because the Fahrenheit scale is not at the ratio level, when it is 100° F it is not twice as warm as when it is 50°F.

Ordinal level data are measured on a scale consisting of discrete categories with an implied rank order as one moves from one category to the next. For example, a student's classification in college is an ordinal variable. The scale is discrete in that it consists of freshman, sophomore, junior, or senior. A given student logically fits into one and only one of the categories, and it does not make sense to be between two categories. To say that you are a junior and three quarters does not really make any sense. However, the scale does have an implied rank order in that you know that anyone in the junior category has more of the thing being measured than anyone in either the freshman or sophomore categories, and less of the thing being measured than anyone in the senior category.

Nominal level data are measured at the simplest level possible. All that is required is a set of discrete categories. Gender, ethnic origin, and brand of toothpaste all represent variables measured at the nominal level.

Take another look at the two groups of variables used to illustrate the difference between discrete and continuous scales. How would you say each variable is measured?

Group 1	Scale Type	Group 2	Scale Type
gender	nominal	age	ratio
classification	ordinal	income	ratio
ethnic origin	nominal	GPA	ratio

If your answers agree with mine then you are probably wrong! While these answers look logical enough, they are based on incomplete information. The answers I have provided are indicative of how these variables are usually measured, but that does not mean it is the only way they are ever measured. Consider income. Usually we would measure it at the ratio level, but what if we measured income as

100,000 or more
75,000–99,999
50,000–74,999
25,000–49,000
below 25,000

or with a more obvious example of

High
Middle
Low

Each of these alternative ways to measure income represents ordinal level data. Many of you may have completed sample surveys at one time or another that contained a question measuring income in a way very similar to one of these. Most variables may be measured at more than one level of data, and keep in mind that the variable itself does not have a characteristic level of data. The level of data is the result of how we choose to measure the particular variable.

Why Does the Level of Data Matter?

Each statistical procedure is based on certain assumptions or requirements. Very often, one of the most important assumptions is that the data being analyzed have been measured at a particular level. Some statistical procedures, such as the computation of percentages, or the mode (a measure of central tendency covered in Chapter 4) may be computed legitimately with data of any level. However, other procedures such as computing the mean (another measure of central tendency, covered in Chapter 4), or correlation and regression (coming soon in Chapters 9 and 10) require at least interval level data.

Some statistical procedures are said to be robust, which means that some of the basic assumptions may be violated without a major loss in meaning of the results. Other statistical procedures are rendered virtually meaningless when the basic assumptions are violated. For example, it makes no sense to compute a mean (or average) for nominal level data. In a group of 100 individuals composed of 30 Germans, 25 Spaniards, 30 Mexicans, and 15 Italians, it would make no sense at all to speak of the mean or average nationality. However, it would make perfect sense to note that 15% of the group was Italian, and that 30% was Mexican.

As a general rule, it is better to collect data at the highest level possible when doing research. For example, collect information on age, income, and so on, at the ratio level if possible. When you begin your statistical analysis you can always convert data measured at a higher level such as ratio to a lower level such as ordinal, but you cannot convert up the scale from ordinal to ratio.

Common Symbols and Mathematics Used in Statistics

The level of mathematical ability required in a basic statistics course is remarkably easy. In fact, there is nothing required in most basic statistics courses that goes beyond what most students master in a first year algebra course in high school. If you can add, subtract, multiply, divide, square a number, and take a square root, you have all of the math skills required to master the material in this text.

What is often confusing to many beginning statistics students is the use of common statistical symbols which have their own unique meaning in statistics. Many of the symbols used in statistics are Greek letters. Let's take a look at a few of the more commonly used statistical symbols and their meaning. Some of these we have already discussed or will discuss shortly, while others will be seen in later chapters.

Commonly Used Statistical Symbols

Symbol	Name	Meaning
Σ	Sigma (uppercase)	To Sum or Add
σ^2	Sigma Squared (lowercase)	Population Variance
σ	Sigma (lowercase)	Population Standard Deviation
μ	Mu (lowercase)	Population Mean
$\bar{X}$	X Bar	Sample Mean
α	Alpha (lowercase)	Level of Significance

You will encounter the uppercase sigma Σ in many statistical formulas. Here are some examples of how sigma and some of the other common statistical symbols are used in statistics. For a hypothetical variable *X*, we may have observed the following values:

$$X\text{: } 8 \quad 8 \quad 9 \quad 10 \quad 12 \quad 12 \quad 14 \quad 15 \quad 16 \quad 20$$

It does not really matter what variable X represents, but suppose in this case X is a measure of educational attainment. As you can see, there are a total of 10 observations, or in the terminology of statistics, our sample size is $n = 10$. The results indicate that educational attainment ranges from a low of 8 years to a high of 20 years.

You will very often encounter a statistical formula that in part looks like this:

$$\sum_{i=1}^{n} X_i$$

This formula would be read as "the sum of the *X* sub *i*, as *i* goes from 1 to *n*." The uppercase sigma means that you should sum or add what is indicated. In this case, you are being asked to sum the values of X. The subscript i to the right of X is just a counter indicating that there are several *X*s. The notation "$i = 1$" below the sigma symbol, and "*n*" above the sigma symbol means that you should begin your summation with the first *X* and continue until you have included the last or *n*th *X*. In other words, all you are being asked to do is add up all the values of *X*.

In our example, that would mean

$$8 + 8 + 9 + 10 + 12 + 12 + 14 + 15 + 16 + 20 = 124$$

Since it is generally expected that we will always add all of the values of *X*, you will usually see the sigma sign without the subscript $i = 1$, and superscript *n* notation, or simply ΣX.

One of the statistical formulas that makes use of the sigma notation is the formula for the mean (or average). It would be written as follows:

$$\bar{X} = \frac{\sum_{i=1}^{n} X_i}{n}$$

You would read the formula as, *X* bar is equal to the sum of the *X*s divided by *n*. The computation is straightforward.

$$\bar{X} = \frac{\sum_{i=1}^{n} X_i}{n} = \frac{124}{10} = 12.4$$

As indicated earlier, since it is usually assumed that all of the values associated with a particular variable will be included in the sum, the formula for a mean will often appear as follows:

$$\bar{X} = \frac{\Sigma X_i}{n}$$

Note that X is not the only symbol used to represent a variable. At times we might use the symbol Y or Z to represent variables. In that case, we might compute the mean for the Y variable by

$$\bar{Y} = \frac{\Sigma Yi}{n}$$

The uppercase sigma sign is also used in two mathematical terms you frequently encounter in statistical formulas:

$$\Sigma X^2 \text{ and } (\Sigma X)^2$$

The first term is read as the sum of *X*s squared. The second term is read as quantity sum of *X*s, squared, or sometimes as sum of *X*s quantity squared. It is easy to confuse these two terms when you first begin to work with them, and to make matters worse, they are often included in the same formula. Once you learn what each means it is easy to tell them apart.

When the exponent 2 is directly on the X it means to square each value of X first, and then add the values. When the exponent is on the parenthesis you should perform whatever mathematical operation is called for inside the parenthesis first, and then square the result. So, in the second term you should add all of the *X*s just as we did before, and then square the total. Keep in mind that the symbol X is just an arbitrary way to refer to some variable. We might just as easily talk about a variable Y or Z. In some applications, when we are working with two variables we will refer to the variables X and Y. Let's look at some hypothetical data below for two variables called X and Y.

(1)	(2)	(3)	(4)
X	X^2	Y	Y^2
8	64	6	36
8	64	5	25
7	49	4	16
5	25	2	4
2	4	0	0
30	206	17	81

Column (1) contains the values for the variable X, and column (3) contains the values for the variable Y. Column (2) contains the values for each X value squared; and, column (4) contains the values for each Y value squared.

Some of the common mathematical operations we might perform, and the way we symbolize them are illustrated below.

Symbol	Result	Meaning
$\Sigma X =$	30	The sum of the Xs
$\Sigma Y =$	17	The sum of the Ys
$\Sigma X^2 =$	206	The sum of the Xs squared
$(\Sigma X)^2 = (30)^2 =$	900	Quantity sum of Xs squared
$\Sigma Y^2 =$	81	The sum of the Ys squared
$(\Sigma Y)^2 = (17)^2 =$	289	Quantity sum of the Ys squared

Let's add one more concept called the sum of the cross products, that is frequently encountered in statistics. Symbolically it is written as

$$\Sigma XY$$

When you encounter the sum of the cross products you are simply expected to multiply the value of the X variable by the value of the Y for each case or element in your sample. Column (3) below represents the cross multiplication of each X value by its corresponding Y value. The sum of the cross products is obtained by simply adding the values in column (3).

(1)	(2)	(3)
X	Y	Y
8	6	48
8	5	40
7	4	28
5	2	10
2	0	0
30	17	126

$$\Sigma XY = 126$$

Note that $\Sigma XY \neq (\Sigma X) \times (\Sigma Y)$

$$126 \neq 30 \times 17$$

$$126 \neq 510$$

Those of you with math anxiety should begin to relax now. If you know how to add, subtract, multiply, divide, square numbers, take square roots, and understand the simple concepts we have just reviewed, then you know how to perform ALL of the mathematical operations required in this textbook. Everything we do from now on will simply build on these basic math skills. The more important issue is to understand statistical analysis.

Introduction to Computer Applications

In this and most subsequent chapters, you will find the following two sections at the end of each chapter: Computer Applications and How to do it. The first provides you with suggested computer exercises keyed to using one of the most widely used statistical packages on the market today, Statistical Package for the Social Sciences (SPSS), and the second provides you with some suggestions on how to complete the exercises. Many of the computer exercises will refer to a real-life data set (the General Social Survey

[GSS]) that is included with the instructor's edition of your text. If you do not have access to SPSS, you may still be able to complete the computer exercises using some other computer statistical package. Your computer exercises for this chapter follow.

Computer Applications

Many of the computer exercises you will encounter in subsequent chapters will require you to use data from a portion of the GSS. The GSS contains data from a survey using a national probability sample of the U.S. population, and has been conducted every 2 years since 1972. Appendix C contains a brief description of the GSS, and a codebook section identifying the variables, their value labels, and distribution in the data set, which is included with the instructor's edition of your text. You will also be expected to use the SPSS, or some similar statistical package to complete the exercises. Appendix D is a guide to using SPSS, and discuses all of the major features that you will need to know how to use to complete the computer exercises. You may also have access to a student manual for SPSS, the actual SPSS user's manual, or the online tutorial. All three of which represent additional sources for assistance.

Your "Computer Applications" exercises for this chapter are

1. Familiarize yourself with the GSS data set.
2. Familiarize yourself with the major features of SPSS.

How to do it

Turn to Appendix C and read it! Then turn to Appendix D and read it!

Summary of Key Points

Knowledge is power, and as you master the material in this book, you will find that a knowledge of statistical analysis will provide you with a certain type of power. You will have the power to examine a set of numbers and know what they mean; the power to analyze data in a meaningful way; and most importantly, the power to present quantitatively derived evidence to support a conclusion. As we continue to live in a time characterized by a genuine explosion of information and data, you will find that the skills you develop through the study of statistical analysis will continue to benefit you long after your college days are over.

Statistics plays a key role in the research process by helping us to analyze the data we have gathered in an effort to answer a fundamental research question. Several important steps and concepts are encountered along the way including the following.

Population of Interest—The population of interest or universe is the entire group that is the subject of your research.

Sample—The sample is a subgroup of the population that is selected in some fashion for direct observation and measurement.

Concepts—Concepts are abstract aspects central to your research effort. Concepts serve as "mental handles" and help us focus on the major research questions which we will investigate.

Proposition—A proposition is a statement of relationship between two concepts. The proposition expresses our expectation of how we think two concepts are related.

Variables—Variables are measurable indicators of concepts.

Hypothesis—A hypothesis is a statement of relationship between two variables. Since variables may be measured we are able to gather data and test a hypothesis.

Attributes—The actual level or state observed for a member of the sample for a particular variable is referred to as an attribute. For the variable gender, the possible attributes are female or male.

Parameters—Parameters are characteristics of a population. They represent the actual true state of a population.
Statistics—Statistics are the characteristics of a particular sample taken from a population. Since many samples are possible, statistics represent estimates of a particular population parameter.
Descriptive Statistics—Descriptive statistics are a set of statistical procedures that describe a set of data with no attempt to generalize the results to any other group.
Inferential Statistics—Inferential statistics are a set of statistical procedures that allow us to generalize the results obtained from an observed sample to a larger population.
Continuous and Discrete Measurement—Variables may be measured in several different ways. Continuous measurement occurs on a scale with an infinite number of points. Discrete measurement occurs on a scale with separate categories.
Ratio Level Data—Ratio level data are measured on a continuous scale with equal appearing intervals and an absolute zero point. Data measured at the ratio level allow us to make ratio type statements of the form: A is twice as much as B.
Interval Level Data—Interval level data are measured on a continuous scale with equal appearing intervals, but lack an absolute zero point. The difference between two intervals at any point along the scale represents the same amount of change in the phenomenon being measured.
Ordinal Level Data—Ordinal level data are measured at the discrete level with separate categories, and an underlying rank order among the categories. Being in a particular category not only tells us the state of the case being observed, but if the case has more or less of what is being measured than cases in adjacent categories.
Nominal Level Data—Nominal level data are measured at the discrete level with separate categories. Nominal level data are the most basic type of measurement possible allowing us to simply categorize our observations.
Independent Variable—In a causal relationship the independent variable is the presumed cause of the dependent variable.
Dependent—Variable in a casual relationship the value of the dependent variable is presumed to be the result or cause of the action of the independent variable.
Discrete Scale—When measuring a variable on a discrete scale the result is categories of the variable. For example the variable age might be measured as young—individuals between the ages of 15 to 29.
Continuous Scale—When measuring a variable on a continuous scale the result is exact values of the variable subject to the level of precision used. For example age might be measured in years and a member of the sample might be 19 years of age.
Scales of Data—Variables may be measured at four levels: The nominal level; the ordinal level (both examples of discrete measurement); the interval level; or, the ratio level (both examples of continuous measurement).

Questions and Problems for Review

1. Define the terms population, sample, statistic, and parameter. How are these terms related?

2. Define the terms concept and variable. How are they related? Identify two concepts and two variables that would be appropriate for a population consisting of single parent families in the United States.

3. Give an example of a concept, and then a variable that would be a good indicator of it. List several possible attributes that might be observed when measuring your variable.

4. Choose a second variable that you think would be related to the first variable you identified. How might the two variables be related?

5. Select a variable that can be measured at more than one level such as ratio, interval, ordinal, or nominal. Give an example of how you could measure the variable at two different levels.

6. Select an article of interest from a scholarly journal in your area of study. Identify the following:
A. Population of interest.
B. Size and method of selection of the sample.
C. The major dependent variable (what is the author trying to explain).
D. The major independent variables (what are the major variables the author thinks explain the dependent variable).
E. The level of measurement for the dependent and independent variables.

7. Identify the following as either concepts, variables, or attributes.
A. occupation
B. $45,900
C. male
D. income
E. plumber
F. $22,500
G. physician
H. gender
I. economic success

8. Using the data below

X	Y
8	12
10	7
14	32
2	4
23	33
15	15
17	22
5	7

9. Calculate the following:
A. ΣX
B. ΣY
C. ΣX^2
D. $(\Sigma X)^2$
E. ΣY^2
F. $(\Sigma Y)^2$
G. $\Sigma X \times \Sigma Y$
H. $\frac{\Sigma X}{n}$
I. $\frac{\Sigma Y}{n}$
J. ΣXY

CHAPTER 2

Sampling

Key Concepts

Population
Sample
Sampling
Probability Samples
Nonprobability Samples
Sampling Distribution

Introduction

As you should recall from Chapter 1, most behavioral science research involves the identification of a **population of interest** consisting of the entire group we wish to study, but we seldom examine the entire population. In most cases research is conducted on a smaller **sample** or subgroup that is drawn from the larger population. **Sampling** is the process of selecting from a population the sample or subgroup that will actually be examined. In this chapter, we will focus on the difference between a population and a sample, the reasons why we choose to examine a sample rather than the entire population when we conduct research, and the process of sampling. We will focus on two broad types of sampling strategies: probability sampling and nonprobability sampling, and we will look at some of the more common sampling techniques of each of those two types. Finally, we will look at the concept of a **sampling distribution**. The sampling distribution is one of the more important concepts in statistical analysis, and like many important concepts in this book, you will find that it will turn up again in some of the later chapters.

Before we begin looking at some of these ideas on sampling, let me remind you of something I asked you to do in the first chapter when I introduced the concepts of population and sample to you. I asked that you imagine that your statistics class consisted of a population of $N = 40$ students, and that we wanted to take samples of size $n = 5$. I was making the point that a single population can have several different samples drawn from it, and I asked you to write down on the first page of your class notes how many different samples of size $n = 5$ you thought we could take from a population of $N = 40$. If you wrote down your guess you might want to refer back to it, and if you did not write down a guess please take a minute to think about it and write it down. We will be coming back to this idea toward the end of the chapter when we discuss the sampling distribution.

What Is Sampling, and Why Do It?

Sampling is the process of selecting a subgroup from a population that we wish to study. Almost all research is based on the examination of a sample rather than the total population for several important reasons. Researchers choose to study a sample because it saves time, money, effort, and in some rare cases because sampling can be more accurate than the study of the entire population. Let's look briefly at each of these reasons for sampling using a national study of the U.S. population as our example.

The population of the United States is approximately 320 million people, and contains over 115 million households. Even if we used households as our **unit of analysis**, you can imagine how much time it would take to contact each and every household in the United States. The federal government actually makes an attempt every 10 years (in years ending in zero) to do just that. Any research project that actually examines the entire population of interest may be referred to as a census. Every 10 years the government conducts the national census of population, in which an effort is made to collect data from every household in the United States. A great amount of time is required to conduct the national census, and in some ways you can think of it as a never ending job. Planning for the next census usually begins as soon as the previous one ends.

Sampling also saves money. The total cost of the national census has risen to the multibillion dollar range, and even less ambitious research projects involving the U.S. population would be quite expensive. Imagine trying to do a simple mail-out mail-back research project in which you sent a brief questionnaire to each of the 115 million or so households in the United States. At current postal rates the cost would be well over 100 million dollars, and that is just for postage! Most national surveys are conducted on a sample of anywhere from 1,000 to 2,000 households, and for a simple mail-out mail-back questionnaire the cost might be $2,500; I think you can appreciate the cost savings from sampling.

Sampling also saves a great deal of effort. Even if you could afford to conduct a national census every time you wished to undertake a research project, the combined effort required just to collect the data would be enormous. Once the data have been collected they must still be coded properly, input into a machine readable format, analyzed, and interpreted. While data analysis and interpretation do not always increase with the size of the sample, the jobs of coding and input certainly do. Do you know how long it would take to input data for 115 million households into a computer data file? (Frankly I don't know either, but I'm pretty sure it would take a long time.)

Finally, in some rare instances, it may be more accurate to sample rather than to examine the entire population. Greater accuracy from a sample compared to a population census is likely to occur in situations where the population is large and rapidly changing. In such situations the time required to complete a population census is so great that it allows the population to change so much that the results from the last stages of the census will refer to a completely different population than the results from the initial stages of the census. Data from a sample of the population can be collected more quickly, and the results from the sample represent a more accurate "snapshot in time" of the population. Although the population itself will continue to change rapidly, we at least know what it was like at the time that the sample was examined.

Sampling Strategies

Probability samples and nonprobability samples are the two broad types of sampling strategies frequently encountered in behavioral science research, and each type may be accomplished by several different methods. A probability sampling strategy is one that to some degree ensures that every element in the population has a chance to be included in the sample. A nonprobability sampling strategy does not ensure that every element in the population has a chance of being selected for the sample. As you might imagine, a probability sampling strategy is more difficult to implement, but the results will be a more valid representation of the source population.

Several examples of probability and nonprobability sampling plans will be discussed below. While both types are important, the probability sampling plans play a more important role in the research process and in statistical analysis. I pointed out in Chapter 1 that statistical procedures can generally be placed into the two categories of descriptive statistics, and inferential statistics. Inferential statistics refers to a group of procedures in which we use information from a sample to infer something about the population from which it was drawn. All inferential statistical procedures assume that a probability sampling plan was used to select the observed sample from the population. If a sample is not representative of the population from which it was drawn, then there is no justification for generalizing the results of the sample back to the population.

Some Basic Sampling Concepts

We need to be familiar with some basic sampling concepts before examining some of the more popular probability and nonprobability sampling methods. Most of the research conducted in the behavioral sciences involves an examination of a population of human beings, but other populations of interest are also examined so the examples illustrated here will be of several types. The concepts of a population and a sample were introduced in Chapter 1, and have been reviewed briefly in this chapter as well. You have seen that when doing research the entire group of interest is called the **population** or universe.

Unfortunately, it is often the case that the true population or universe can never be completely identified. The population from which the sample is actually drawn is sometimes referred to as the **sampling population**, and it will often deviate to some extent from the true population of interest. Suppose we wanted to conduct some research on all full-time undergraduate students at your college or university. The true population of interest is composed of all students who meet the criteria of being: undergraduates, and taking enough courses to be classified as full-time students. To draw the sample we would need to work from a list or data base in which these students are identified. That list or data base will comprise the sampling population, and it is unlikely that it is ever a completely accurate reflection of the true population of full-time undergraduate students. At any given time, there will be some full-time undergraduate students who are not on the list, perhaps due to an oversight on someone's part, or perhaps they have been admitted after the list was obtained, or for any number of other reasons. Also at any given time, there will be students on the list of full-time undergraduate students who should not be there, some because they have left the school, some because they are no longer taking a full-time load of courses, and some for other reasons. Small differences between the true population of interest and the sampling population from which the sample is actually drawn will not cause a problem, but major differences represent a threat to the validity of the research.

Two other terms to be familiar with in the process of sampling are the unit of analysis and the **sampling unit**. The unit of analysis in research refers to the person or thing from which we will actually collect data. In behavioral science research the individual is frequently the unit of analysis. Other units of analysis are possible, and are frequently used. We might just as well be doing research on communities, and our unit of analysis might be cities or metropolitan areas. We might also conduct research on prisons, and our unit of analysis would be the prison. The variables in our research would reflect characteristics of the prison such as its size, its location, the type of facility (local, state, or federal), the level of security, the sex of the inmates, and so on.

The sampling unit refers to the units that are actually selected from the population for inclusion in the sample. Oftentimes the sampling unit and the unit of analysis are the same thing. In doing research at your college or university we might work with a list of students to draw our sample so the sampling unit would be students, and we would eventually collect data from the students selected so the unit of analysis would be students. But there are occasions where the sampling unit and the unit of analysis are different. For example, instead of using a list of students to draw our sample we might use a list of courses offered in the current academic term and obtain a sample by selecting a certain number of the courses being offered. Once the courses have been selected we would collect data from the students in the selected courses. The sampling unit in that case would be the courses offered in the current academic term, but the unit of analysis would be the individual students enrolled in the courses. Many national studies are conducted in a manner where the sampling unit and the unit of analysis are different. The sampling unit may be the household or the telephone number of the household, and the unit of analysis will be an individual within the household.

Nonprobability Sampling Methods

Nonprobability sampling methods are generally easier to apply than are probability methods, but they are not always representative of the population from which they are drawn. The results from a nonprobability sample may not be generalized to the larger population from which it was drawn, but that does not mean that the observations obtained from the sample are of no use. Nonprobability sampling plans are more useful the more homogeneous the population is and less useful when population is more heterogeneous. One frequent use of nonprobability sampling plans is in pretesting the data collection instruments or procedures to be used in a larger research project. In pretesting, the emphasis is not on the actual results from the sample, but is on the process of obtaining the results. In a pretest of the methodology, we are more concerned with the following issues: Are the questions in a questionnaire clear? Are the directions clear and easy to follow? Do the members of the research staff understand what to do? Another frequent use of nonprobability sampling plans is situations that require rapid results.

There are many types of nonprobability sampling methods, and we will cover just a few of the ones that are more frequently used in behavioral science research. **Accidental sampling** is one of the most frequently used nonprobability sampling methods. Accidental sampling is accomplished by recruiting anyone you can find to become a member of your sample until the desired sample size is obtained. Local television news crews often claim to have taken a "random sample of opinion" when they take their cameras out on the street and film the first four or five people who are willing to speak with them. As you will see in the next section, such a strategy is very far removed from a true random sample, instead what they are doing is accidental sampling.

A second nonprobability sampling strategy is referred to as captive sampling. **Captive sampling** is accomplished by using any group for the sample over which the researcher has temporary control. College professors are frequent users of the captive sampling method when they bring various questionnaires into the classroom for their students to complete.

Quota sampling is a nonprobability sampling plan that is similar to accidental sampling except that the researcher is looking for individuals with particular characteristics among those who are selected for the sample. For example, as a class project you might conduct some research on undergraduates at your college or university. With quota sampling, each of you would be assigned the responsibility of finding a certain number of individuals for the sample who satisfied various criteria. One person might be expected to interview five male business majors, another person might be expected to interview five female science majors, another might be expected to interview five commuter students, and so on.

Spatial sampling and **systematic sampling** are two nonprobability sampling methods that are frequently used to sample individuals in public settings. Spatial sampling is often used to sample individuals in large crowds such as concerts or festivals of some type. Spatial sampling is accomplished by having members of the research team spread out at roughly equal intervals at the perimeter of the crowd. The team will take a specified number of steps into the crowd, stop, and then attempt to interview the first person to their right. If that person refuses, they will continue to ask individuals in the immediate vicinity until an interview is obtained. They will then take the specified number of steps further into the crowd, stop, and attempt to interview the first person to their left. The pattern of advancing into the crowd and attempting to interview individuals alternately to the right and left continues until they reach the far side of the crowd. When completed, the members of the research team have taken a sample of individuals throughout the entire crowd as opposed to collecting data from individuals in some isolated section.

Systematic sampling is also used for sampling individuals in public settings, although in a setting that is different than a large crowd. Systematic sampling is often used in situations where there is a great deal of pedestrian traffic such as a retail mall. Interviewers are positioned in high traffic areas, and are instructed to allow a certain number of individuals to pass (let's call that number "*k*") and then to attempt to interview the *k*th + 1 person who walks by. If that person refuses they continue to solicit individuals for an interview until an interview is completed, then they allow "*k*" individuals to pass by before attempting to complete another interview. Neither spatial sampling nor systematic sampling will result in a representative sample, but by having the research team proceed in stages into the crowd, or by having the interviewers allow "*k*" individuals to pass before attempting another interview does accomplish something very important. It prevents the research team from allowing their own personal biases (some of which will be subconscious) from determining the composition of the sample. Each of us is more comfortable with certain types of people than we are with others, and if we are simply instructed to go to a location and come back with 20 completed interviews we are likely to approach the first 20 people with which we happen to feel comfortable. That may make our job easier, but it introduces an unnecessary bias into the sample.

Two final special nonprobability sampling methods are **judgmental sampling** and **snowball sampling**. Judgmental sampling is essentially a handpicked sample composed of members of a population known to be particularly knowledgeable with respect to the issues under investigation. The sample is selected based on either the researcher's own expertise as to whom in the population is particularly knowledgeable, or on the basis of suggestions from a local informant. Snowball sampling takes its name from the analogy of what happens when a snowball becomes larger and larger after it is sent rolling down a hill. With snowball sampling the researcher begins with a few knowledgeable individuals from a population, and then asks them for suggestions as to whom to interview next. If every person can provide two additional names, it does not take very long before the sample becomes quite large.

Probability Sampling Methods

Probability sampling methods are all based on the premise that, to one degree or another, all elements of the population will have a chance to be included in the sample. Again, we will just cover a few of the more frequently used methods, and you might notice that the sampling methods we have already examined represent the nonprobability alternative of the probability plans we will examine below.

The most frequently used probability sampling method, and the one that most of the others is patterned after, is called **simple random sampling**. Simple random sampling is accomplished by first obtaining as complete as possible a listing of the population of interest, then the desired number of individuals for the sample is selected on a "random" basis. Random does not mean selection in a haphazard or accidental basis, but rather in the context of research, random means without bias. The selection should be made such that every element in the population has an "equal" chance of being selected for the sample.

There are several different methods of achieving a random selection. Increasingly, a random sample is being accomplished through the use of various computer programs that are designed to select cases at random from a given population. The computer method is particularly useful when your population of interest is already in a machine readable form. The most basic method of achieving a random selection is the random draw where all of the elements of the population are represented in a container of some type and the number desired for the sample are drawn "at random" from the container one at a time. It is at this point that you can see why I put the word equal in quotation marks above when I said that a random sample was one in which all elements of the population had an equal chance of being selected. Suppose we were going to draw a random sample of $n = 5$ students from a class with a population size of $N = 40$, and we were going to accomplish this by the random draw method by putting all of your identification cards in a box from which we would draw a total of 5 for the sample. The first person drawn had a 1 out of 40 chance of being selected; the second person drawn had a 1 out of 39 chance of being selected; and so on to the last person drawn who had a 1 out of 36 chance of being selected. As you can see, the probability of being selected is not exactly equal, but in most instances the differences in the probability of being selected as you move from the first person chosen to the last person chosen are so small as to be inconsequential. There are those who suggest that it is possible to keep the probabilities exactly equal by, in this case, returning the selected identification card back to the box after the identity of the person selected for the sample has been recorded. Returning the identification card to the box does not really keep the probabilities of being selected equal because you would not select the same individual twice for the sample. If you were to select an identification card that had been selected previously you would ignore it the second time, so in effect, after the first person has been selected the box would contain 39 identification cards that are eligible for selection and 1 card that is just taking up space. The probability for the second person entering the sample would still be 1 out of 39 even though there are a total of 40 cards in the box. Yet, there are those who think they are accomplishing something by returning the selected cases back to the box. The logic is a little like buying a six pack (of juice of course), drinking one and putting the empty can back in the refrigerator. You have five cans of juice and an empty can—try and drink it!

There are a couple of other methods of obtaining a random selection that are used when the population of interest is in the form of a written list. Each member of the population would be given a code number such as 1, 2, 3, …, 40 in the case of our classroom example. The sample would be obtained by using a **random number table**. A random number table is a table of random numbers. It is just a series of digits selected at random, usually by a computer, and printed out on as many pages as necessary. The table is usually organized in a fashion to make it easy to read, but there is no underlying structure or meaning to the presentation of the numbers. A representation of a typical random number table is presented below with column numbers and row numbers added to make it easier to read.

	1	2	3	4
1	6 8 3 0 0	3 3 2 0 9	5 9 0 1 1	2 8 6 9 4
2	4 8 2 2 1	5 9 2 1 1	8 8 6 3 2	4 2 1 9 7
3	8 4 7 4 0	2 4 7 9 6	7 5 4 2 6	0 8 5 3 1

Figure 2.1 A Facsimile of a Random Number Table

The random number table would continue on for several pages with row after row, and column after column of random digits printed on the pages. A random number table is used by selecting a particular place to begin, and then moving through the table in a systematic fashion selecting numbers corresponding to the code numbers of the members of the population. Note that once you select a place to begin you move through in a systematic fashion, and not by skipping about from place to place. In our example, each row contains four blocks of five numbers each. Our population contains a total of 40 individuals, and we wish to select a total of five for the sample, so we will need to work with groups of two digits as we move through the random number table. Let's suppose we decided to begin in column 1, row 1, and we are going to use the last two digits in each group of numbers. From our starting point we will move down the first column to the end, and then move to the second column looking at the last two digits and moving down that column, and so on until we have selected the five cases for the sample. The table is represented again below with the last two digits underlined.

The first two digits we encounter are "0 0." These two digits would be ignored because we began our coding sequence with the number 1 (or you could think of it as "01"). The next two digits are "2 1," so the person coded as 21 would be the first case selected for the sample. Moving down the column, the next to digits are "4 0," so the person coded as 40 would be the second person selected for the sample. Moving to the top of the next column we see the digits "0 9," so the person coded as 9 would be selected for the sample. Moving down the column we see the digits "1 1," meaning that the person coded as 11 would be selected for the sample. The next two digits we see are "9 6," which will be ignored because our population only has 40 elements in it. Moving to the top of the next column we see another two digit sequence of "1 1." The person coded as 11 has already been selected for the sample, so the second instance of "1 1" will be ignored, and we will move on until we find another eligible two digit sequence that will allow us to complete the sample. The next two digits of "3 2" indicate that the person coded as 32 should be included, and that will complete our sample of $n = 5$ with the individuals coded as follows having been selected for the sample: 21, 40, 9, 11, and 32.

Many of you will have access to a method similar to the random number table that you can use for selecting a random sample. If you have a calculator that goes beyond the basics, you might find that it has a built-in function that generates a sequence of random digits for you. It is usually a second function key, and the key may be marked "rnd #," "ran #," or something similar. Not all calculators have that particular function, but if yours does you might activate that function several times and record the sequence of digits that is displayed (sometimes the display is in decimal form, but you can ignore the decimal point). Essentially, your calculator is generating random numbers that you could use to select a true random sample in place of the other methods that we have discussed above.

Stratified random sampling is very similar to simple random sampling, but it takes advantage of the fact that populations often have certain characteristics that differentiate some groups from others. These sets of characteristics are referred to as strata. Geologists often speak of rock strata that are just layers of certain types of rock-piled layer upon layer. You probably have seen evidence of rock strata when traveling on highways that pass through mountainous areas where you can see the exposed face of a hillside that has been cut to make room for the highway. Populations can be thought of as having strata

	1	2	3	4
1	6 8 3 <u>0 0</u>	3 3 2 <u>0 9</u>	5 9 0 <u>1 1</u>	2 8 6 <u>9 4</u>
2	4 8 2 <u>2 1</u>	5 9 2 <u>1 1</u>	8 8 6 <u>3 2</u>	4 2 1 <u>9 7</u>
3	8 4 7 <u>4 0</u>	2 4 7 <u>9 6</u>	7 5 4 <u>2 6</u>	0 8 5 <u>3 1</u>

Figure 2.2 A Facsimile of a Random Number Table with Selected Digits Underlined

as well, and stratified random sampling is accomplished by taking a set of random samples, one from each stratum. For example, an undergraduate college population can be thought of as a single group, undergraduate students, but there are also various subgroups or strata in the population that might be relevant to a particular piece of research. Some of the more common ways we might think of a population of undergraduates as being stratified would include: sex, classification, major, age, residential or commuter, and race just to name a few.

A stratified random sample is a little more difficult to conduct than a simple random sample because more information is required. We not only need to be able to identify the population of interest, but we must also be able to classify each member of the population with respect to the stratification characteristics that are relevant for our research. Once that is done we may begin to sample randomly from within each of the strata. Stratified random sampling may be done on a proportional or a nonproportional basis depending upon the nature of the research to be conducted. With proportional stratified random sampling the proportion of the population in each stratum is maintained in the sample. For example, if sex and classification were two stratification variables deemed important for a particular piece of research, and 20% of the population consisted of sophomore males then 20% of the sample would consist of sophomore males. With the proportional approach, the same proportions found in the population would be found in the sample.

Nonproportional stratified random sampling results in the over-sampling of some groups in the population and the under-sampling of others. Nonproportional stratified random sampling is done when the data from some groups in the sample are considered more relevant than data from others. If we were conducting research on college undergraduates with respect to career plans, or their evaluation of the quality of the academic program we would probably want to use a nonproportional stratified random sampling plan, and we would want to over-sample juniors and seniors. Juniors and seniors are closer to graduation and are more likely to have firmer career plans than are freshmen or sophomores. Similarly, juniors and seniors have had more exposure to the academic program, and we might have reason to value their input more than the input from students who have had less exposure to the program. Other types of research questions might make a proportional sampling plan, the better choice. Both methods are legitimate, and the choice should be based on the overall objectives of the research project being undertaken.

Systematic sampling can also be used as a probability type of sampling method. Typically, one will begin with a listing of the population, and then will generate the sample by selecting every "*k*th" name on the list. "*k*" is sometimes referred to as the sampling interval, and it is related to the probability of a member of the population being selected for the sample. For example, if we had a population with a total of $N = 800$ persons, and we wanted to select 10% of the population for the sample then we would need to select 1 name out of every 10 on the list. In other cases we might know how many individuals we wanted in the sample, and would use that information to calculate "*k*." If we wanted a total sample size of $n = 50$ from our list of $N = 800$ names, we would find "*k*" by dividing the population size by the desired sample size. In this case: $800/50 = 16$, which would indicate that we should select 1 name out of every 16 on the list. Systematic sampling is an acceptable alternative to simple random sampling as long as there is no bias in the construction of the listing of the population.

The probability sampling methods we have discussed so far have one thing in common: they are only feasible with relatively small populations. In most cases, it is not even remotely feasible to try to draw a simple random sample of really large populations such as that of the United States. The current population of the United States is approximately 320 million persons, and even if we wanted to forgo a sample of individuals and draw a sample of households the task would still be Herculean. You can imagine the difficulty of identifying the addresses of over 100 million households in the United States, writing them down on pieces of paper, finding a container big enough to hold them all, and then selecting the few thousand or so that are usually examined for a national sample. Given the difficulty of dealing with large populations inherent in the more traditional methods, researchers interested in taking a probability sample of large populations, such as that of the United States, usually rely on a method called cluster sampling.

Cluster sampling is based on the fact that the 115 million or so individual households in the United States are tied to larger and larger geographic areas. The process of cluster sampling begins not by selecting individual households, but by selecting geographic areas where households are located. Beginning with geographic areas has two advantages over households: first, there are fewer of them; and second, they are much easier to identify. The geographic unit most frequently chosen as a beginning point is the "county." The county is an ideal unit with which to begin because there are a manageable number of them (about 3,300 in the United States), it's easy to get a list of them all, and the census results indicate how many households are in each county. Identifying the number of households in each county is important because if the final sample is to be representative, a county's chance of being selected should be proportional to how many households it contains. For example, a county with five times the number of households as another should have five times the chances of being selected in the first stage of the sampling process. Once a certain number of counties has been selected it is likely that the process will continue with another random draw of smaller geographic units. The process continues with the selection of smaller and smaller geographic units until we arrive at a random selection of geographic areas that are small enough to allow us to work with households. In most cases the stopping point is the geographic area called a "census block."

Census blocks are just geographic areas defined by the U.S. Bureau of the Census, which are bounded by physical features. The simplest example is what we would think of as a block in a city: an area bounded by four streets. Features other than streets may serve as boundaries such as railroad tracks, a river or stream, a highway, or any number of other things as long as they are physical features that can be seen. Census blocks are geographic areas that are small enough that once a sample of them has been obtained field representatives can compile a list of the actual housing units within each one. At this stage, we now have a listing of housing units from which a true probability sample of all households in the United States may be selected.

The Sampling Distribution

Up to this point, we have discussed the process of sampling, and various methods of probability and nonprobability sampling. We are going to conclude the chapter with a discussion of the **sampling distribution**. The sampling distribution is a very important concept that will surface in several of the later chapters. Before I define the sampling distribution and discuss some of its characteristics, it might be useful to illustrate the sampling process in general to emphasize the distinction between population parameters, sample statistics, and the variability that is inherent in generalizations based on the results from sampling.

How Much Change Is in Your Pocket?

Suppose we have a rather small population of $N = 5$ individuals, and each person has a certain amount of pocket change. The five individuals are listed below with the amount of change in his or her pocket indicated below the name.

Alice	Bob	Carol	David	Eve
$0.65	$0.80	$0.45	$0.15	$0.50

Figure 2.3 A Population of Five Individuals

In Chapter 1, we saw that characteristics of populations are called parameters, and that parameters represent exact values or "true" values for the entire population. In our small population of $N = 5$ we have data for two variables, the name of the individual, and the amount of change in each person's pocket. The only parameter we can measure is the mean (or average) amount of change in the pockets of the members of the population. You also may recall from Chapter 1 that we use the Greek letter lowercase Mu (μ) to represent a population mean. We can calculate μ by simply adding the total amount of change in each person's pocket in the population, and then dividing by the total number of individuals in the population.

$$\mu = \frac{.65 + .80 + .45 + 1.5 + .50}{5} = \frac{2.55}{5} = .51$$

So, the mean amount of change in the pockets of the population is 51 cents. This population characteristic we have measured represents an exact and true value because we have examined every element of the population.

We also saw in Chapter 1 that characteristics of samples are called statistics, and that we could think of statistics as estimates of the population parameters. Below, we are going to put all five members of the population in a box, and take a simple random sample of $n = 2$. We will then calculate the mean amount of pocket change in our sample of $n = 2$, and compare it to the true value represented by the population mean μ. (You may remember from Chapter 1 that sample statistics are represented by English letters, and that a sample mean is represented by $\bar{x}$.)

We have taken a simple random sample of two individuals from the population, and we have calculated the sample mean to be 55 cents. Fifty-five cents is the exact mean for our sample comprised of Alice and Carol, who happen to be the two members of the population chosen for the sample, but .55 is not exactly equal to the population mean that we calculated earlier to be .51. Still, in the absence of any other information we could use the results of our observation of the sample of $n = 2$ to estimate what the population mean equals. We would not be exactly correct, but the results are relatively close.

Of course, Alice and Carol is not the only possible sample of size $n = 2$ that could be drawn from our population of $N = 5$. Their mean amount of pocket change was relatively close to the population mean, but some possible samples of size $n = 2$ will not be as close. There are two particular samples that will probably not be very close at all to the true population mean. One sample would be the two members of the population with the most and the second most amount of pocket change (Alice and Bob, whose mean is 72.5 cents), and the other sample would be the two members of the population with the least and the second least amount of pocket change (David and Carol whose mean is 30 cents).

So far, we have identified three possible samples of size $n = 2$ from the population. The first sample mean was reasonably close to the true population mean, but the other two were not as close and did not

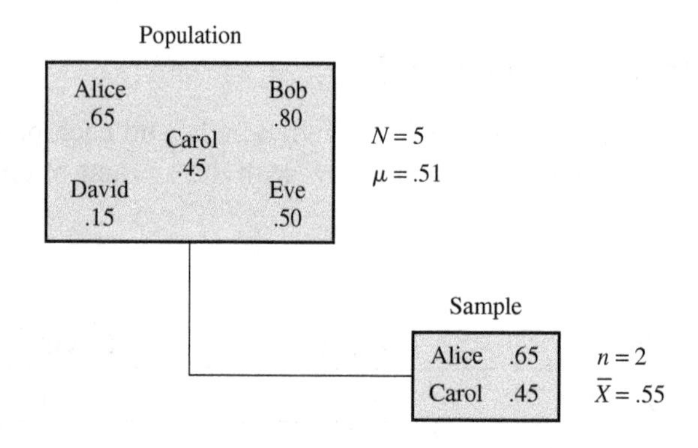

Figure 2.4 A Random Sample of Size $n = 2$ Drawn from the Population of $N = 5$

give us as good an estimate of the population parameter. What about all of the other possible samples of size $n = 2$ that could be drawn from the population. Below, I will identify all of the possible samples of size $n = 2$ that can be drawn from our population of size $N = 5$. To save space, I will abbreviate the names of the population members by using the first letter of their name.

	Alice	**Bob**	**Carol**	**David**	**Eve**
Population Member's Pocket Change	\$0.65	\$0.80	\$0.45	\$0.15	\$0.50

Sample and Mean of All Possible Samples of Size n = 2

Sample	Mean	Sample	Mean	Sample	Mean	Sample	Mean
A B	\$0.725						
A C	0.55	B C	\$0.625				
A D	0.40	B D	0.475	C D	\$0.30		
A E	0.575	B E	0.50	C E	0.47	D E	\$0.325

Figure 2.5 All Possible Samples of Size $n = 2$ Drawn from the Population of $N = 5$

As you can see, there are 10 possible samples of size $n = 2$ that can be drawn from a population of size $N = 5$. Some of the 10 samples provide us with means that are reasonably close to the true population mean of 51 cents, but others are not very close. All 10 sample means are presented below from low to high along with the true population mean of 51 cents.

$$.30\ .325\ .40\ .475\ .475\,|\,.55\ .575\ .625\ .65\ .725$$
$$\uparrow \mu = .51$$

Notice that the 10 sample means are distributed symmetrically around the true population mean of 51 cents, that is, there are five of the possible sample means that are less than the true population mean, and there are five of the sample means that are greater than the true population mean. There is one other very important feature to notice about the 10 sample means we were able to draw from the population. What do you think would happen if we were to calculate a mean value for all 10 of the sample means? Note that we are not going to compute an individual sample mean, we are going to compute the mean of all 10 means (or in other words, we are going to average all 10 of the averages). How close do you think the mean of all the sample means will come to the true population mean of 51 cents? The mean of all 10 sample means is computed below:

$$\frac{.30+.325+.40+.475+.475+.55+.575+.625+.65+.725}{10} = \frac{5.10}{10} = .51$$

The mean of all 10 possible sample means of size $n = 2$ drawn from our population of $N = 5$ is 51 cents, which is exactly equal to the population mean from which the samples were drawn. Even though some of the possible sample drawn from a population will be less than the true population mean, and some of the possible sample means drawn from a population will be greater than the true population mean, the mean of all of the possible sample means drawn from a population will always be equal to the population mean from which the samples were drawn.

What Is the Sampling Distribution?

The previous discussion has not only served to reinforce some of the concepts we discussed in Chapter 1 concerning population parameters and sample statistics, but it has also introduced you to the concept of a sampling distribution. The sampling distribution is the set of all possible samples of size "*n*" taken from a population. Normally, when we conduct research we define a population of interest, and then draw a single sample from the population for observation. The particular sample that we select comes from the sampling distribution of all samples of that particular size that theoretically could be drawn from the population.

In the actual practice of research we can never know which sample we have obtained, but we can know quite a bit about the set of samples from which it came (the sampling distribution). The example I have used here with a very small population of $N = 5$, and a small sample size of $n = 2$ is useful for illustrating the general idea of a sampling distribution. When the population is larger and the sample size is larger a very important principle in statistics begins to apply—the **Central Limit Theorem**—that gives us some very precise information regarding the nature of the sampling distribution. One very important characteristic of the sampling distribution that you have already seen is that the mean of all of the sample means will equal the population mean from which the samples were drawn. A second important characteristic (again, when the population and sample are larger than what I have demonstrated here) is that the set of samples in the sampling distribution takes on a particular shape or pattern. Not only will half of the samples be greater than the true population mean and half be less than the true population mean (as was the case above), but all of the samples will fall into the same general shape or pattern. We will see what that particular shape looks like in the next chapter, but let's look at two more examples of a sampling distribution taken from our small population of $N = 5$.

The Sampling Distribution When $n = 3$ and $n = 4$.

How many possible samples of size $n = 3$ can we draw from our same small population of $N = 5$ people? It turns out that we can draw 10 different samples of size $n = 3$ from our population. The samples and the mean amount of pocket change are listed below.

	Alice	Bob	Carol	David	Eve
Population Member's Pocket Change	$0.65	$0.80	$0.45	$0.15	$0.50

Sample and Mean of All Possible Samples of Size $n = 3$

Sample	Mean	Sample	Mean	Sample	Mean
A B C	$0.6333	B C D	$0.4667	C D E	$0.3667
A B D	0.5333	B C E	0.5833		
A B E	0.65C	B D E	0.4833		
A C D	0.4167				
A C E	0.5333				
A D E	0.4333				

Figure 2.6 All Possible Samples of Size $n = 3$ Drawn from the Population of $N = 5$

The 10 possible sample means of size $n = 3$ are presented below in order from low to high along with the true population mean of .51.

$$.3667\ .4167\ .4333\ .4667\ .4833\ |\ .5333\ \ .5333\ .5833\ .6333\ .65$$
$$\uparrow \mu = .51$$

There are several things that you should notice about the 10 possible sample means of size $n = 3$ drawn from our population. First, just as before when we saw all possible samples of size $n = 2$, the samples of size $n = 3$ are also distributed symmetrically around the true population mean of 51 cents. Five of the possible sample means are less than the true population mean, and five of the sample means that are greater than the true population mean. Second, if we were to compute the mean of all 10 sample means of size $n = 3$ we would again find that the mean of all of the samples in the sampling distribution is equal to the population mean from which they were drawn.(The sum of the 10 sample means is equal to $5.0999, and the mean is equal to .50999. The small difference between the actual mean of the sampling distribution and the population mean of .51 is due to rounding error.) Third, notice that some of the 10 sample means are closer to the true population mean than others, but the most extreme sample results (the very highest and the very lowest) are closer to the true population mean than the most extreme results were when we took all possible samples of size $n = 2$. In other words, when the sample size gets larger the results of the sample get better. Even the worst case samples become closer to the true population mean. We will see that in action one more time by selecting all possible samples of size $n = 4$.

We were able to draw 10 different samples from our population of $N = 5$ when our sample size was $n = 2$, or $n = 3$. When the sample size is $n = 4$ we will only be able to draw a total of five samples. The samples along with the mean amount of pocket change for each one are represented below.

The five possible sample means of size $n = 4$ are presented below in order from low to high along with the true population mean of .51.

$$.4375\ .475\ |\ .5125\ .525\ .60$$
$$\uparrow \mu = .51$$

	Alice	**Bob**	**Carol**	**David**	**Eve**
Population Member's Pocket Change	$0.65	$0.80	$0.45	$0.15	$0.50

Sample and Mean of All Possible Samples of Size *n* = 4

Sample	**Mean**	**Sample**	**Mean**
A B C D	$0.5125	B C D E	$0.475
A B C E	0.60		
A B D E	0.525		
A C D E	0.4375		

Figure 2.7 All Possible Samples of Size $n = 4$ Drawn from the Population of $N = 5$

We do not see a perfect symmetrical distribution of the sample means around the true population mean of .51 because there are only five possible samples of size $n = 4$ that can be drawn from the population. However, we do see that there are two sample means less than the true population mean, two sample means that are greater than the true population mean, and one that is almost equal to the true population mean. Also notice that as the sample size increased to $n = 4$, the value of all of the possible sample means becomes closer to the true population mean. Finally, note that the mean of the five possible sample means of size $n = 4$ is equal to the population mean of $\mu = .51$.

How Many Sample Means Can Be Drawn from a Given Population?

You may have noticed that even with the simple examples above it was rather tedious to try to identify every possible sample of a particular size from a given population. What if we had a larger population of maybe $N = 40$, and we wanted to know how many samples of size $n = 5$ we could draw from it (by the way, if you haven't written your guess down . . .)? In mathematics this is referred to as a combination problem: how many unique combinations of size 5 can we create from a group of 40 elements? Or sometimes the problem is stated as "how many combinations of 40 things can we create 5 at a time?"

As you might guess, there is a formula to compute the solution to combination problems. The general formula to compute the number of combinations of n things selected r at a time is as follows:

$$\mathrm{C}\binom{n}{r} = \frac{n!}{r!*(n-r)!}$$

where n represents the size of the pool and r represents the number of items selected each time.

Note that I am following the conventional notation by using a lowercase "*n*" as the size of the population of elements, and a lower case "*r*" as the number to be selected. To be consistent with our population and sample examples, "*n*" represents the population size, and "*r*" represents the sample size.

The exclamation point indicates that we must calculate a factorial product. You may recall the use of factorials from previous mathematics courses. The use of factorials is rather simple, for example, 3! is equal to 3*2*1 or 6. The value of 4! is equal to 4*3*2*1 or 24. Some of you may have calculators that will compute the product of a factorial. The key may be marked as "X!," or perhaps "N!," and a few of you may have calculators marked "nCr," which will solve the entire combination problem. If you have a combination problem key on your calculator you probably have a similar key marked "nPr." The "nPr" key is used to compute the number of permutations of "*n*" things taken "*r*" at a time. A permutation differs from a combination in that the order of elements does not matter in a combination. For example, the outcomes:

A B C, A C B, B A C, B C A, C A B, C B A

All represent a single combination (the letters "A" "B" and "C" in various orders), but they represent six different permutations. When computing the number of possible samples that may be drawn from a population, we are interested in the number of possible combinations. The sample "Alice" and "Bob" is the same as the sample of "Bob" and "Alice"; the order does not matter.

In our first sampling example we wanted the number of combinations of $n = 5$ things taken $r = 2$ at a time, and we were able to list a total of 10 outcomes. By formula we would solve:

$$\mathrm{C}\binom{5}{2} = \frac{5!}{2!(5-2)!}$$

$$C\binom{5}{2} = \frac{5!}{2!*(5-2)!}$$

$$= \frac{5*4*3*2*1}{2*1 \;*\; 3*2*1} = \frac{120}{12} = 10$$

In the second example, we wanted the number of combinations of $n = 5$ things taken $r = 3$ at a time, and we again were able to list a total of 10 outcomes. By formula we would solve:

$$C\binom{5}{3} = \frac{5!}{3!*(5-3)!}$$

$$= \frac{5*4*3*2*1}{3*2*1 \;*\; 2*1} = \frac{120}{12} = 10$$

And finally in our last example, we wanted the number of combinations of $n = 5$ things taken $r = 4$ at a time, and we were able to list a total of 5 outcomes. By formula we would solve:

$$C\binom{5}{4} = \frac{5!}{4!*(5-4)!}$$

$$= \frac{5*4*3*2*1}{4*3*2*1 \;*\; 1} = \frac{120}{24} = 5$$

What about the classroom example? How many possible samples of size $n = 5$ can we draw from a population of $N = 40$? By formula we would solve:

$$C\binom{40}{5} = \frac{40!}{5!*(40-5)!}$$

or

$$= \frac{40!}{5*4*3*2*1 \;*\; 35!}$$

At this point, we may simplify matters a little. The term in the parenthesis gave us 35! in the denominator. By beginning to expand the numerator we will also have 35! at some point, and we can then cancel those two terms out.

$$= \frac{40*39*38*37*36*\cancel{35!}}{5*4*3*2*1*\cancel{35!}}$$

The result becomes

$$= \frac{78,960,960}{120} = 658,008$$

From a population as small as $N = 40$, there are 658,008 different samples of size $n = 5$ can be drawn! How close was your guess? Most of the guesses from students in my classes are not very close. Frequent guesses are 8 (40 divided by 5); 25 (the square of 5); 200 (40 times 5); and, 1,600 (the square of 40).

Remember that when we conduct research we normally only draw a single sample from the population. As you have seen, in any given sampling distribution some of the samples are better than others (i.e., some sample means are closer to the true population mean than others). When we draw our single sample do we get a good one, or do we get a bad one? The fact is that we can never be sure, but we do know a couple of things. First, we know that if we took all possible samples and calculated the mean of the entire sample means the result would equal the population mean from which the samples were drawn. Second, we know that as the population and sample get larger the sampling distribution will take on a particular shape. These two facts about the sampling distribution are very important, and in many ways serve as the foundation for many of the statistical tests that we will examine in later chapters. There is one final fact about the sampling distribution that I will point out now, and will demonstrate in Chapter 6. You have seen that some of the samples in the sampling distribution are better than others, but once the sample size gets to a particular level even the worst possible sample you could draw from the population is actually going to be a pretty good estimate of the true population mean. Generally speaking, once a sample size reaches the range of $n = 1200$ to $n = 1500$ (assuming that the sample is drawn in a probability fashion) even the "bad" samples in the sampling distribution are actually going to be pretty good.

Calculate the Odds of Winning the Lottery!

One additional application of combination problems is in calculating certain types of lottery results. Many states have adopted lottery games as a method of raising revenue as an alternative to taxes. There are a variety of lottery games, but most states have a "lotto" type game in which players pick a group of six numbers from a larger pool. If the six numbers that you select match the six numbers drawn at random from the pool you will win the grand prize. Your odds of winning are actually based on a combination problem. You can think of the large pool of numbers as the "population" and the six numbers that you select as a particular "sample" from that population. You have chosen one particular outcome from the population, so your odds of winning depend on the total number of possible outcomes of six numbers that can be drawn from the pool of numbers. States vary in the size of the pool of numbers from which you will choose your six numbers, but let's assume that a typical pool will consist of 45 numbers. At that point your odds of winning can be calculated by solving the following combination problem.

$$C\binom{45}{6} = \frac{45!}{6!*(45-6)!} \text{ or } \frac{45!}{6!*\ 39!}$$

Again at this point we may simplify matters a little. The term in the parenthesis gave us 39! in the denominator. By beginning to expand the numerator we will also have 39! at some point, and we can then cancel those two terms out.

$$= \frac{45*44*43*42*41*40*\cancel{39!}}{6*5*4*3*2*1*\cancel{39!}}$$

The result becomes:

$$= \frac{5{,}864{,}443{,}200}{720} = 8{,}145{,}060$$

So there are 8,145,060 possible combinations of six numbers selected from a pool of 45. Your odds of matching all six numbers would be 1 out of 8,145,060 or a probability of .000000123. Those are very poor odds, and you actually have a greater probability of being struck by lightning than winning the typical state lottery. What keeps people playing is the realization that no matter how great the odds are someone is going to win.

Computer Applications

Figure 2.3 presents a population of five individuals consisting of their name and amount of pocket change.

1. Create a data set consisting of the name of each individual in column one, and the amount of pocket change in column two.
2. Use the random sampling feature to select two successive samples of size $n = 3$ from the population of all individuals.
3. Compare your two selected samples with the list of all possible samples of size $n = 3$ presented in figure 2.6. Which two of the samples did Statistical Package for the Social Sciences (SPSS) select for you?

How to do it

Review Appendix D as necessary to create the data file. Remember that to input the name data you must first define the variable for column one as a string variable (use the "data define" option under the "Data" procedure). Obtain your random samples by using the "select cases" option under the "Data" procedure. Be sure you do not delete the unselected cases, or you will have to re-input the data.

Summary of Key Points

Most research in the behavioral sciences does not involve the direct examination of the entire population of interest. Instead, a smaller subgroup of the population is selected for observation, and often the results observed for the sample are generalized back to the population from which it was selected. Sampling is the process of selecting from the population the small subgroup that will actually be observed. Sampling methods include both probability and nonprobability types. A probability sampling method should be selected if the results from the sample are to be generalized back to the population.

Population (or Population of Interest)—The total group to which a research project is directed.
Sampling Population—The population from which a sample will actually be drawn.
Sample—The subgroup for observation which is selected from the population of interest.
Probability Sampling—A group of sampling methods in which all members of the population have a chance of being selected for the sample. Major methods of probability sampling include simple random sampling, stratified random sampling, systematic sampling, and cluster sampling.
Nonprobability Sampling—A group of sampling methods in which the sample is not necessarily representative of the population from which it is drawn. Major methods of nonprobability sampling include accidental sampling, captive sampling, quota sampling, spatial sampling, systematic sampling, judgmental sampling, and snowball sampling.
Sampling Distribution—The set of all possible samples of size "n" taken from a particular population.
Sampling—The process of selecting from a population the sample or subgroup from which we will actually collect data for analysis.
Unit of analysis—The unit of analysis in research refers to the person or thing from which we will actually collect data. In behavioral science research the individual is frequently the unit of analysis.

Sampling unit—The sampling unit refers to the units that are actually selected from the population for inclusion in the sample. For example the sampling unit might be the household and the sample would consist of all members of each household that was selected.

Accidental sampling—A nonprobability sampling plan in which the sample consists of anyone you can find to become a member of your sample until the desired sample size is obtained.

Captive sampling—A nonprobability sampling plan in which any group over which the researcher has temporary control becomes the sample.

Quota sampling—A nonprobability sampling plan similar to accidental sampling except that the researcher is looking for individuals with particular characteristics among those who are selected for the sample.

Spatial sampling—A nonprobability sampling plan in which members of the research team spread out at roughly equal intervals at the perimeter of the crowd and then a specified number of steps into the crowd, stop, and then attempt to interview the first person to their right. If that person refuses, they will continue to ask individuals in the immediate vicinity until an interview is obtained. The process is repeated alternating attempted interviews to the left and right until the desired sample size is obtained.

Systematic sampling—A nonprobability sampling plan often used in situations where there is a great deal of pedestrian traffic such as a retail mall. Interviewers are positioned in high traffic areas, and are instructed to allow a certain number of individuals to pass and then attempt to interview an individual for the sample. The process is repeated until the desired sample size is obtained.

Judgmental sampling—A nonprobability sampling plan consisting of a hand-picked sample comprised of members of a population known to be particularly knowledgeable with respect to the issues under investigation. The sample is selected based on either the researcher's own expertise as to whom in the population is particularly knowledgeable, or on the basis of suggestions from a local informant.

Snowball sampling—A nonprobabiltiy sampling plan that takes its name from the analogy of what happens when a snowball becomes larger and larger after it is sent rolling down a hill. With snowball sampling the researcher begins with a few knowledgeable individuals from a population, and those interviewed for suggestions as to whom to interview next.

Simple random sampling—A probability sampling plan in which members of the sample are obtained on a random basis (i.e. without any bias) from a complete list of the population of interest.

Random number table.—A random number table is a series of digits selected at random, usually by a computer, and printed out on as many pages as necessary. It is one method of selecting a simple random sample.

Stratified random sampling—A probability sampling method very similar to simple random sampling, but it takes advantage of the fact that populations often have certain characteristics that differentiate some groups from others. These sets of characteristics are referred to as strata and in a stratified random sample elements for the sample are selected from each stratum.

Systematic sampling—A probability type of sampling method when the list from which the sample is drawn contains all of the members of the population. Typically, one will begin with a listing of the population, and then will generate the sample by selecting every "kth" name on the list. "k" is sometimes referred to as the sampling interval, and it is related to the probability of a member of the population being selected for the sample i.e. if the sample should equal ten percent of the population then every tenth name will be selected.

Central Limit Theorem—A theorem stating that for a population with a mean μ and a variance σ^2 the sampling distribution (the set of all possible samples of sample size n taken from the population) becomes normal as the sample size increases with a mean (of the sampling distribution) equal to the mean of the population from which the samples were drawn.

Questions and Problems for Review

1. How does a population parameter differ from a sample statistic?

2. Which sampling method is most appropriate if a researcher wants a probability sample taken from a population of college faculty, and the researcher wants to be sure that the sample will contain both male and female faculty members?

3. If an academic department at a college contains a total of 10 faculty members, how many different samples of size $n = 4$ could be drawn from the population?

4. Given the identity and age of the following small population of $N = 5$ individuals, create the sampling distribution of samples of size $n = 2$ (identify every possible sample of size 2). Adam T. (15), Beth K. (48), Kyle B. (10), Sara B. (10), Sherry R. (62)

5. Compute the mean of each of the samples from the sampling distribution above, and verify that the mean of all of the samples is equal to the population mean from which they were drawn.

6. A professor is interested in measuring student attitudes regarding the quality of the food service at your college, and uses a sample of her students in a marketing class. What type of sampling method is being used? Can the results of the research legitimately be generalized to the entire student body? Why or why not?

7. What type of probability sampling plan is most often used for the study of very large populations? Is simple random sampling a logical alternative? Why or why not?

8. Identify the nonprobability sampling methods that are equivalent to the following probability methods:
   a. simple random sampling
   b. stratified random sampling
   c. cluster sampling
   d. systematic sampling

9. How does judgmental sampling differ from snowball sampling?

10. Which two types of nonprobability sampling plans are often used to sample large crowds in public places? In what type of situation is each the more appropriate choice?

Part II Descriptive Statistics

Part II Descriptive Statistics consists of Chapters 3–5. In this part of the text, you will begin to see how statisticians reduce a set of data to a form that is more easily interpreted. Chapter 3 includes a discussion of frequency distributions, and several major types of graphing techniques. Chapter 4 includes measures of central tendency such as the mode, the median, and the mean. Part II ends with Chapter 5 on measures of variation dealing with several statistical techniques that indicate how much difference or "variation" there is in a set of scores.

CHAPTER 3

Data Reduction: Frequency Distributions and the Graphic Display of Data

Key Concepts

Frequency Distributions	Percentage	Graphic Display of Data
Upper and Lower Limits	Proportions	Normal Distribution
Interval Midpoint	Percentiles	Skewed Distributions

Introduction

The first two chapters have served to introduce you to a variety of basic concepts that are fundamental to the study of statistics, and to clarify the distinction between a population of interest in a research project and the sample which is the group from which data are obtained. Once we progress in the research process to the point that we have collected data from the sample, we are faced with the task of making sense of it all. The focus in this chapter is on **data reduction**. Data reduction refers to several different procedures for reducing or organizing data into a form that will aid in interpretation. One major method of data reduction is to create a **frequency distribution**. A second major method is to present the data in some type of graphic display. We will begin with a discussion of frequency distributions for a single variable, and then continue with a discussion of major types of graphic displays.

The Construction of Frequency Distributions

Frequency distributions are used to organize a collection of raw data into a form that makes the data easier to interpret. We will examine two types of frequency distributions that will differ by the size of the interval we use to group the variable being presented. Frequency distributions with an interval size "i" = 1 are used to report the number of observations in a set of data at exact values of the variable that has been measured; we will refer to frequency distributions with an interval size of $i = 1$ as "simple frequency distributions." Frequency distributions with an interval size greater than one, $i > 1$, report the number of observations in a set of data within a given range of values of the variable that has been measured; frequency distributions with an interval size of $i > 1$ will be referred to as "frequency distributions." A set of raw data is presented below for a sample of $n = 25$ individuals. The variable "age" has been measured, and we will present the data in a simple frequency distribution ($i = 1$), and then in a frequency distribution with an interval size of $i = 5$. After you have seen what each type of frequency distribution looks like, we will discuss the procedure for constructing each type.

Raw Data on Age for a Sample of $n = 25$

21 38 30 18 24 21 39 28 25 27
39 18 19 37 34 25 19 24 21 37
19 28 37 34 28

Age	f	Cf
39	2	25
38	1	23
37	3	22
34	2	19
30	1	17
28	3	16
27	1	13
25	2	12
24	2	10
21	3	8
19	3	5
18	2	2
	$\Sigma f = 25$	

Figure 3.1 Presentation of the Variable Age in a Simple Frequency Distribution ($i = 1$)

The presentation of the raw data in Figure 3.1 can give you some idea of why we might want to create a frequency distribution. We have data on the ages of $n = 25$ individuals, but it is difficult to get a sense of what the data indicate when they are presented in an unorganized fashion. For example, it takes some time to find the oldest age in the data, the youngest age in the data, and to have a good sense of how the data are distributed between those two extremes. In Figure 3.1, we have presented the same data in a simple frequency distribution with an interval size of $i = 1$. Most frequency distributions are generally comprised the three columns seen here, but other columns presenting information such as the relative percentage of observations in each interval, or the cumulative percentage of observations at each interval may also be included. Let's examine the purpose of each of the three columns presented above in turn.

The first column contains the values of the variable that has been measured; in this case, the variable is age. Frequency distributions where the interval size $i = 1$ report exactly each of the observed values of the variable that has been measured beginning with the smallest value at the bottom of the frequency distribution and ending with the highest value at the top of the frequency distribution.

The second column is labeled "*f*," and contains the number of observations or frequencies associated with each of the observed values of the variable. The uppercase sigma at the bottom of the column indicates the sum of the frequencies, which in this case is 25. The sum of the frequencies is always the same as the sample size "*n*". We began with $n = 25$ observations in the raw data, and we still have a total of 25 observations in the simple frequency distribution. The difference is that we have ordered all of the observed values from low to high in the first column, and we have saved space by indicating the number of observations associated with each value in the frequency column.

The final column is labeled "Cf," and it contains the cumulative frequency information. The cumulative frequency column provides us with a running total of the number of frequencies at any given point in the distribution plus all of the frequencies below that point. For example, the value of 17 in the cumulative frequency column associated with the observed age of 30 indicates that there are a total of 17 individuals at age 30 or younger. The value of 10 in the cumulative frequency column associated with the age of 24 indicates that there are a total of 10 individuals at age 24 or younger. Note that we do not provide a sum for the cumulative frequency column.

The cumulative frequency column is constructed by transferring the frequency value associated with the smallest interval directly across to the cumulative frequency column. (In this example, the smallest interval is the age 18, and the frequency value is "2".) The next value in the cumulative frequency column as you move up the frequency distribution is obtained by adding the value in the frequency column to the value in the cumulative frequency column just below it. For example, the appropriate value in the cumulative frequency column associated with the age = 19 entry is obtained by adding the value in the frequency column associated with age = 19 (in this case, there are three frequencies.) to the value in the cumulative frequency column just below the age = 19 entry. (In this case, the value is 2.) The three frequencies in the age = 19 category added to the two cumulative frequencies in the age category just below the age = 19 category equal a total of five cumulative frequencies. The remainder of the entries in the cumulative frequency column is obtained in a similar fashion. The final entry at the top of the frequency distribution (the age = 39 category) should always equal the sum of the frequencies.

What are the Advantages of a Simple Frequency Distribution?

If you compare the raw data with the same data presented in a simple frequency format in Figure 3.1, you should see some major differences. With just a quick examination of the data in the simple frequency distribution, you can see that the youngest age is 18, the oldest age is 39, and all of the observed values of age in between the two extremes. You can also see that there are a total of 25 observations, and how those observations are allocated among the various ages of the sample. One final advantage of a simple frequency distribution is that we are able to achieve all of this convenience without any loss of precision.

We began with 25 observations when the raw data were presented; we still have 25 observations in the simple frequency distribution, and most important, we know the exact value of the 25 observations. Unfortunately, this level of precision is not maintained when we move to a frequency distribution where the interval size is larger than 1, such as the case in Figure 3.2.

Frequency Distributions Where $i > 1$

Frequency distributions with an interval size larger than $i = 1$ are used when the range of observed values for a variable in the sample becomes large. Notice that when we moved from the raw data to the simple frequency distribution, we were able to list all of the observed values by using a total of 12 intervals. (There were 12 age values that we had to identify to allocate all 25 cases in our sample.) The effectiveness of a simple frequency distribution begins to diminish as the number of observed values becomes larger. At some point, it becomes more effective to use larger intervals to group the observed values of the variable so that the total number of intervals required to represent the entire distribution stays at a manageable number. There are no hard and fast rules as to the number of intervals one should have in a frequency distribution, but a maximum of somewhere between 10 and 15 intervals usually works quite well.

Age	f	Cf
35–39	6	25
30–34	3	19
25–29	6	16
20–24	5	10
15–19	5	5
	$\Sigma f = 25$	

Figure 3.2 Presentation of the Variable Age in a Frequency Distribution ($i = 5$)

Figure 3.2 presents the same raw data that we began with in a frequency distribution with intervals of size $i = 5$. We still account for all $n = 25$ individuals in the sample, but instead of reporting the number of frequencies at exact ages, we now report the number of frequencies associated with 5-year age intervals. The presentation of the data in the frequency distribution is otherwise identical to that of the simple frequency distribution. The first column contains the 5-year age intervals for the variable we have measured. The youngest age interval is at the bottom of the distribution, and the highest age interval is at the top of the distribution. The second column contains the frequencies associated with each age interval, and the sum of the frequencies is presented at the bottom of the column. The third column contains the cumulative frequency information, which is again a running total of frequencies as we move up the distribution.

Construction of a frequency distribution with an interval size larger than $i = 1$ is slightly more involved than the simple frequency distribution. The following guidelines can serve as a model on the construction of a frequency distribution.

Guidelines for the Construction of a Frequency Distribution

1. Identify the smallest and largest observed values in the distribution, and compute the difference between the two scores by subtracting the smaller score from the larger score.

2. Determine either the number of desired intervals you will use in the frequency distribution (i.e., the number of categories you want to have), or the size of the desired intervals (i.e., how large each category or interval will be). The choice is the result of a value judgment. You may decide that you want the frequency distribution to have 10 intervals, or you may decide that you want each interval to be of size $i = 4$, or $i = 5$, or so on.
3. If you know how large you want each interval to be, you are ready to move to step 4. If you decide you want a certain number of intervals in the frequency distribution, you will need to calculate the approximate size of each interval by dividing the difference between the largest and smallest scores (obtained in step 1) by the number of desired intervals. For example, if the difference between the largest and smallest scores is 85, and you know that you want a frequency distribution with 10 intervals, then the approximate size of each interval would be 8.5.
4. Select a final interval size that is of a logical size. To some extent, "logical size" is another value judgment, but generally logical-sized intervals consist of the single digits through 5, and then multiples of 5, 10, 25, 50, 100, and so on. You can certainly construct a frequency distribution with multiples of other sizes, but you will find that intervals, for example, of size 17, 28, or 136 will be much more difficult to work with. In the example above, the approximate interval of 8.5 would be rounded to either 5, or 10.
5. Construct the first interval (the one at the bottom of the distribution) so that it meets the following two criteria. The first interval should contain the smallest observed value in the data, and the interval should begin with a multiple of the interval size. By beginning the first interval with a multiple of the interval size, you will find that it is much easier to construct the remaining intervals.
6. Construct the remaining intervals using the final interval size until you have the interval that will contain the largest observed value in the data.

An Example of Constructing a Frequency Distribution

The following example will progress through the six guidelines for the construction of a frequency distribution. The raw data below will serve as the basis for the example.

Raw Data on Income (in thousands) for a Sample of $n = 40$

80	45	30	22	76	49	26	54	65	18
16	34	78	83	32	43	46	29	22	83
57	62	38	72	20	71	55	41	26	30
19	33	27	51	46	72	32	31	28	60

1. The smallest and largest observed values in the data are 16 and 83, with a difference of 67.
2. I have arbitrarily decided to present the data in eight intervals.
3. The approximate interval size will be the difference between the smallest and largest values (67) divided by the number of desired intervals (8); $67 \div 8 = 8.375$.
4. The approximate interval size of 8.375 will be rounded to a logical interval size of 10. (We could also have used 5, but that interval size would result in more intervals than I would like.)
5. The first interval is constructed so that it contains the smallest score, and begins with a multiple of the interval size. The smallest value in the raw data is 16, so our first interval should contain the value 16, and begin with a number that is evenly divisible by our interval size of 10. Therefore, the starting point of the interval should be 10. Take care in using the proper ending point for the interval. We want the intervals to be of size $i = 10$, and we know that it should begin with the value 10. The proper interval is not 10–20, as many might think. If you write out the numbers from 10 through 20 you will find that there are 11 numbers, which is equivalent to an interval size of $i = 11$. The proper beginning and ending points of the first interval are 10–19.

6. Beginning the first interval with a multiple of the interval size makes the construction of the remaining intervals very easy. The second interval will begin with 20, the third interval will begin with 30, and so on through to the last interval that will be 80–89. We know to stop at that point because the highest observed score in the data is 83.

The final frequency distribution will take the following form (Figure 3.3).

Income	f	Cf
$80–89	3	40
70–79	5	37
60–69	3	32
50–59	4	29
40–49	6	25
30–39	8	19
20–29	8	11
10–19	3	3
	$\Sigma f = 40$	

Figure 3.3 Presentation of the Variable Income (in thousands) in a Frequency Distribution ($i = 10$)

The advantages in interpretation of the data by presentation in the frequency distribution can be seen by comparison with the presentation of the data in the disorganized raw form. It is much easier to see the values over which the data range in the frequency distribution, for example, the highest and lowest income categories can be found at a glance. It is also much easier to see how the data are distributed among the various income intervals. For example, almost half of all the incomes are 39 thousand dollars or less. (Note the cumulative frequency value of 19 in the Cf column associated with the income interval 30–39.)

The one disadvantage of using a frequency distribution where the interval size is larger than one is that we have lost some precision in the data. With the raw data, we know that there are three individuals with incomes in the 60 thousand dollar range, and we know that those incomes are exactly: 60, 62, and 65 thousand. Once the data have been placed in the frequency distribution, we only know that there are three individuals whose incomes are somewhere between 60 and 69 thousand, but we no longer know exactly what they are. There will be times when we want both the convenience of using a frequency distribution, and the ability to make calculations on the data that require an exact value for the observations. In those cases, we will be making an assumption as to the exact value associated with the frequencies in a particular interval based on some other characteristics of a frequency distribution that will be examined later in the chapter.

Frequency Distributions May Be Created with Any Type of Data

Our previous examples have created frequency distributions with ratio level data, but frequency distributions may be created with data at any level of measurement including those of the interval, ordinal, and nominal levels. A frequency distribution with interval level data would appear identical to that of our previous examples with ratio level data. A frequency distribution with ordinal level data would be constructed in much the same way as one with interval or ratio level data. The smallest category would appear at the bottom of the frequency distribution, the intermediate categories would appear in order increasing by size as you move up the distribution, and the largest category would appear at the top of the distribution. The number of observations at each category of the measured variable would be placed in the frequency column with a sum at the bottom of the column. The cumulative frequency column would provide a running total by category as you move up the distribution. A frequency distribution with ordinal level data might appear as follows (Figure 3.4).

Nominal level data may also be presented in a frequency distribution, but the procedure varies slightly from what is used with higher order data. Nominal level data consist only of named categories, and there is no underlying dimension of measurement as you move from one category to another. Consequently, the order of presentation of the various categories or intervals is completely arbitrary; there is no logical top or bottom of the distribution. Also, the cumulative frequency column should not be included because without an underlying dimension of measurement the concept of accumulating frequencies as we move up the distribution does not apply. Consider the example below.

The frequency distribution in Figure 3.5 with nominal level data lists various categories of color preference, and the number of individuals who prefer each category. A cumulative frequency column would not be logical in this case because the order in which the colors are listed does not reflect any underlying level of measurement. To say that there are a total of 55 individuals who prefer "red or below" in the distribution would not make any sense. The color presented below "red" could easily be any of the other three colors in the distribution; in this case, it just happens to be "orange." In our example with ordinal level data in Figure 3.4, we can have a logical sense of cumulative frequency. There is a definite logic in progressing from one class to another in the presentation of the data, and the statement that "there are a total of 105 students at the junior level or less," has definite meaning.

Class	f	**Cf**
Senior	30	135
Junior	45	105
Sophomore	25	60
Freshman	35	35
	$\Sigma f = 135$	

Figure 3.4 A Frequency Distribution with Ordinal Level Data

Color Preference	f
Green	19
Blue	16
Red	21
Orange	34
	$\Sigma f = 90$

Figure 3.5 A Frequency Distribution with Nominal Level Data

Midpoints and Limits in a Frequency Distribution

We saw earlier that moving from raw data to a frequency distribution with an interval size larger than $i = 1$ resulted in some loss of precision in the data. There are times when we wish to have the convenience of a frequency distribution, but still want to be able to perform calculations with the data requiring greater precision than is possible with intervals larger than $i = 1$. The solution to this dilemma is to make certain assumptions about the data, which we are able to do in frequency distributions presenting interval or ratio level data. Frequency distributions of this type have some important characteristics called **interval midpoints**, **upper limits**, and **lower limits** for each interval. We will examine these characteristics below as we develop a frequency distribution with intervals equal to $i = 5$ for the raw data presented below.

Raw Data on Hours of Television Watched per Week by Preschoolers for a Sample of $n = 30$

62	53	94	52	82	68	57	69	60	54
78	59	84	82	59	55	61	76	62	57
59	88	72	86	63	76	75	80	92	64

The data on hours of television watched per week by preschoolers for a sample of $n = 30$, children will be presented in a frequency distribution with an interval of size $i = 5$. We will follow the same general guidelines presented in Figure 3.3 for the construction of a frequency distribution, but the task will be made easier because we have arbitrarily decided on the desired interval size of $i = 5$. We begin by identifying the smallest and largest scores that in this case are 52 and 94. Knowing that we want to construct intervals of size $i = 5$, we are ready to construct the first interval such that it contains the smallest score and begins with a multiple of the interval size. The first interval should, therefore, be "50–54." Even though the smallest score is "52," we want to begin the first interval with a number that is evenly divisible by our interval size of $i = 5$. Doing so makes the construction of the remaining intervals much easier. By identifying the highest observed score, "94," we also know the last interval should be "90–94." At this point, we are ready to construct the remaining intervals, account for the frequencies in each interval, and to construct the cumulative frequency column all of which are presented in Figure 3.6.

The intervals in a frequency distribution based on interval or ratio level data have interval midpoints, upper limits, and lower limits. An interval's midpoint is simply the numerical value in the center of the

Television Viewed	f	Cf
90–94	2	30
85–89	2	28
80–84	4	26
75–79	4	22
70–74	1	18
65–69	2	17
60–64	6	15
55–59	6	9
50–54	3	3
	$\Sigma f = 30$	

Figure 3.6 Presentation of the Variable Television Viewed (in hours) in a Frequency Distribution ($i = 5$)

interval. If we were to take any of the intervals in the frequency distribution in Figure 3.6 and write out the entire interval, the result would appear as follows:

60–64 = 60 61 62 63 64
↑

As you can see, the interval midpoint (or the point in the middle) is the value "62." An easy way to find the midpoint of any interval is to add the beginning and ending points that you see for the interval, and to divide the result by "2". In this example, we would add 60 + 64 to get a total of 124, and that sum divided by 2 results in the midpoint of "62". The midpoints of the other intervals may be obtained in a similar fashion.

Interval midpoints are not always whole numbers, and any even-sized interval will have a midpoint that is a fractional number. The intervals in Figure 3.3 are of size $i = 10$, and we could find the midpoints in the same way as we have above: we could write out the interval and find the point in the middle, or we could add the beginning and ending points and divide the sum by "2". For example, the interval 40–49 can be represented as follows:

40–49 = 40 41 42 43 44 45 46 47 48 49
↑

The midpoint of the interval is the value "44.5," which we could also obtain by 40 + 49 = 89, and 89 ÷ 2 = 44.5. When you compute the midpoints for the intervals in a frequency distribution, you will find that the midpoints will increase or decrease by the same amount as the interval size. For example, if the interval size is $i = 10$, then the midpoints will increase or decrease by "10," such as: 44.5, 54.5, 64.5, and so on.

Each of these intervals also has an upper limit and a lower limit. The interval limits that you see, such as 40–49, or 50–54 are sometimes referred to as the "apparent limits," because they represent the limits where the intervals appear to begin and appear to end. However, each of the intervals has an upper limit and a lower limit, both of which are sometimes referred to as the real limits because they represent the points where the intervals of the frequency distribution really begin and end. The upper and lower limits are the points midway between where one interval appears to end and the next appears to begin. For example, given the two intervals "50–54," and "55–59," one interval appears to end at "54," and the next appears to begin at "55" The upper limit of the 50–54 interval would be midway between the numbers "54," and "55," so the upper limit is "54.5." The same value, "54.5," would also be the lower limit of the "55–59" interval. Intervals at the bottom and the top of the distribution are treated as if the intervals continued on in the same pattern. For example, the upper limit of the interval "90–94" in the frequency distribution in Figure 3.6 would be "94.5" even though there is no interval above it. Similarly, the interval "50–54" would have a lower limit of "49.5" even though there is no interval below it. The intervals from the frequency distribution in Figure 3.6 are presented below in Figure 3.7 along with their lower limits, upper limits, and midpoints.

Upper and lower limits have several uses, the most obvious of which is to guide the placement of observed values of a variable that fall between two adjacent intervals. For example, an observed value of "59.2," should be placed in the interval "55–59," whereas an observed value of "59.7," should be placed in the interval "60–64." You might notice that there is some overlap between the upper and lower limits of the intervals, in fact, the upper limit of one interval is equal to the lower limit of the adjacent higher interval. The overlap between intervals can sometimes lead to confusion when a variable is observed to have a value equal to an overlapping value. Which is the proper interval for an observed value of "74.5" that represents the upper limit of the "70–74" interval, and the lower limit of the "75–79" interval? Actually, an observed value of "74.5" can properly be placed in either of the two intervals of which it is a part. What is important in a case like this is that you are consistent in the way you place similar overlapping values. Consistency is usually achieved by adopting a decision rule to guide the placement of overlapping values. Any number of

Lower Limit	Interval	Upper Limit	Midpoint
89.5	90–94	94.5	92
84.5	85–89	89.5	87
79.5	80–84	84.5	82
74.5	75–79	79.5	77
69.5	70–74	74.5	72
64.5	65–69	69.5	67
59.5	60–64	64.5	62
54.5	55–59	59.5	57
49.5	50–54	54.5	52

Figure 3.7 Lower Limits, Upper Limits, and Midpoints of the Intervals from Figure 3.6

decision rules are possible, for example, always round up to the higher interval, always round down to the lower interval, always round to the even number, always round to the odd number, take turns, and so on. What is important is that your decision rule provides a consistent policy that does not introduce any bias in the data.

One final point on the construction of intervals is the use of an open interval. All of the examples we have seen up to this point have involved closed intervals, that is, each interval had a definite beginning point and a definite ending point. An **open interval** is used when a set of data has one or two extreme observed values at either end, but the remaining observed values fall into a more compact range. An open interval in the form of "95+," or "<50" can be used in the frequency distribution to avoid a series of closed intervals without any observed values. For example, if the raw data for Figure 3.6 had an additional observation of "32," it would require the addition of four more closed intervals, three of which would have zero frequencies, to accommodate the one extreme observation of "32." A useful alternative would be to use an open interval at the bottom of the distribution in the form of "<50," with one frequency, and we could avoid the insertion of several closed intervals without any frequencies.

Proportions, Percentiles, Deciles, and Quartiles

Whether data are presented in a raw form or in a frequency distribution, we are often interested in describing how the data are distributed around certain points. We have seen the need to identify the smallest and largest scores in a distribution as a first step in constructing a frequency distribution, but there are other points within a distribution that provide useful information. For example, we may want to know what score or value in a distribution has half of all the other scores below it, or we may want to know what score marks the highest 10% of the distribution. We usually refer to these groupings of scores as percentiles, deciles, or quartiles. We will examine each of these concepts after briefly mentioning the difference between a proportion and a percentage.

We saw earlier that the cumulative frequency column provided the number of scores below a given interval in the frequency distribution. Often we will want to express that number of scores at or below a given point in the distribution as the percentage of all scores, or as the proportion of all scores. The computation of both percentages and proportions should be well familiar to you, but we will review them briefly as a foundation for the upcoming discussion of percentiles.

Percentiles and proportions are very similar concepts, and actually only differ by a matter of scale. Percentages range from zero to one hundred, and proportions range from zero to one. A proportion may be converted to a percentage by multiplying the value by 100, and a percentage may be converted to a proportion by dividing the value by 100. The proportion .75 is equivalent to the percentage 75% (Figure 3.8).

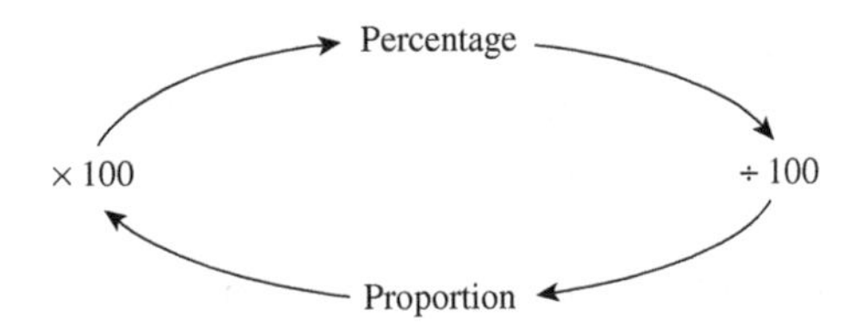

Figure 3.8 Percentages and Proportions

Television Viewed	*f*	Cf	Proportion at or below Interval
90–94	2	30	30/30 = 1.00
85–89	2	28	28/30 = .93
80–84	4	26	26/30 = .87
75–79	4	22	22/30 = .73
70–74	1	18	18/30 = .60
65–69	2	17	17/30 = .57
60–64	6	15	15/30 = .50
55–59	6	9	9/30 = .30
50–54	3	3	3/30 = .10
	$\Sigma f = 30$		

Figure 3.9 Computation of Proportion of Total Frequencies below each Interval

We will use the data from the frequency distribution originally presented in Figure 3.6 to compute the proportion of all scores at or below each of the intervals in the frequency distribution. For example, the cumulative frequency column indicates that there are a total of 18 scores with a weekly television viewing time of "70–74" hours or less. The proportion of scores with a weekly television viewing time of "70–74" hours or less may be found by dividing the number at that point or lower, a total of 18, by the total number of individuals in the sample ($n = 30$). The result is .60, and if we wanted to express the amount as a percentage we would multiply the proportion .60 by 100 resulting in 60%.

The percentage of all scores below a given point in a distribution is sometimes referred to as the score's percentile rank. We can then define a **percentile** as the point in a distribution that has a given percentage of all scores equal to or below it. For example, the 60th percentile, sometimes written as P60, is the score in the distribution that has 60% of all of the other scores at or below it. We can determine from inspection what the 60th percentile would be in the frequency distribution above, although the answer may not be as obvious as it may seem at first glance. We have already computed the proportion of scores below each interval based on the values in the cumulative frequency column. The interval "70–74" has .60 or 60% of all scores below it, so the question is what single value represents the 60th percentile. Keep in mind two things. First, the computation of the proportion of all scores below each of the intervals in the frequency distribution above is based on the number of scores in each interval and all of those below it, so the 60th percentile is the value at the very end of the "70–74" interval. Second, the value "74" is not the end point of the "70–74" interval. The "true" end point of the "70–74" interval is its upper limit, which is "74.5."

We will be working with a variety of percentile, and percentile related problems in this section, and these types of problems involve an element of common sense. You have seen that a frequency distribution has a bottom and a top, and that the lower scores are at the bottom and the higher scores are at the top. A percentile is a value in the distribution (a number in one of the intervals) that has a certain percentage of the other scores at or below it. You may not be able to tell exactly what score represents the 95th percentile in the frequency distribution in Figure 3.9 just by looking at it, but you certainly should know that it will be a value very close to the top of the distribution. A particular score must be very high in the distribution if it is going to have 95% of all of the other scores at or below it. Similarly, the 50th percentile should be a score in the middle of the distribution, and the 10th percentile should correspond to a score toward the bottom of the distribution. As we begin some of the computational problems, you might want to look at the distribution and estimate what the approximate answer should be. Not all of the statistical applications we will examine in the remainder of this text will have a common sense element to them, but when there is one you should use it to your advantage.

The Relationship between Percentiles, Deciles, and Quartiles

Most applications in behavioral science research will look at key points within a frequency distribution as percentiles, but deciles and quartiles are occasionally used as well, with quartiles being used quite often in educational research. Percentiles, deciles, and quartiles are all similar ways to describe various groupings of scores within a distribution. Percentiles break a distribution into 100 units, deciles break a distribution into 10 units, and quartiles break a distribution into 4 units. Just as percentiles correspond to percentage units, deciles correspond to groupings of 10% points, and quartiles correspond to groupings of 25% points.

Deciles and quartiles follow the same logic as percentiles. The higher the decile or quartile value, the higher you are to the top of the distribution. The problems we will examine next will pertain to percentile calculations, but you can easily find any decile or quartile by just solving for its percentile equivalent. For example, the ninth decile is the point in a distribution with 90% of all of the other scores at or below it, so the ninth decile is exactly the same as the 90th percentile. Similarly, the third quartile is the point in the distribution with three quarters or 75% of all of the other scores at or below it, so the third quartile is exactly the same point as the 75th percentile. Figure 3.10 illustrates the relationship among percentiles, deciles, and quartiles.

Percentile	Decile	Quartile
100th	10th	4th
90th	9th	
80th	8th	
75th		3rd
70th	7th	
60th	6th	
50th	5th	2nd
40th	4th	
30th	3rd	
25th		1st
20th	2nd	
10th	1st	

Figure 3.10 The Relationship among Percentiles, Deciles, and Quartiles

Finding the Score Corresponding to a Given Percentile

We will be solving two types of problems relating to percentiles. The first problem is one in which we are given a particular percentile, and want to know what score defines it. For example, given a particular frequency distribution we might want to find the score that represents the 40th percentile (P40).

The second type of problem we will examine in the next section can be thought of as the reverse procedure. We will be given a particular score in the frequency distribution, and will want to find the percentile it represents.

Finding the score in a distribution that represents a particular percentile is relatively simple. We will use the frequency distribution on hours of television viewing to illustrate the procedure by finding the score that represents the 40th percentile.

Television Viewed	f	Cf
90–94	2	30
85–89	2	28
80–84	4	26
75–79	4	22
70–74	1	18
65–69	2	17
60–64	6	15
55–59	6	9
50–54	3	3
	$\Sigma f = 30$	

The first step is to find the number of scores that will be below the given percentile (F_P). To do this, we convert the desired percentile to a proportion—the 40th percentile is the proportion .40—and then multiply the proportion by the total number of frequencies contained in the distribution.

$$F_P = .40 \times 30 = 12$$

At this point, we do not know what the 40th percentile is, but we do know that it is the point in the distribution that has a total of 12 frequencies (40% of all scores) below it. The next step is to use the cumulative frequency column to find which of the intervals in the frequency distribution contains the 40th percentile. We will begin at the bottom of the distribution and move up the cumulative frequency column until we encounter the first interval with at least 12 frequencies below it. The value "9" in the cumulative frequency column at the "55–59" interval indicates that there are a total of nine frequencies from the end of that interval to the bottom of the distribution. Moving up one more interval to the "60–64" interval, we see the value "15" in the cumulative frequency column, which indicates that there are a total of 15 scores from the end of that interval to the bottom of the distribution, so now we know that somewhere in the "60–64" interval we will find the score that has 12 of the total scores below it. Once we have identified the interval in the frequency distribution containing the desired percentile, we can use the following formula to compute the value corresponding to the percentile.

Formula for finding a score associated with a given percentile is as follows:

$$P_x = LL + \left[\left(\frac{F_P - CF_{below}}{f_{int}}\right) \times i\right]$$

where

P_x = the desired percentile
F_P = the number of frequencies below the desired percentile
CF_{below} = the value from the cumulative frequency column for the interval just below F_P

LL = the lower limit of the interval containing F_P
f_{int} = the number of frequencies in the F_P interval
i = the interval size.

The frequency distribution is rewritten below with the interval determined by the value of F_P (12), and the value of CF_{below} highlighted in bold:

Television Viewed	f	Cf	
90–94	2	30	
85–89	2	28	
80–84	4	26	
75–79	4	22	
70–74	1	18	
65–69	2	17	
60–64	6	15	<—(F_P = 12 is in this interval)
55–59	6	**9**	<—(CF_{below} = 9)
50–54	3	3	
	$\Sigma f = 30$		

To find the 40th percentile for the frequency distribution, we simply substitute into the formula as follows:

$$
\begin{aligned}
P_{40} &= 59.5 + \left[\left(\frac{12-9}{6}\right) \times 5\right] \\
&= 59.5 + \left[\left(\frac{3}{6}\right)\right] \times 5 \\
&= 59.5 + (.5 \times 5) \\
&= 59.5 + 2.5 = 62
\end{aligned}
$$

So the score 62, representing 62 hours per week of television viewing, is the 40th percentile. Another way to state the result is that 40% of the sample watched 62 hours or less of television per week.

We will summarize the steps in finding a given percentile below, and then use the income data from the frequency distribution appearing earlier in Figure 3.3 to provide a second example of finding a percentile.

Steps in Finding a Percentile

1. Convert the desired percentile to a proportion, and multiply the result by the sum of the frequencies (the sample size) to find F_P (the number of frequencies below the desired percentile).
2. Locate the interval in the frequency distribution containing the desired percentile by moving up the cumulative frequency column to the first interval containing the number of frequencies below the desired interval.
3. Identify the following values: the lower limit of the interval containing the desired percentile (LL), the frequencies in the interval containing the desired percentile (f_{int}), the value in the cumulative frequency column below the interval containing the desired percentile (CF_{below}), and the interval size (i).
4. Substitute the appropriate values into the formula, and complete the computation.

We will use the frequency distribution below presenting data on income to find the 90th percentile. As I indicated earlier, there is some common sense to problems such as this. We know the 90th percentile is going to be the income level that has 90% of all of the frequencies in the distribution below it, so it will be an income quite high in the distribution. We may not be able to look at the distribution and know exactly what the value of the 90th percentile is, but we know it definitely will be in the top half, and will probably be in the top one or two intervals.

We begin the computations by finding how many of the total frequencies will be below the 90th percentile (F_P) by multiplying the value ".90" times the total number of frequencies ($n = 40$). Then using the cumulative frequency column, we will be able to identify the interval containing the 90th percentile.

$$F_P = .90 \times 40 = 36$$

Income	f	Cf	
80–89	3	40	
70–79	5	37	<—($F_P = 36$ is in this interval)
60–69	3	**32**	<—($CF_{below} = 32$)
50–59	4	29	
40–49	6	25	
30–39	8	19	
20–29	8	11	
10–19	3	3	
	$\Sigma f = 40$		

Once we have identified the interval containing the 90th percentile, we are ready to substitute the appropriate values into the formula. (Note that in this case, the interval size $i = 10$.)

$$\begin{aligned} P_{90} &= 69.5 + \frac{36 - 32}{5} \times 10 \\ &= 69.5 + \frac{4}{5} \times 10 \\ &= 69.5 + .80 \times 10 \\ &= 69.5 + 8 = 77.5 \end{aligned}$$

The resulting value of "77.5" represents the 90th percentile, or in other words, 90% of the sample earns 77.5 thousand dollars or less.

Finding the Percentile Corresponding to a Given Score

Computing the percentile corresponding to a given score in a distribution is a process that is very similar to that of finding the score associated with a given percentile. The elements in the two formulae are the same, and moving from one problem to another primarily involves a simple algebraic manipulation of the terms in the formula to allow us to solve for the percentile once we are given the score. The formula is presented below, and note that the formula results in a proportional value rather than a percentile. We will need to multiply the final result by 100 to convert the proportion to a percentile if desired.

Formula for finding a percentile associated with a given score:

$$P_x = \frac{CF_{Below} + \left[\left(\frac{X - LL}{i}\right) \times f_{int}\right]}{\Sigma f}$$

where

P_x = the desired proportion
CF_{below} = the value from the cumulative frequency column for the interval just below F_P
X = the value of the score that is given
LL = the lower limit of the interval containing X
f_{int} = the number of frequencies in the interval containing X
i = the interval size
Σf = the sum of the frequencies (the sample size).

A typical application of finding a percentile associated with a given score can be illustrated with the same frequency distribution of income with which we have been working. For example, we might want to know what percentile is associated with the income level of 42 thousand dollars. The first step is to identify the interval containing the given score of "42." Once that has been accomplished, we simply substitute the appropriate values, familiar to us from our previous percentile problems, into the formula. Once again you might want to take advantage of the common sense nature of percentile problems, and try to estimate what the final answer will be by inspecting the frequency distribution. The value "42" is close to the middle of the distribution, so the percentile associated with 42 thousand dollars should be somewhere around the 50th percentile.

Income	f	Cf	
80–89	3	40	
70–79	5	37	
60–69	3	32	
50–59	4	29	
40–49	6	25	<—(X = 42 is in this interval)
30–39	8	**19**	<—(CF_{below} = 19)
20–29	8	11	
10–19	3	3	
	$\Sigma f = 40$		

$$P_x = \frac{19 + \left[\left(\frac{42 - 39.5}{10}\right) \times 6\right]}{40}$$

$$= \frac{19 + \left[\left(\frac{2.5}{10}\right) \times 6\right]}{40}$$

$$= \frac{19 + (.25 \times 6)}{40}$$

$$= \frac{19 + 1.50}{40}$$

$$= \frac{20.50}{40}$$

$$= .5125$$

The proportion of scores in the sample at or below 42 thousand dollars is .5125, and if we wanted to report it as a percentile we would multiply the value by 100 resulting in the 51.25th percentile. The result is very close to what we expected given the location of 42 thousand in the frequency distribution. Our next example of computing the percentile associated with a given score will use the frequency distribution of television viewing that we used previously. In this case, we will find the percentile of someone who watched "88" hours of television per week. The value "88" is very close to the top of the frequency distribution, so we should expect a final answer close to the 90th percentile. (Note that the interval size in this frequency distribution is $i = 5$.)

Television Viewed	f	Cf	
90–94	2	30	
85–89	2	28	<—($X = 88$ is in this interval)
80–84	4	**26**	<—($CF_{below} = 26$)
75–79	4	22	
70–74	1	18	
65–69	2	17	
60–64	6	15	
55–59	6	9	
50–54	3	3	
	$\Sigma f = 30$		

$$\begin{aligned} P_x &= \frac{26 + (88 - 84.5)/5 \times 2}{30} \\ &= \frac{26 + (3.5)/5 \times 2}{30} \\ &= \frac{26 + .7 \times 2}{30} \\ &= \frac{26 + 1.4}{30} \\ &= \frac{27.4}{30} \\ &= .913 \end{aligned}$$

The proportion of scores in the sample viewing 88 or fewer hours of television per week is .913, and if we wanted to report that as a percentile, we would report the 91.3rd percentile. Again, the answer is close to what we had expected by looking at the relative position of 88 in the frequency distribution.

Up to this point, we have focused on the frequency distribution and its characteristics as a method of data reduction. Certainly, the frequency distribution is one very effective way to reduce a large amount of data into a form that is more easily interpreted; however, it is not the only method at our disposal. Graphic display of data is also an effective way to present information, and graphic presentations are becoming a more important component of both professional and popular press presentations of information. The next major section of this chapter deals with the underlying logic of graphic presentation, and provides examples of some of the more popular types of graphic displays.

The Graphic Display of Data

A graphic display of data can be a very effective way of conveying a sense of understanding to a broad audience in a relatively short period of time. Being able to "see" what the data look like can be a real advantage; however, graphic displays have their limitations just as frequency distributions do. In fact, graphic displays may even be more limited if they are used as substitutes for other types of presentations rather than being used as accompaniments. A "picture" of how the data appear in combination with a more detailed presentation is ideal, but a graphic display alone does not begin to provide the type of precision required in behavioral science research.

We will examine four types of popular graphing techniques in this section. The first three, the **bar graph**, the **histogram**, and the **frequency polygon** are all similar in their presentation of data. The final graphing method, the **ogive chart**, presents data in a slightly different fashion, and is seen less seldom than the other three.

The Bar Graph, the Histogram, and the Frequency Polygon

The bar graph, the histogram, and the frequency polygon all present data in a similar fashion, and are all based on the same underlying logic. All three graphing techniques present data on a coordinate system similar to the Cartesian coordinate system consisting of a horizontal and a vertical axis, which you may remember from earlier courses in mathematics. The horizontal axis is used to present values of the measured variable with values increasing in size as you move from left to right. The vertical axis is used to present the number of frequencies associated with each observed value of the measured variable with values increasing in size as you move from the bottom to the top of the axis.

Each of the graphing techniques may be applied to raw data or to data in a frequency distribution, but graphing raw data becomes less practical as the sample size increases. We will present examples using the frequency distributions on income, and television viewing that have been used previously in the chapter. The horizontal axis presents values of the measured variable, and it is permissible to label the values along the horizontal axis with the actual interval being presented, or if space is limited, the values may be labeled with the interval midpoint. The scale used to present frequencies along the vertical axis should be adjusted depending on the number of frequencies associated with each interval. For example, if there are no more than five or six frequencies in any given interval in the frequency distribution, then each mark along the vertical axis should represent one additional frequency. However, with larger sample sizes, it is likely that some intervals will have a large number of frequencies, and it might be more appropriate to have each mark along the vertical axis represent multiple frequencies.

The first illustration of graphing is a bar graph depicting the income data. The horizontal axis is labeled with the midpoint of each income interval, and each mark along the vertical axis represents two frequencies to have a final product with a proper scale. The data are presented again below, and then graphed in Figure 3.11.

Income	f	Cf
80–89	3	40
70–79	5	37
60–69	3	32
50–59	4	29
40–49	6	25
30–39	8	19
20–29	8	11
10–19	3	3
	$\sum f = 40$	

The income data are presented in Figure 3.12 in the form of a histogram. The general form of presentation is identical with the measured variable on the horizontal axis, and the frequencies presented on the vertical axis (Figure 3.12).

There is a great deal of similarity between the bar graph and the histogram. At first glance, the only difference might appear to be the width of the bars used to represent the number of frequencies in each income interval, and the fact that the bars in the histogram touch each other whereas those in the bar graph do not. In one sense, those are the only differences, but what is important is what is implied by the fact that the bars in the histogram are attached to each other. We have discussed different types of data in this chapter as well as the earlier ones. Some data, nominal level data for example, are categorical with no underlying level of measurement. The other types of data are either measured on a continuous scale, as in the case of interval and ratio level data, or are categorical but with an underlying level of measurement such as ordinal level data. The underlying level of measurement implying movement from one category to the next that is present in ordinal, interval, and ratio level data are compatible with the presentation in the histogram. Just as one income level flows smoothly to the next income level, the bars in the histogram flow smoothly to each other. Consequently, the histogram would not be the appropriate choice for nominal level data because the categories are strictly discrete units with no underlying level of measurement. The bar graph is the appropriate choice for nominal level data, but may be used with data measured at the other levels as well.

The third graphing technique we will examine is called the frequency polygon. The frequency polygon shares the same logic as the bar graph and histogram, and may be familiar to you as a "line graph." The only difference between the frequency polygon and the histogram is that a line is used to connect points representing the number of frequencies at each interval midpoint in the graph rather than the

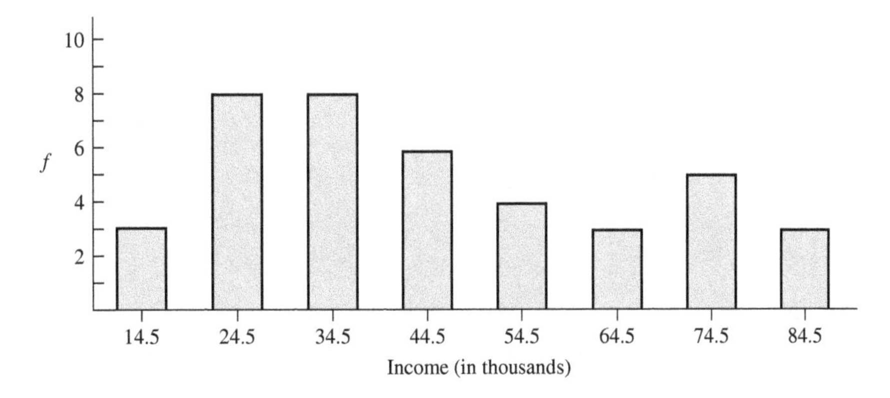

Figure 3.11 Bar Graph of Income Data

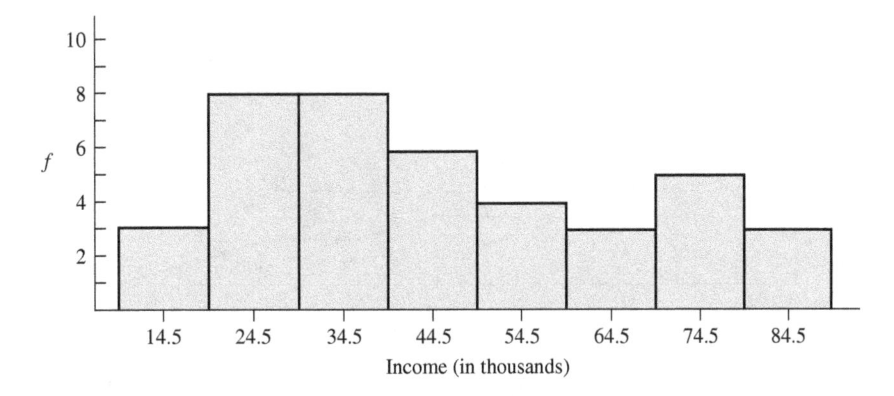

Figure 3.12 Histogram of Income Data

adjacent bars used in the histogram. Frequency polygons are often used when presenting several data sets on a single graph. Presenting multiple data sets with a bar graph or a histogram can result in a chart that is harder to interpret. When multiple data sets are presented in the frequency polygon, one often uses different symbols to represent each data set presented, and includes a legend at the bottom of the graph to identify each symbol used. We will illustrate the frequency polygon using the weekly television viewing data with the addition of data for two older age groups, teens and adults. All three data sets will be presented on the same frequency polygon (Figures 3.13 and 3.14).

A histogram, or bar graph could have been used to present the television viewing data, but the resulting graph would have been more difficult to interpret. Bar graphs, histograms, and frequency polygons are not the only graphing techniques available, but they are three of the more popular. Graphic presentation of data is becoming more popular, and most computer statistical analysis packages, computer spreadsheet packages, and even some computer word processing packages have very impressive graphing capabilities. A traditional limitation to some of the more elaborate computer generated graphic presentations is that what could be seen on the computer's video display did not translate well onto the printed page from a single color printer; however, recent advances in color printers for personal computers have given us a great deal more flexibility.

The final example of graphing a single variable we will examine is called the ogive chart. The ogive chart differs slightly from that of the first three that we have seen in that the ogive presents cumulative

Television Viewed	**Preschool** f	**Teen** f	**Adult** f
90–94	2	0	1
85–89	2	1	0
80–84	4	2	1
75–79	4	2	1
70–74	1	3	2
65–69	2	3	6
60–64	6	7	4
55–59	6	7	8
50–54	3	4	7
	$\Sigma f = 30$	$\Sigma f = 29$	$\Sigma f = 30$

Figure 3.13 Preschool, Teenage, and Adult Television Viewing per Week (in hours)

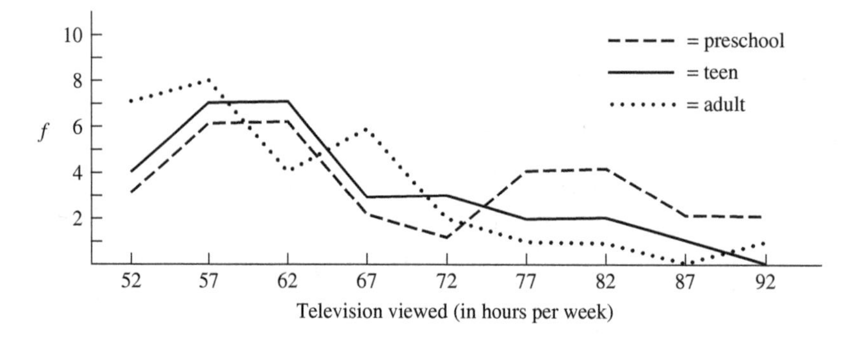

Figure 3.14 Frequency Polygon of Weekly Television Viewing by Age

frequencies on the vertical axis instead of simple frequencies. The ogive chart is useful for presenting a cumulative experience of some type over time. For example, we might have data on employee mistakes made during a 7-day training period. A frequency polygon would present the number of errors made day-by-day, but the ogive chart will present the cumulative number of errors day-by-day. The resulting graph can provide some useful insight into the progress being made. Data representing training errors for the ogive chart are presented below.

Training Errors during a 7-Day Training Period

Day	f	Cf
7	0	13
6	1	13
5	1	12
4	2	11
3	2	9
2	3	7
1	4	4
	$\Sigma f = 13$	

What does the ogive chart tell you about the employee's progress during training? Does this seem like an employee you would want to keep, or would you be better off replacing this employee? Consider the following four ogive charts depicting cumulative errors over a 7-day training period. Which do you think represents the most promising employee, and which represents the least promising? (Figure 3.16)

Comparing the pattern from the first ogive chart to the other four may help you interpret the results. Which employee is most promising, and which is the least promising? You might want to think about what the cumulative frequency column would have to look like to generate each of the four charts if you have a problem seeing any particular pattern. Employees "C" and "D" are not very good prospects. Employee "D's" chart exhibits a progressive straight line, which would be the result of a steady pattern of errors occurring day after day. Employee "D" is not getting any worse over time, but there is no evidence of progress either. Employee "C" is even worse. The upturn in the slope of the curve after the first 3 days of training indicates that more errors are being made every day as time goes on.

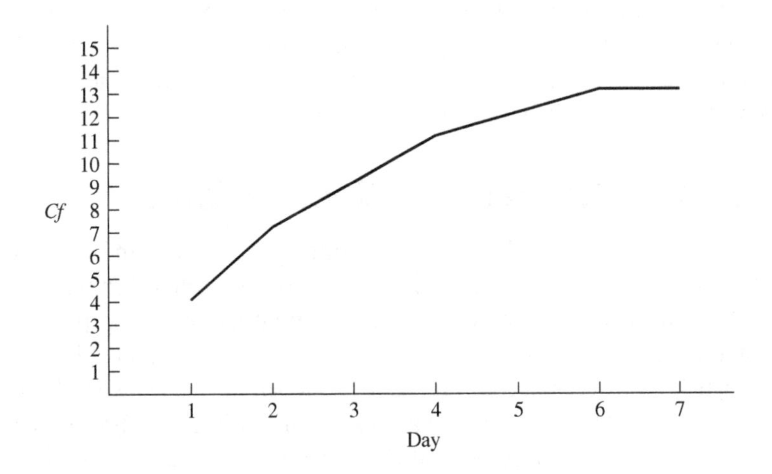

Figure 3.15 Ogive Chart of Cumulative Errors

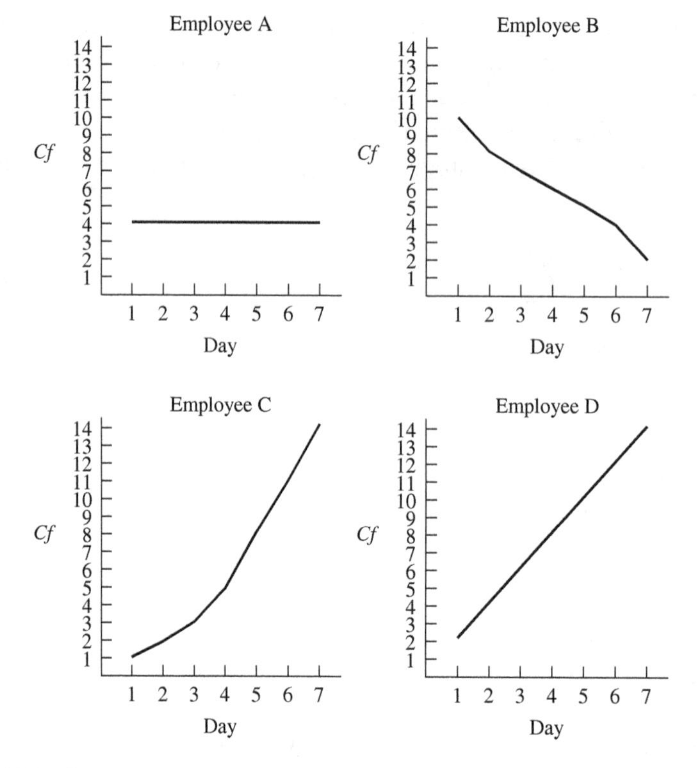

Figure 3.16 Four Hypothetical Ogive Charts

What about employee "B?" The ogive chart for employee "B" represents quite an accomplishment. To generate the pattern seen for employee "B," the values in the cumulative frequency column would have to decrease over time. Once an error has been made, it remains (there is no extra credit in life.), so the pattern of a negative slope for employee "B" is not possible in an ogive chart. Employee "A" clearly represents the most promising employee. The straight horizontal line would be the result of someone who made all of his or her errors on the first day of training, and then did not make another error. How does our original case compare with the other four? The original example actually represents a promising employee, and one that is certainly better than "C," or "D." The pattern in the original ogive chart is sometimes referred to as the "classic learning curve." The steep slope in the beginning, which then tails-off or flattens out toward the end is indicative of how we often learn. That type of curve is indicative of someone who made many errors in the beginning, but then begins to catch on and makes fewer and fewer errors toward the end of the training period. You can see that this was the case if you examine the daily frequency of errors presented in the frequency distribution.

Some Common Graphic Patterns Seen in Data

Any variable that we measure can be graphed, and the resulting picture can indicate a great deal about the distribution of the data. There are a few patterns that tend to occur quite often when the distribution of a variable is graphed. We will illustrate some of the more commonly occurring patterns with the use of a frequency polygon, and then indicate what the pattern tells us about the data (Figure 3.17).

Each of the two distributions depicted in Figure 3.17 are said to be skewed. A distribution is said to be skewed when a large proportion of the observations occur at either the bottom or the top of the distribution, and then tail-off toward the other end of the distribution. The nature of the skew is named for the direction in which the distribution tails-off. For example, a distribution in which most of the cases

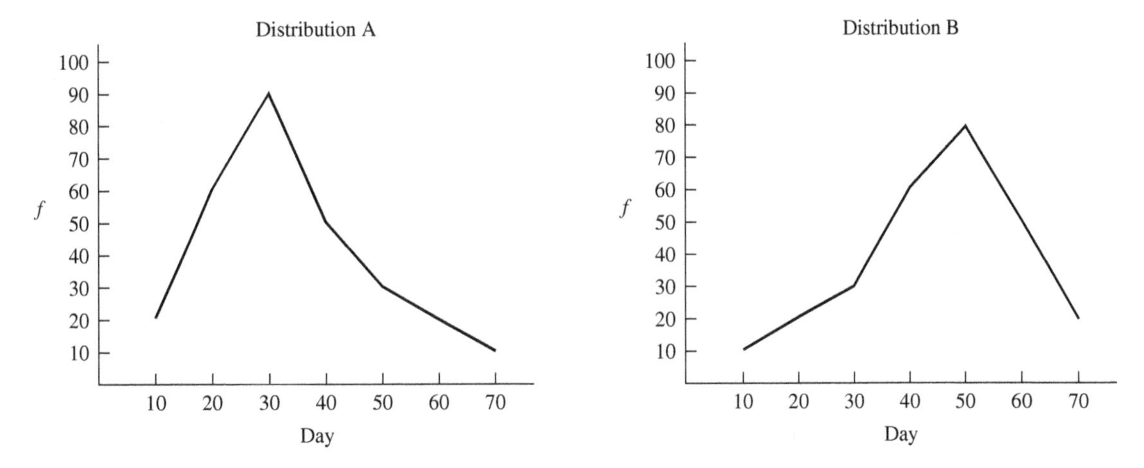

Figure 3.17 Some Typical Data Patterns—Skewed Distributions

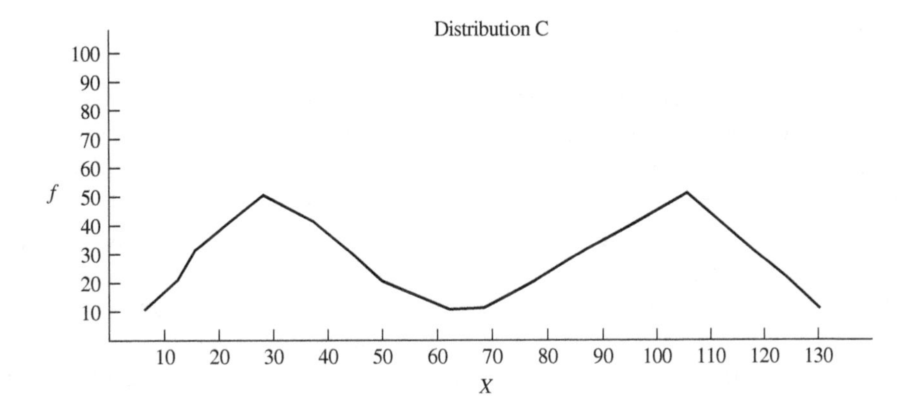

Figure 3.18 A Typical Data Pattern—A Bimodal Distribution

are found toward the bottom of the distribution (where the negative or low values are present), and then tails-off toward the right (where the positive or higher values are present) is said to be skewed to the right, or to be positively skewed. Distribution "A" is skewed to the right, or has a positive skew. Similarly, a distribution in which most of the cases are found toward the top of the distribution (where the positive or high values are present), and then tails-off toward the left (where the negative or lower values are present) is said to be skewed to the left, or to be negatively skewed. Distribution "B" is skewed to the left, or has a negative skew (Figure 3.18).

Distribution "C" has two main concentrations of observations, the first group peaks at about the value of $X = 30$, and the second group peaks at about the value of $X = 110$. The point in a distribution of scores where the highest concentration of observations is found is called the mode. Distribution "C," like many other distributions you may encounter, has two modes, and as such is referred to as being bimodal. Examination grades in certain courses are often bimodal. You will find a grouping of students in the upper grade range, another grouping of students in the lower grade range, and almost no one in between in the middle grade range. One last distribution shape that is commonly seen is represented in Figure 3.19.

The shape of distribution "D" in Figure 3.19 goes by a number of names, and many of you may be familiar with it as the "bell-shaped curve." In statistics, the shape of distribution "D" is referred to as a **normal distribution**. You might notice that a normal distribution is symmetrical, and that the

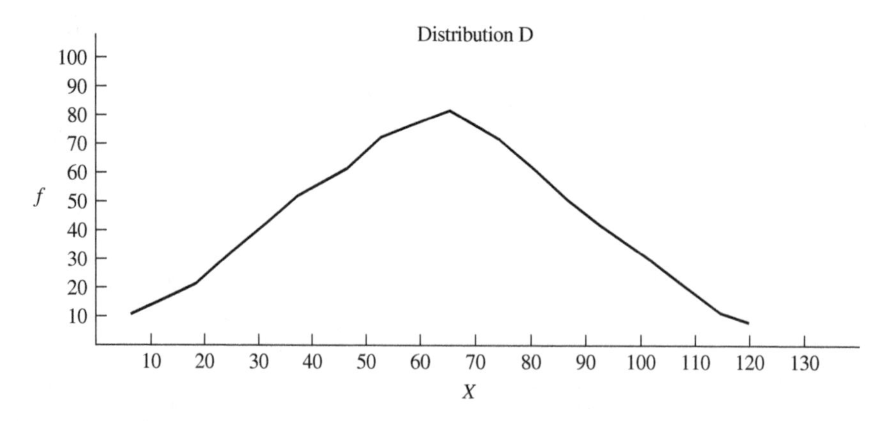

Figure 3.19 A Typical Data Pattern—A Normal Distribution

mode appears at the center of the distribution. Not only does the mode appear at the center of a normal distribution, but the mean (the average), and the median are located at the center as well. (The mode, the mean, and the median are all measures of central tendency that we will examine in more detail in Chapter 4.)

Many of the variables we encounter in behavioral science research resemble a normal distribution, and for this and other reasons the normal distribution is an important concept. We will devote an entire chapter to the normal distribution, and you will find that the normal distribution will surface later in the text as well. In one sense, you have already encountered a normal distribution. In Chapter 2 on sampling, we examined the concept of a sampling distribution, the distribution of all possible sample means of a particular size drawn from a population. I mentioned several important characteristics of the sampling distribution, one of which was that the sampling distribution took on a particular shape once the population and sample size became large. The shape a sampling distribution takes is that of a normal distribution. We will be examining this and other important characteristics of normal distributions in Chapter 6.

Computer Applications

1. Enter the raw data on age used to generate the frequency distribution in Figure 3.1.
2. Generate a frequency distribution of the data using Statistical Package for the Social Sciences (SPSS), be sure to specify a descending order for the data presentation.
3. Recode the data (into a new variable called "CAT" for category) using the data ranges presented in Figure 3.2.
4. Generate a frequency distribution for the recoded variable.
5. Compare the SPSS frequency distributions with those in the text. Are the frequency counts the same? What additional types of summary information does SPSS include in its frequency distributions?
6. Select several variables from the General Social Survey (GSS) data set and create a bar graph, frequency polygon, and histogram from the data. Which graphic presentation seems to better fit the variables you have chosen?

How to do it

Enter the age data for Figure 3.1 into column one of a new data file. Generate your first frequency distribution by using the "Frequencies" option under "Summarize" for the "Statistics" procedure. Be sure to use the "Format" option to specify a descending order for the frequency distribution.

Use the "Recode" (into a different variable) option under the "Transform" procedure to create the new variable called "CAT." Highlight the variable to be recoded and move it to the selected box by clicking on the directional arrow. Specify the name for the new variable in the "Output Variable" box, and click on the "Change" button. Click on the "Old and new Values" button to specify the data ranges: for example, enter 15 through 19 as the old value range, and then enter 1 as the new value range. Be sure to then click on the "Add" button to make the change effective. Complete the recode by specifying the additional data ranges and their new values (click on "Add" after each one). Complete the recode by clicking on "Continue" and then "OK." You should see the new variable on the data editor screen.

Use the "Define Variable" "Labels" option under the "Data" procedure to create value labels for the variable "CAT." Enter the value (e.g., "1") and the label you want associated with it (e.g., "15–19"). Be sure to click on the "Add" button after specifying each label.

Now repeat the frequencies procedure in SPSS for the new variable "CAT" and compare the output with Figure 3.2.

Click on the "Graphs" procedure and select a graphic format (bar, line, or histogram). Click on "Define" to select a variable, and then generate the graphic by clicking on "OK." Compare the results for a single variable with several different graphic displays (i.e., a bar graph and then a line graph). When creating a histogram click on the "Display Normal Curve" option. How closely does the graph of the variable you selected resemble a normal distribution?

Summary of Key Points

Data reduction refers to methods of transforming raw data into a more easily interpreted form. One major method of data reduction is the frequency distribution, which may be used with intervals of size $i = 1$, or with larger intervals when the distribution contains a broader range of data. We have also examined some of the more popular graphing techniques, which use graphic display as a method of reducing data to a more manageable form. Finally we have seen several types of distributions such as the normal distribution and **skewed distributions**, which tend to be characteristic of many of the variables measured in behavioral science research. Many of the topics covered in this chapter will surface again in later chapters as we begin moving into the more computational aspects of statistical analysis.

Frequency Distribution—A method of data reduction in which raw data are organized into ordered categories of various sizes.

Lower and Upper Limits—The actual beginning and ending points of an interval in a frequency distribution as opposed to the apparent limits, which are the points where an interval appears to begin and end.

Interval Midpoints—The center point of an interval in a frequency distribution. Many computations involving frequency distributions assume that the frequencies in a particular interval are equal to the size of the interval midpoint.

Percentiles—A percentile is the percentage of a distribution equal to or below a given point. The 75th percentile is the score in a distribution with 75% of all scores in the distribution at or below it.

Bar Graph—A common graphing technique in which the height of bars displayed on an axis represent the number of frequencies in the category being represented by the bar.

Histogram—A graphing technique similar to a bar graph except that the bars in a histogram are contiguous representing the continuous nature of the underlying variable. Histograms are appropriate for data of the ordinal, interval, or ratio level.

Frequency Polygon—A common graphing technique in which a line is used to connect points representing the number of frequencies at each point in a distribution.

Ogive Chart—A graphing technique that is used to display cumulative frequencies rather than simple frequencies.

Normal Distribution—A distribution of scores that is symmetrical around a center point where the mean, the mode, and the median of the distribution are found. A normal distribution resembles the shape of a bell.
Skewed Distribution—An asymmetrical distribution in which a large concentration of scores is found at one end of the distribution.
Data Reduction—Techniques to organize or "reduce" raw data (individual scores) to a form that is easier to interpret. Examples include frequency distributions and a variety of graphing techniques.
Open Interval—An interval at the beginning and/or end in a frequency distribution that has an indefinite beginning or ending point. For example the variable age might be presented with intervals of size $i = 5$ such as: 15–19; 20–24; 25–29, and so on (examples of closed intervals of size i = 5), or an open interval may be used such as < 20 (an open interval for all subjects of age less than 20), 20–24, 25–29.
Frequency Distribution—A data reduction technique where raw data are organized into groups of a particular range.
Normal Distribution—A theoretical distribution for a variable created by W. S. Gossett in which the mean, mode, and median are all at the same point with half the distribution at or above that central point and half of the distribution below that central point. The distribution tails off symmetrically from the central point in a shape often described as a "bell shaped" curve.
Skewed Distribution—A distribution which is not normal in shape but rather has a major cluster of scores at one end or the other with the remaining scores forming an elongated tail toward the right (positive skew) or the left (negative skew).
Bimodal Distribution—A distribution with two clusters of scores near each end of the distribution. For example a set of exam grades might have a cluster of grades at the low end (the D and F range) and a second cluster of scores at the high end (the B and A range) with not too many scores in the center (the C range).

Questions and Problems for Review

1. Using the raw data below, create a frequency distribution with an interval size $i = 1$. Plot the data as a frequency polygon, and interpret the shape of the distribution.

Data:	12	9	5	9	5	11	5	6	7	12
	5	8	6	8	6	10	7	7	5	2

2. Which type of graphic display is most appropriate for the frequency distribution below? Graph the data using the most appropriate choice.

Major	f
Math	20
English	15
History	25
Psychology	45
Art History	10
	$\Sigma f = 115$

3. Create a frequency distribution with an interval size $i = 10$ with the following data. Identify each interval's midpoint, lower limit, and upper limit.

Data:										
	77	82	90	75	87	99	102	112	92	88
	124	145	134	76	97	144	138	99	100	122
	91	76	133	149	87	79	91	80	107	105

4. Using the frequency distribution below, find the following percentiles: 25th, 50th, 75th, 95th.

Age	*f*	Cf
90–99	15	326
80–89	30	311
70–79	52	281
60–69	24	229
50–59	45	205
40–49	46	160
30–39	24	114
20–29	35	90
10–19	20	55
0–9	35	35
	$\Sigma f = 326$	

5. Using the frequency distribution in problem 4 above, compute the percentile defined by the following observed ages: 65, 42, 18.

6. Data are presented below for the number of errors made by a trainee during a 7-day training period. Create an ogive chart of the cumulative frequencies, and assess the employee's potential.

Day	*f*	Cf
7	5	30
6	4	25
5	4	21
4	3	17
3	5	14
2	4	9
1	5	5
	$\Sigma f = 30$	

7. Create a histogram for the data below, and describe the resulting distribution.

I.Q.	f	Cf
140–149	3	163
130–139	6	160
120–129	12	154
110–119	32	142
100–109	54	110
90–99	25	56
80–89	18	31
70–79	8	13
60–69	5	5
	$\Sigma f = 163$	

8. Create a frequency distribution with an interval size $i = 5$ with the following data. In addition to the frequency, and cumulative frequency columns, also provide columns to indicate each intervals relative percentage, and cumulative percentage.

Data:										
	27	32	30	25	27	39	42	52	32	28
	44	25	54	26	37	64	58	62	50	61
	49	36	38	49	47	57	39	40	37	55

9. Using the frequency distribution you create for the data in problem 8, compute the percentile defined by the values: 30, 40, 50, 60.

10. Create a histogram using the relative percentage of the total cases in each interval for the frequency distribution you create for the data in problem 8. Compare the resulting histogram with one using the actual number of cases in each interval.

CHAPTER 4

Measures of Central Tendency

Key Concepts

Central Tendency
Mode
Bimodal
Trimodal
Median
Mean
Grand Mean

Introduction

The data reduction techniques discussed in Chapter 3 are ideal ways to begin the process of taking a large amount of data and reducing it to a form that is easier to interpret. For example, a frequency distribution is a good way to indicate the range of a set of numbers, and how the values of a particular variable are distributed across that range. Measures of central tendency take this process one step further by giving us a single number intended to describe the central point of the distribution. A measure of central tendency is the answer to the question "What is the single number that best describes the typical value in this set of data?"

In this chapter, we examine three common measures of central tendency: the **mode**, the **median**, and the **mean**. Examples are provided for several types of data presentations because we cannot always control the form in which data are presented to us. The mode and mean are examined for three different types of data presentation: raw data consisting of individual scores; simple frequency distributions where the interval size is equal to 1; and, frequency distributions where the interval size is greater than 1. The median is examined for raw data, and for frequency distributions where the interval size is greater than 1.

Mode

The mode is the simplest measure of central tendency, and is defined as the most frequently observed value of a measured variable in a set of data. Let's look at a simple example using raw data (a group of individual scores). Suppose we have measured the variable age for a group of $n = 9$ elementary school children. We might observe a distribution of scores such as the following:

X: 6, 6, 8, 8, 8, 8, 11, 12, 12

We have a total of nine scores representing children ranging from age 6 to age 12. The mode is simply the value of the variable that occurs most frequently. In this case, the mode is equal to 8, because there are more observations at age 8 (a total of four children) than any other age. Keep in mind that the mode is a value of the measured variable, in this case 8. It is not 4 that just represents the number of observations at the mode.

Determining the mode in a simple frequency distribution is just as easy. Again, we might be measuring the variable age for a group of elementary school children. In this case, we have a sample size of $n = 20$, which we have displayed in the frequency distribution below.

X	f	Cf
12	7	20
10	3	13
8	3	10
7	5	7
6	2	2
	$\Sigma f = 20$	

We have a total of 20 observations that range from two students at age 6 to seven students at age 12. The mode is the value of the most frequently observed value of our measured variable. In this case, the mode is equal to 12. The mode is equal to 12 because there are more observations at that value of the variable (seven of them) than there are at any other value of X. Also note that sometimes the mode is not a very good measure of central tendency, if by central tendency we mean a value of X that tends to be toward the center of the distribution. In this case, for example, the mode is at the extreme top of the distribution. However, the mode still represents what is "typical" of the distribution in that more observations were at a value of 12 years of age than any other.

The final case we will examine in determining the mode is for a frequency distribution. As our sample size increases, or as the range of the values of the measured variable increases, we might move to a frequency distribution with an interval size greater than 1, as in the example below. In this case, we might have measured the number of times a child has been absent from school for a group of $n = 75$ children.

X	f	Cf
15–17	9	75
12–14	16	66
9–11	20	50
6–8	16	30
3–5	10	14
0–2	4	4
	$\sum f = 75$	

There are a total of 20 observations for the interval 9–11, so that tells us where the mode is located, but how do we report it? This is one of those occasions in statistics where some flexibility is allowed. For a frequency distribution, we may report the mode as the entire interval in which it is found, in this case we would report 9–11; or, we may report the midpoint of the interval as the mode. So if we prefer to report a single value we could simply say the mode is 10 (the midpoint of the 9–11 interval). Also note in this example that the mode is located toward the center of the distribution.

There is one other unusual situation you may encounter when reporting a mode for a particular distribution. Consider the frequency distribution below.

X	f	Cf
80–89	4	165
70–79	8	161
60–69	12	153
50–59	20	141
40–49	36	121
30–39	29	85
20–29	36	56
10–19	12	20
0–9	8	8
	$\sum f = 165$	

We have a total of $n = 165$ observations ranging from the interval 0–9 to the interval 80–89. We know that we can report the mode as either the entire interval of our measured variable X, or we can report the midpoint. But in this situation we have two intervals that each has a total of 36 observations. This is an example of a distribution that is said to be **bimodal**. The distribution has two modes, and we should report them both. We may report the intervals 20–29, and 40–49 as the modes, or we may report the two midpoints 24.5, and 44.5 as the modes.

Some distributions may even have three modes, in which case the distribution is said to be **trimodal**. It is possible that some distributions will have an even larger number of scores or intervals with the same high number of observations, but it would be unusual to report more than three modes. If four or more scores or intervals are tied with the same large number of observations, it is usually better to simply say that the distribution does not have a mode.

The mode is not only the simplest measure of central tendency, it is also a measure that may be applied to data of any level. The examples we have looked at up to now have represented ratio level data. However, we can just as easily, and just as meaningfully, report a mode for nominal or ordinal level data. Consider the frequency distribution below representing classification.

Classification	f	Cf
Senior	74	400
Junior	98	326
Sophomore	120	228
Freshman	108	108
	$\Sigma f = 400$	

In this distribution of $n = 400$ students by classification, the mode is sophomore because there are more observations at that category than any other.

Median

The median is the second measure of central tendency that we will examine. The median may not be as familiar to many people as the mean, but the median can provide a great deal of valuable information about a distribution. The median is defined as the value of a measured variable that divides the distribution in half. Just as the median of a highway is a strip of grass or concrete that has half of the highway on one side, and half of the highway on the other side, the median of a distribution is the value that will have half of the distribution above it, and half of the distribution below it.

Calculating the median for raw data is rather simple, but the procedure varies slightly depending on whether you have an even number of observations or an odd number of observations.

The raw data case where the sample size is odd

When the sample size is an odd number, the median will equal a value that is actually represented in the distribution of scores. Consider the following group of $n = 9$ scores.

X: 12 13 9 24 35 21 16 27 8

To find the median, we must first arrange the scores in order from the lowest to the highest. The distribution is rewritten in numerical order below, and labeled as x_1 to x_9.

X_i:	x_1	x_2	x_3	x_4	x_5	x_6	x_7	x_8	x_9
X:	8	9	12	13	16	21	24	27	35

The median is the value of X that has half of the other scores greater than it, and half of the other scores less than it. Since there are a total of $n = 9$ scores, the median is the fifth value of X, or in this case the median is equal to 16. The median is 16 because the value 16 has half of the other scores above it (observations x_6 to x_9), and half of the other scores below it (observations x_1 to x_4).

A simple way to determine which observation of X is the median when you have an odd number of scores is to let the value of the median be represented by X_i

$$\text{where } i = \frac{n+1}{2}$$

In this example $n = 9$, so

$$i = \frac{n+1}{2} = \frac{9+1}{2} = \frac{10}{2} = 5$$

The median is equal to the fifth value of *X*. In this distribution, once the *X*s have been arranged in order, x_5 is equal to the value 16.

We have examined relatively simple examples to save space, but suppose we had a total of $n = 135$ values of *X*. Which value of *X* would be the median?

$$i = \frac{n+1}{2} = \frac{135+1}{2} = \frac{136}{2} = 68$$

The median would be equal to the 68th value of *X* in the ordered distribution.

The raw data case where the sample size is even

When there is an even number of observations in the distribution, the median usually will not equal an actual observed value of *X*, because there is no value of *X* that falls in the exact middle of the distribution. The median will equal the midpoint of the two most center values of *X* in the ordered distribution (if the two most center values of *X* happen to be equal, then the median will equal an observed value in a distribution with an even sample size). In the case of an even number of observations the value "*i*" is equal to the total sample size divided by 2, and the two most center values of *X* will be the *i*th value and the *i*th+1 value of *X*. Consider the distribution of $n = 10$ scores below (which have already been arranged in order from lowest to highest).

X_i:	x_1	x_2	x_3	x_4	x_5	x_6	x_7	x_8	x_9	x_{10}
X:	20	23	37	43	57	63	66	67	102	114

We must first find the value of *i*, where given an even number of observations:

$$i = \frac{n}{2} = \frac{10}{2} = 5$$

The median of the distribution of scores will equal

$$\frac{x_i + x_{i+1}}{2} = \frac{x_5 + x_6}{2}$$

the midpoint between the fifth and sixth score. In the distribution of $n = 10$ scores, the median is equal to

$$\frac{57 + 63}{2} = \frac{120}{2} = 60$$

Which two scores would we average to find the median if our sample size was $n = 228$? The value of *i* would be 228/2 or 114, so the median would be the midpoint between the 114th score and the 115th score.

Calculating the median for a frequency distribution

Calculating the median for a frequency distribution is somewhat more involved. Consider the frequency distribution below.

X	f	Cf
25–29	20	125
20–24	22	105
15–19	28	83
10–14	20	55
5–9	25	35
0–4	10	10
	$\Sigma f = 125$	

The frequency distribution contains a total of $n = 125$ scores that range from the 0–4 interval up to the 25–29 interval. The median will be a score, or value of X, that has half of the frequencies above it, and half of the frequencies below it. So to find the median, we must find the value of X that has 50% of the scores on either side of it. While the process for finding the median in a frequency distribution is a little more involved than it is for a set of raw data, the good news is that you already know how to do it! The median in a frequency distribution is just another name for the 50th percentile, so you can find the median just as you found any other percentile in Chapter 3.

Recall the procedure for finding a given percentile. The first step is to find the number of scores that will be below the given percentile (F_P). To do this, we convert the desired percentile to a proportion—the 50th percentile is the proportion .50—and then multiply the proportion by the number of frequencies contained in the distribution.

$$F_P = .50 \times 125 = 62.5$$

Then the general formula for finding a given percentile is given by the equation below.

$$P_x = LL + \left[\left(\frac{F_P - CF_{below}}{f_{int}} \right) \times i \right]$$

where

P_x = the desired percentile
CF_{below} = the value from the cumulative frequency column for the interval just below F_P
LL = the lower limit of the interval containing F_P
f_{int} = the number of frequencies in the F_P interval
i = the interval size.

The frequency distribution is rewritten below with the interval determined by the value of F_P (62.5), and the value of CF_{below} highlighted in bold:

X	*f*	Cf	
25–29	20	125	
20–24	22	105	
15–19	28	83	<--(F_P = 62.5 is in this interval)
10–14	20	**55**	<--(CF_{below} = 55)
5–9	25	35	
0–4	10	10	
	$\Sigma f = 125$		

To find the median for the frequency distribution, we simply substitute into the formula as follows:

$$P_{50} = 14.5 + \frac{62.5 - 55}{28} \times 5$$

$$P_{50} = 14.5 + \frac{7.5}{28} \times 5$$

$$P_{50} = 14.5 + .268 \times 5$$

$$P_{50} = 14.5 + 1.34 = 15.84$$

The median of the frequency distribution is 15.84.

Consider a second example consisting of a set of $n = 40$ examination scores arranged in a frequency distribution.

X	*f*	Cf
90–99	9	40
80–89	10	31
70–79	12	21
60–69	6	9
50–59	2	3
40–49	1	1
	$\Sigma f = 40$	

The frequency distribution ranges from the 40–49 interval up to the 90–99 interval, with most of the students scoring above a grade of 70. To find the median we must first find the number of scores that will be below the median.

$$F_P = .50 \times 40 = 20$$

We know now that the median will be the value of X that has a total of 20 frequencies below it. Using the formula for finding a percentile yields the following (Note that our interval size in this example is $i = 10$):

$$P_{50} = 69.5 + \frac{20-9}{12} \times 10$$

$$P_{50} = 69.5 + \frac{11}{12} \times 10$$

$$P_{50} = 69.5 + .917 \times 10$$

$$P_{50} = 69.5 + 9.17 = 78.67$$

The median score in the frequency distribution is 78.67, which indicates that anyone earning that grade on the exam was in the middle of the distribution.

To Find the Median

When n is odd: $i = \frac{n+1}{2}$	The median is the value of raw data the ith X.
When n is even: $i = \frac{n}{2}$	The median is the value of raw data $\frac{X_i + X_{i+1}}{2}$
For a frequency distribution	The median is the value of the 50th percentile.

Up to this point, you have seen applications of the median for raw data for both even size samples and odd size samples, and for a frequency distribution. The median still may not seem very familiar to you, but you have probably encountered reports of median values more often than you might think. What types of data are often reported as median values? Very often the median is reported for variables that can have a wide range of values, and in particular a small number of values at one extreme end of a distribution. Two examples that you will often encounter are data on income, and data on the sales price of houses.

Consider two simple distributions presented below, where the mean and median are each reported.

X_1:	33	35	45	67	78	85	92	95	100
X_2:	33	35	45	67	78	85	92	95	1,270

	Sample Size	Median	Mean
Distribution 1	9	78	70
Distribution 2	9	78	200

Each distribution has nine observations, and they each have the same median equal to 78. In fact, the two distributions are identical with the exception of the last observation. However, the difference in that single observation is enough to increase the mean of the second distribution by 130 points. Note that the effect of the extreme score in the second distribution results in a mean value that is over twice as large as the second largest value of the distribution (95). The mean, which is intended to be a measure of central tendency, is not really close to any of the observations in the distribution. The median, in contrast, remains stable at a value of 78 for each of the distributions, even when a large value is encountered in distribution 2.

Why is the median so often used as a measure of central tendency for income data and housing data? One major reason is that income and housing data often contain some extreme values at the high end of those distributions. Most households in the United States earn at least $7,500 but below $200,000 (Current median household income in the United States is approximately $50,000. An income of $12,000 would put your household at the 10th percentile, while a household income of approximately $190,000 would put your household at the 95th percentile.) Yet, there are some extreme cases on the upper end of the income scale. You can imagine what would happen to the mean income value for a group of individuals if an extreme case were to be included.

A similar situation might arise in the case of the sales price of houses. Most private homes in the United States sell from between $25,000 and $500,000. The median sales price of existing homes in the United States is currently around $200,000 (the median sales price of new homes would be higher). Yet I am sure you are aware of homes that have sold for much more, and recently several single family homes have sold for just over one hundred million dollars! Again, the effect of extreme values on the upper end of the distribution is to create an upward bias on the mean, but the median will remain stable in those cases.

Mean

The mean is the third measure of central tendency that we will examine, and it is the one familiar to most people, although most people know the mean by a more common name—the average. Calculating a mean for a set of raw data is a simple process. The formula for the mean is as follows:

$$\overline{X} = \frac{\sum_{i=1}^{n} X_i}{n}$$

where $\sum X$ is the sum of all of the *X*s in the distribution (recall from Chapter 1 that the "$i = 1$ to n" notation means that we begin with the first X and continue to the nth or last one); and, n is the sample size.

We are using the symbol $\overline{X}$ (X bar) in the formula suggesting that we are computing a mean for a sample, but the formula would be the same when computing a mean for a population which we symbolize with the Greek letter μ (mu). To compute the mean, we simply sum the values for a measured variable, and divide the total by the number of observations. In the following example with raw data we have $n = 7$ observations of a variable labeled $\overline{X}$.

$$\overline{X}: 24\ 56\ 75\ 32\ 23\ 19\ 30$$

Summing the values of $\overline{X}$ results in a total of 259, and then dividing the total by 7, the number of observations, results in the mean value of 37.

$$\overline{X} = \frac{\sum_{i=1}^{n} X_i}{n} = \frac{259}{7} = 37$$

A mean may be calculated for data in a simple frequency distribution ($i = 1$), and for data in a frequency distribution ($i > 1$) with a simple modification in the basic formula. We will look at each of these situations in turn. Consider the following simple frequency distribution.

X	f
11	1
9	5
8	4
5	3
3	3
	$\sum f = 16$

In this simple frequency distribution, we have a total of $n = 16$ scores that range from 3 to 11. We do not need cumulative frequency information to compute the mean, so that column has been omitted. Our formula for the mean must be modified slightly to account for the fact that we have multiple observations of the various values of X. The modified formula becomes

$$\overline{X} = \frac{\sum_{i=1}^{n} fX_i}{n}$$

where $\sum fX_i$ represents the sum of each value of X multiplied by its frequency of occurrence in the frequency distribution; and,
n is the sample size (recall that $n = \sum f$).

The first step is to multiply each observed value of X by the corresponding number of observations as indicated by the frequency column. The result of this multiplication is placed in the fX column:

X	f	fX
11	1	11
9	5	45
8	4	32
5	3	15
3	3	9
	$\sum f = 16$	$\sum fX = 112$

We can complete the calculation for the mean by dividing the $\sum fX$ of 112 by the number of observations, $n = 16$ as follows:

$$\overline{X} = \frac{\sum_{i=1}^{n} fX_i}{n} = \frac{112}{16} = 7$$

Keep in mind that the total sample size represented in our simple frequency distribution is 16. The observations are presented in five categories (3, 5, 8, 9, and 11), but we still have a total of 16 observations.

It is a common mistake to confuse the number of categories in a simple frequency distribution with the total sample size. To do so will result in an incorrect computation of the mean. In this case, the mean would have been 22.4 if we had mistakenly divided the ΣfX (112) by 5 instead of 16.

Some statistical calculations reflect a certain amount of common sense, and the mean is one of them. You should be able to look at a distribution of numbers and make a reasonable approximation of what the mean will be. One thing for certain is that the mean must be within the range of numbers in the distribution. In our distribution, the numbers range from 3 to 11. Common sense should tell us that there is no way the mean value for a set of numbers ranging from 3 to 11 will equal 22.4. One way to avoid mistakes such as this is to concentrate on understanding what each statistical procedure is intended to indicate rather than just focusing on how to plug numbers into the formula and generate an answer.

The mean may also be calculated for a frequency distribution where the interval size i is greater than 1. We use a formula for the mean in this case very similar to the one used with a simple frequency distribution.

$$\bar{X} = \frac{\sum_{i=1}^{n} fX_i}{n}$$

where $\Sigma X'i$ represents the sum of the midpoint (X') for each interval multiplied by the number of frequencies in the interval; and, n is the sample size

The only difference is that in this case X' (read as X prime) refers to the midpoint of the interval. As you saw in Chapter 3, constructing a frequency distribution with intervals greater than 1 benefits us by making it easier to interpret a large set of raw data, but we lose some precision of measurement in the process. We may see that there are a total of five observations in the interval 20–24, but we no longer know exactly what they are. Still, there are times when we must regain the original precision of our measurement represented by the raw data, or find a way to approximate it. In calculating a mean for data in a frequency distribution, we approximate precise measurement by assuming that all of the observations within a given interval are located at the midpoint of the interval.

Consider the frequency distribution below.

X	f	Cf
45–49	6	50
40–44	8	44
35–39	12	36
30–34	10	24
25–29	9	14
20–24	5	5
	$\Sigma f = 50$	

We have a total of $\Sigma f = 50$ scores ranging from five scores in the 20–24 interval to six scores in the 45–49 interval. To calculate the mean we either must know the exact value of each score, or make an assumption about the exact value of each score. In this case, we will assume that each score is equal to the midpoint of the interval in which it is found. Recall that an easy method of calculating the interval midpoint is to add the apparent upper and lower limit of each interval and divide by 2. We will not need the cumulative frequency column to calculate the mean, so let's rewrite the frequency distribution replacing the cumulative frequency column with a new column containing each interval's midpoint represented by X'. Also note that the formula for calculating the mean for a frequency distribution requires us to multiply the interval midpoint (X') by the number of observations found in each interval (f), so we will create a column for the resulting product of fX'.

X	f	X'	fX'
45–49	6	47	282
40–44	8	42	336
35–39	12	37	444
30–34	10	32	320
25–29	9	27	243
20–24	5	22	110
	$\Sigma f = 50$		$\Sigma fx' = 1{,}735$

All that is required to finish the calculation of the mean for the frequency distribution is to divide the $\Sigma fX'$ by the sum of the frequencies.

$$\bar{X} = \frac{\sum_{i=1}^{n} fX_i'}{n} = \frac{1735}{50} = 34.7$$

The mean for the frequency distribution is equal to 34.7, and again this is a result that should be somewhat intuitive. Looking at the range of the values represented by the frequency distribution, and the number of frequencies in each interval, you should expect a mean or "average" value somewhere close to the center of the distribution. After all, the mean is a measure of central tendency. If the distribution were skewed in some fashion, for example, with a large number of frequencies toward the upper end, then the mean would be a little higher. But you should certainly observe a mean value within the limits of the distribution.

Data at the interval level or higher are required to legitimately calculate the mean. Using integers to substitute for nominal or ordinal level categories do not allow us to calculate a true mean. For example, coding males as (1), and females as (2) will not transform the nominal categories of gender into interval level data.

Gender	f	fX
(2) Female	75	150
(1) Male	50	50
	$\Sigma f = 125$	$\Sigma X = 200$

$$\bar{X} = \frac{\sum_{i=1}^{n} fX_i}{n} \neq \frac{200}{125} \neq 1.6$$

We cannot say that the mean gender is equal to 1.6. The most appropriate way to summarize data such as these at the nominal level is to report percentages.

The Mode, Median, and Mean Compared

The mode, median, and mean are all suitable as measures of central tendency. They each provide us with a single statistic that tells us something about what is typical of a given distribution. The principal advantage of the mode is that it may be computed for any level of data, and is the only measure of central tendency that may be legitimately computed for nominal and ordinal level data.

The mean and median each require at least data at the interval level. The median is an informative measure of central tendency, which is not subject to the presence of one or two extreme scores in a distribution. The mean is the most familiar of the measures of central tendency, although as we have seen, it is subject to bias from extreme values. The mean has one additional advantage. The mean is much more likely to be used in conjunction with other statistical procedures. For example, the entire concept of variance covered in Chapter 5 involves the examination of variation in a distribution as measured from the mean. In addition, the key statistical tests of significant difference covered in Chapter 11 are based on a comparison of the difference in the means of a population and a sample, or the difference in the means of two samples.

The Grand Mean

The **Grand Mean** is the final concept covered in this chapter. The grand mean differs from the sample or population mean that we have previously examined. The mean that is reported for a population or a sample is a mean of individual scores. The grand mean arrives at the same information, but it is based on a mean of means. Consider the example below where hypothetical data on grade point average for a group of students is presented by classification.

Class	Mean GPA	f	Cf
Senior	3.20	50	230
Junior	3.05	75	180
Sophomore	2.85	45	105
Freshman	2.75	60	60
		$\Sigma f = 230$	

The frequency distribution indicates the grade point average by classification for each of the four groups represented, but we do not have the grade point average for the total group of 230 students.

Calculating the grand mean would be simple if each group had the same number of individuals in it. For a situation similar to this one we could simply add the four individual grade point average (GPA) values, and then divide by 4 (the number of groups represented). In this example, which is more likely to be typical of what you would encounter, the four groups have different sample sizes so we will have to weight each of the four group means accordingly. Each group's mean GPA will be weighted by multiplying the group's GPA by the number of individuals in the group. The weighted group means will be summed, and then divided by the total sample size (note that we do not simply divide the weighted group means by the number of groups).

The frequency distribution is rewritten below with a column containing the weighted mean GPA in place of the cumulative frequency column which is not necessary in this procedure.

Class	Mean GPA	f	GPA $\times f$
Senior	3.20	50	160.00
Junior	3.05	75	228.75
Sophomore	2.85	45	128.25
Freshman	2.75	60	165.00
		$\Sigma f = 230$	ΣGPA $\times f = 682.00$

The grand mean, symbolized by $\bar{X}_T$ (for total mean) may be computed by the following formula:

$$\bar{X}_T = \frac{\sum_{i=1}^{k} f_i \bar{X}_i}{\Sigma f}$$

where $\Sigma f_i \bar{x}_i$ is the sum of each group mean multiplied by its sample size and Σ is the total sample size.

The number of groups included in the frequency distribution is represented by the letter k (in this case four). Substituting the appropriate values into the formula yields:

$$\bar{X}_T = \frac{\sum_{i=1}^{k} f_i \bar{X}_i}{\Sigma f} = \frac{682.0}{230} = 2.97$$

The resulting value of 2.97 represents the GPA of the entire group of $n = 230$ students.

Consider another example below where mean salaries are provided by gender, but we do not have a mean salary for the entire workforce.

Group	Mean Salary	f	Salary $\times f$
Male	83,500	50	4,175,000
Female	85,300	40	3,412,000
		$\Sigma f = 90$	ΣSalary $\times f = 7{,}587{,}000$

Substituting the appropriate values into the formula yields:

$$\bar{X}_T = \frac{\sum_{i=1}^{k} f_i \bar{X}_i}{\Sigma f} = \frac{7587000}{90} = 84300$$

The mean income of all employees is $84,300. What does the amount 7,587,000 obtained in the intermediate step indicate? Would this number ever be of any significance, or is it just a step in the process of obtaining the grand mean? In the previous example, the sum of each group's GPA multiplied by the number of frequencies in the group yielded 682.00, which has little if any additional significance beyond being an intermediate calculation toward the grand mean. However, in this example, the value of 7,587,000 might have some additional significance. In this case, it represents the value of the total annual payroll.

Making a consistent effort to understand the process is the key to developing a firm foundation in statistics. You can get by with knowing how to plug the numbers into a formula, but the power of statistical analysis can only be developed through understanding the process, and seeing the linkages between concepts. In Chapter 5, we will examine the concept of variation in a distribution. One key aspect of the next chapter deals with the idea of variation, or distance of each observation, from the mean.

Summary of Key Points

Measures of central tendency provide a summary statistic intended to tell us what value in a distribution of scores is typical of the distribution.

Mode—The mode is the simplest measure of central tendency representing the most frequently observed value of a variable.
Median—The median is the midpoint in a distribution of scores. Half of the scores will be above the median, and half of the scores will be below the median.
Mean—The mean is the arithmetic average of a set of scores. The mean is more sensitive to extreme scores at either end of a distribution than the median.
Grand Mean—The grand mean is a mean calculated from a set of means rather than from a set of individual scores.
Bimodal—A distribution that has two modes.
Trimodal—A distribution that has three modes.

Questions and Problems for Review

1. Distinguish between the three major measures of central tendency.

2. Calculate the mode, mean, and median for the raw data below:

   X: 23 45 32 12 11 43 23 43 45 268 67 34

3. Would the mean or the median be the better measure of central tendency for the distribution in problem 2? Why?

4. Calculate the mean for the simple frequency distribution below:

X	f	Cf
18	9	44
15	10	35
13	8	25
11	11	17
10	6	6
	$\Sigma f = 44$	

5. Calculate the mode, mean, and median for the frequency distribution below:

X	f	Cf
35–39	8	75
30–34	11	67
25–29	18	56
20–24	16	38
15–19	12	22
10–14	10	10
	$\Sigma f = 75$	

6. Calculate the overall GPA (grand mean) for the students represented below from the information provided:

Major	f	GPA
Economics	8	3.12
English	15	2.85
Mathematics	10	3.25
Psychology	15	3.35
Sociology	12	3.50
	$\Sigma f = 60$	

7. Why do the sociology majors have the highest GPA for any of the groups represented in problem 6?

8. Select an article of interest from a scholarly journal in your area of study. Which measure of central tendency is reported more often? Is more than one measure of central tendency reported?

9. Compute the mean, mode, and median for the raw data below:

X: 20 23 26 28 29 30 33 33 34 36 38 39 39 40 40 40 42 43 45 47 48 48 49 50 52 54 55 56 56 57 59 60 62 63 65 66 67 67 68 68

10. Create a frequency distribution for the data in problem 9 with an interval size $i = 10$ (review the guidelines for creating a frequency distribution in Chapter 3 if necessary), and calculate the mean, mode, and median for the data in the frequency distribution. How do the values of the mean, mode, and median for the frequency distribution differ when compared to the raw data calculations?

CHAPTER 5

Measures of Variation

Key Concepts

Deviation from Mean
Range
Interquartile Range
Semi-Interquartile Range
Variance
Standard Deviation
Explaining Variance

Introduction

Measures of variation refer to a group of statistics that is intended to provide us with information on how a set of scores are distributed. An examination of measures of variation is a logical extension of any description of a data set using the measures of central tendency that we examined in the previous chapter. Consider a case where there are two sections of a course in statistics, and you are told that each section is taught by the same professor, each section has an enrollment of 15 students, and that the mean, and median score on a recent examination is 80 in both sections of the course. Without any additional information you would be tempted to conclude that the performance of the students in the two sections of the course is reasonably similar. As a matter of fact, all of the information up to this point would suggest that the performance of the students in the two sections is identical.

Now suppose that you are shown the actual performance of each student on the examination in both sections of the course (see Figure 5.1).

Clearly the performance of the students in the two sections is radically different. The score of 80 is not only the mean of section 1, but also a score that seems to be more representative of the performance of the entire class. While not all of the students scored 80, more were at that score

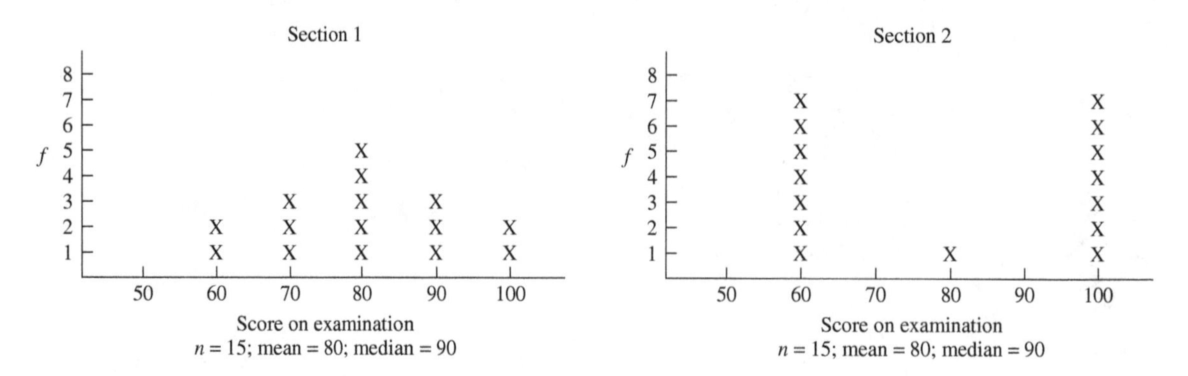

Figure 5.1 Graph showing distributions for the two sections

than any other, and the number of students scoring higher or lower than 80 falls off the further the scores deviate from 80. The performance of the students in section 2 is far different even though the distribution has the same mean and median score as the first distribution. In section 2 the mean of 80 is not at all representative of the typical performance. In fact, only one student earned a score at the mean. Seven students earned perfect scores of 100, while the remaining seven students earned a very low score of only 60.

Just as the mean is a single number that is designed to tell us where the central point of a distribution of scores is located, measures of variation are single numbers that are designed to tell us how the individual scores are distributed. By examining both a measure of central tendency such as the mean, and an appropriate measure of variation, we will be able to know not only where the central point of a distribution is located, but also if it tends to look more like the distribution of scores in section 1 or if it looks more like the distribution of scores in section 2.

The measures of variation examined in this chapter can be divided into two groups. The first group of statistics measures variation in a distribution in terms of the distance from the smaller scores to the higher scores. Included in this group of measures of variation is the range, which is a simple measure of the variation in a distribution computed by examining the distance from the smallest score to the largest score. Also included in this group of statistics are the interquartile range (IQR), and the semi-interquartile range (SIQR). These latter two measures of variation are often used in educational research. The second group of statistics measures variation in terms of a summary measure of each score's **deviation from the mean**. The two statistics of this type that we will examine are the variance, and the standard deviation. The measures of variation based on deviation from the mean tend to be more useful, and are fundamental concepts in behavioral science research.

The Range, Interquartile Range, and Semi-interquartile Range

Range

The simplest measure of variation in a distribution of data is the **range**. The range is defined as the distance from the smallest observed score to the largest observed score in a set of data. For raw data, the range may be computed by subtracting the lower limit of the smallest observed score from the upper limit of the largest observed score. Consider the set of raw data below:

X: 2 4 5 5 7 8 10 12 15 18 21

The range is $(21.5 - 1.5) = 20$.

A simple alternative is to subtract the smallest observed score from the largest observed score, and then add 1 additional unit of measurement to compensate for the upper and lower limits. In this case, the range is

$$(21 - 2) + 1 = 20$$

In either case, the set of data ranges over 20 units, just as if you began reading a book on page 2 and continued through page 21 you would have read a total of 20 pages.

Keep in mind when using the alternative method of subtracting the smallest observed score from the largest observed score that we add 1 *unit of measurement*, not just 1. For example, if we had the following data on income:

Income: 12,000 14,000 18,000 35,000 46,000 58,000

The range would be computed as:

$$(58{,}000 - 12{,}000) + 1{,}000 = 47{,}000$$

It would not be correct to compute the range as:

$$(58{,}000 - 12{,}000) + 1 = 46{,}001$$

The computation is similar for data organized in a simple or full frequency distribution as illustrated below.

X	f	Cf
12	7	20
10	3	13
8	3	10
7	5	7
6	2	2
	$\Sigma f = 20$	

In a simple frequency distribution where the interval size $i = 1$, we do not lose any precision in measurement. In this case the range would be computed as

$$(12.5 - 5.5) = 7$$

or, alternatively:

$$(12 - 6) + 1 = 7.$$

In a frequency distribution where the interval size $i > 1$, we follow the same general procedure, but use the lower limit of the smallest interval, and the upper limit of the largest interval as our parameters for the computation of the range. For the data below:

X	f	Cf
25–29	9	75
20–24	16	66
15–19	20	50
10–14	16	30
5–9	10	14
0–4	4	4
	$\Sigma f = 75$	

the range would be computed as: [29.5 – (–0.5)] = 30
or, alternatively (29 – 0) + 1 = 30.

Interquartile range

The **interquartile range** (IQR) is defined as the distance from the 75th percentile to the 25th percentile in a set of data. In the previous chapter on central tendency we saw that a few extreme scores at one end of the distribution can bias a measure such as the mean. The same situation is true for a measure of variation such as the range. A few extreme scores at one or the other end of a distribution will affect the size of the range. The IQR is an alternative measure of variation that eliminates the effect of the extreme scores in a distribution by reporting the range between the 75th and 25th percentiles. In effect, the IQR represents the range of the middle 50% of the distribution, and ignores the top 25% and bottom 25% of the data that may be subject to extreme scores.

Computing the IQR is as simple as subtracting the 25th percentile from the 75th percentile:

$$\text{IQR} = \text{P75} - \text{P25}$$

The formula for finding a particular percentile from Chapter 3 is provided below, along with the frequency distribution previously used to illustrate the procedure. To compute the IQR we will need to find both the 25th percentile and the 75th percentile.

X	f	Cf
25–29	20	125
20–24	22	105
15–19	28	83
10–14	20	55
5–9	25	35
0–4	10	10
	$\Sigma f = 125$	

The general formula for finding a given percentile is given by the equation below.

$$\text{P}_x = \text{LL} + \left[\left(\frac{F_\text{P} - \text{CF}_\text{below}}{f_\text{int}}\right) \times i\right]$$

where

P_x = the desired percentile
F_P = the number of frequencies below the desired percentile
CF_below = the value from the cumulative frequency column for the interval just below F_P
LL = the lower limit of the interval containing F_P
f_int = the number of frequencies in the F_P interval
i = the interval size.

For the 25th percentile:

$$F_\text{P} = (.25) \times (125) = 31.25$$

For the 75th percentile:

$$F_\text{P} = (.75) \times (125) = 93.75$$

The intervals containing the 25th and 75th percentiles are indicated below:

X	f	Cf	
25–29	20	125	
20–24	22	105	<--(F_P= 93.75 is in this interval)
15–19	28	**83**	<--(CF_{below} = 83)
10–14	20	55	
5–9	25	35	<--(F_P= 31.25 is in this interval)
0–4	10	**10**	<--(CF_{below} = 10)
	$\Sigma f = 125$		

To find the 25th percentile we simply substitute appropriate values into the formula as follows:

$$P_{25} = 4.5 + \left[\left(\frac{31.25 - 10}{25}\right) \times 5\right]$$

$$P_{25} = \left[\left(4.5 + \frac{21.25}{25}\right) \times 5\right]$$

$$P_{25} = 4.5 + (.85 \times 5)$$

$$P_{25} = 4.5 + 4.25 = 8.75$$

To find the 75th percentile we simply substitute appropriate values into the formula as follows:

$$P_{75} = \left[\left(19.5 + \frac{93.75 - 83}{22}\right) \times 5\right]$$

$$P_{75} = 19.5 + \left[\left(\frac{10.75}{22}\right) \times 5\right]$$

$$P_{75} = 19.5 + (.49 \times 5)$$

$$P_{75} = 19.5 + 2.45 = 21.95$$

Having computed both the 25th and the 75th percentiles, we may now compute the IQR:

$$IQR = 21.95 - 8.75 = 13.2$$

The resulting value of 13.2 for the IQR indicates that there is a range of 13.2 points for the middle 50% of the distribution. The advantage of the IQR over the simple range is that any bias that might result from a few extremely high scores or a few extremely low scores (or both) has been eliminated.

Semi-interquartile range

The final range based measure of variation presented in this chapter is the **semi-interquartile range**. The concept "interquartile range" suggests a measure of variation based on a quartile, or 25% of the distribution; however, the IQR actually represents the range of the middle 50% of the distribution. The SIQR is an alternative to the IQR that comes closer to representing a "quartile or 25%" size range in the distribution. The SIQR is simply the IQR divided by 2.

$$\text{SIQR} = \frac{\text{IQR}}{2}$$

In the case of our previous example, the SIQR is

$$\text{SIQR} = \frac{13.2}{2} = 6.6$$

The IQR and the SIQR are widely used in education research where there always seems to be one or two students at each extreme of the distribution.

The Variance and the Standard Deviation

Variance

The **variance** and the **standard deviation** are two measures of variation that are based on the concept of deviation from the mean. For any distribution of scores measured on a continuous scale we can compute a mean, and then measure the distance of each score from the mean. For example, the set of 6 scores presented below have a mean equal to 8.

X
13
11
9
7
5
3

We may then define deviation from the mean (di) as the distance of each score from the mean, or

$$\text{d}i = X - \bar{X}$$

Using our set of 6 scores, we may then calculate the deviation from the mean for each score.

X	$X - \bar{X} = \text{d}i$
13	(13 – 8) = 5
11	(11 – 8) = 3
9	(9 – 8) = 1
7	(7 – 8) = –1
5	(5 – 8) = –3
3	(3 – 8) = –5

One way we might construct a summary measure of variation in a distribution of scores is to compute the average deviation of each score from the score's mean. To do this, we would simple sum the individual deviations from the mean which we have just calculated, and then divide by the number of observations we have.

$$\frac{\Sigma di}{n} = \frac{5+3+1+(-1)+(-3)+(-5)}{6}$$

$$\frac{\Sigma di}{n} = \frac{0}{6} = 0$$

It may seem quite logical to construct a measure of variance by calculating the average deviation from the mean for a set of scores, but there is one small problem. The sum of the deviations from the mean for all distributions is always the same thing—0.

$$\Sigma di = 0$$

One solution to this problem is to base our measure of variance on the *squared deviation* from the mean. By squaring the result of $(X - \bar{X})$, we will eliminate the negative numbers, and prevent the negative deviations and positive deviations from canceling each other out.

Applying this strategy to our original distribution will give us the following result:

X	$X - \bar{X}$	$(X - \bar{X})^2$
13	$(13 - 8) = 5$	$(5)^2 = 25$
11	$(11 - 8) = 3$	$(3)^2 = 9$
9	$(9 - 8) = 1$	$(1)^2 = 1$
7	$(7 - 8) = -1$	$(-1)^2 = 1$
5	$(5 - 8) = -3$	$(-3)^2 = 9$
3	$(3 - 8) = -5$	$(-5)^2 = 25$

Now if we want to construct a measure of variation that gives us a single number representing the average variation of each score in a distribution, we can use the mean (or average) of the squared deviations of each score from the distribution mean. We need only sum the squared deviations from the mean, and then divide by the number of observations.

$$\frac{\Sigma(X - \bar{X})^2}{n} = \frac{25+9+1+1+9+25}{6}$$

$$\frac{\Sigma(X - \bar{X})^2}{n} = \frac{70}{6} = 11.67$$

The resulting value indicates that the mean squared deviation of each score from the distribution mean is 11.67; or stated differently, on average, the distance squared of each score from the mean is 11.67 units.

The statistic we call the **variance** represents the mean squared deviation from the mean for a set of data. The logical formula for the variance is simply:

$$\text{Variance} = \frac{\Sigma(X - \bar{X})^2}{n}$$

where

X = each value of X in the distribution
$\bar{X}$ = the mean of the distribution
n = the sample size.

What does the variance really tell us?

Recall the situation from earlier in the chapter where we had two distributions representing the performance of two sections of a class on an exam with the same number of observations, the same mean, and the same median. Yet, we could see by simple inspection that the two distributions were very different. The variance of the distribution will tell us how representative the mean is of each of the scores in the distribution. The closer each individual score is to the mean the smaller the variance will be. If each score is at the mean in a distribution the variance will equal zero, indicating that there is no variation from the mean across the entire distribution. The farther each of the individual scores is from the mean the greater the variance will be, indicating that the mean is not as typical of the individual scores in the distribution.

Examine the three simple distributions below.

A	B	C
10	10	8
8	10	6
6	6	6
4	2	6
2	2	4

Each distribution contains 5 observations, and each distribution has a mean equal to 6. Yet the observations differ with respect to how much the individual scores vary from the mean. Since the variance represents the average squared deviation from the mean, which distribution would you expect to have the greatest variance? Which distribution should have the smallest variance?

The mean value of 6 seems to be most typical of the scores in distribution C, so it should have the smallest variance. The scores in distribution B are much farther from the mean value of 6, so it should have the largest variance. The scores in distribution A appear somewhat in between, and should have a variance between that of distribution B and distribution C.

Let's calculate the variance for each distribution below:

A	$(X-\bar{X})$	$(X-\bar{X})^2$	B	$(X-\bar{X})$	$(X-\bar{X})^2$	C	$(X-\bar{X})$	$(X-\bar{X})^2$
10	4	16	10	4	16	8	2	4
8	2	4	10	4	16	6	0	0
6	0	0	6	0	0	6	0	0
4	–2	4	2	–4	16	6	0	0
2	–4	16	2	–4	16	4	–2	4
	$\Sigma(X-\bar{X})^2 = 40$			$\Sigma(X-\bar{X})^2 = 64$			$\Sigma(X-\bar{X})^2 = 8$	
Variance:		$\frac{40}{5} = 8$			$\frac{64}{5} = 12.8$			$\frac{8}{5} = 1.6$

As suspected, the variance in distribution C is the smallest. For distribution C the average squared deviation from the mean is 1.6 units. Distribution B has the largest variance with an average squared deviation of 12.8 units. Distribution A is in between the two with an average squared deviation from the mean of 8 units.

It is important to realize that the size of the variance does not have any special underlying standard interpretation. The value of the variance does not have a special meaning like your blood pressure, where you know that you are in reasonably good condition with a systolic blood pressure of 120 and a diastolic pressure of 80. There is no normal or abnormal range for the variance. What the variance is telling you is simply what the squared distance is from the typical or average score to the mean. Variance, along with the mean, can then allow you to have an idea of what a particular distribution might look like, and will allow you to judge how well the mean serves as a measure of central tendency.

Consider the case of the three simple distributions we just examined. Typically, information for such distributions would not provide individual scores, but would be presented in summary form as follows:

	Distribution		
Summary Statistics	A	B	C
Sample Size:	5	5	5
Mean:	6	6	6
Variance:	8.0	12.8	1.6

Even though we do not have the individual scores available, we can reach some fairly accurate conclusions about what each of these distributions would look like. For example, we can see that all three distributions are of the same size ($n = 5$), and that all three have the same mean ($\bar{X} = 6$). The fact that the variance for distribution C is only 1.6 indicates that most of the individual scores in the distribution should be very close to the value of 6. After all, a distribution where every score is equal to the mean will have a variance of zero (remember that variance can never be negative). Similarly, we would assume that the individual scores in distribution B must be much more diverse or spread out around the mean since the variance is so much larger.

A computational formula for variance

Up to this point we have utilized a logical formula for the variance that is useful for demonstrating how the variance is computed, but requires us to go through some unnecessary steps. A computational formula may be used, which simplifies the calculations, especially when a larger data set is involved. The computational formula presented below may look a little more difficult at first, but with a little experience using it you will likely find it to be much easier.

$$\text{Variance} = \frac{\Sigma X^2 - \frac{(\Sigma X)^2}{n}}{n}$$

where

ΣX^2 = the sum of the Xs squared
$(\Sigma X)^2$ = the quantity, sum of Xs squared
n = the sample size.

Notice that the computational formula for variance contains both the sum of Xs squared term, and the quantity sum of Xs squared term.

The term in the numerator: $\Sigma X^2 - \frac{(\Sigma X)^2}{n}$

represents the sum of the squared deviations from the mean that was formally written as

$$\Sigma(X - \bar{X})^2$$

This term is also sometimes referred to as simply the sum of squares, and will play a role in several statistical procedures that will be examined in later chapters.

Let's demonstrate that the computational formula for the variance will provide the same results that we previously obtained for our three distributions.

A		B		C	
X	X^2	X	X^2	X	X^2
10	100	10	100	8	64
8	64	10	100	6	36
6	36	6	36	6	36
4	16	2	4	6	36
2	4	2	4	4	16
$\Sigma X = 30$	$\Sigma X^2 = 220$	$\Sigma X = 30$	$\Sigma X^2 = 244$	$\Sigma X = 30$	$\Sigma X^2 = 188$
$(\Sigma X)^2 = 900$		$(\Sigma X)^2 = 900$		$(\Sigma X)^2 = 900$	

Variance Computation		
Distribution A	Distribution B	Distribution C
$\dfrac{220-\dfrac{900}{5}}{5}$	$\dfrac{244-\dfrac{900}{5}}{5}$	$\dfrac{188-\dfrac{900}{5}}{5}$
$\dfrac{220-180}{5}$	$\dfrac{244-180}{5}$	$\dfrac{188-180}{5}$
$\dfrac{40}{5}$	$\dfrac{64}{5}$	$\dfrac{8}{5}$
8	12.8	1.6

In each case we obtain the same result for the variance with the computational formula that we previously obtained with the logical formula. Remember that the variance will never be negative. If your calculation of the variance results in a negative number, you can be sure that you have made an error somewhere. Confusing the ΣX^2 with the $(\Sigma X)^2$ in the computational formula is a common mistake that will result in a negative number, as is neglecting to divide $(\Sigma X)^2$ by n before subtracting the result from ΣX^2.

Some important terminology and symbols for variance

At this point we need to introduce some terminology, and appropriate symbols for the variance. There are three situations where we might want to calculate the variance. The logical formula and its computational alternative that we have been using are appropriate for two of the three situations.

Recall from Chapter 1 the distinction between a population representing the entire collection of units of interest in a research project, and a sample that is the smaller subgroup that we select for actual observations. The variance for a population is represented by the lower case Greek letter sigma squared, or σ^2. The population variance σ^2 may be calculated using the formula that we have worked with up to this point. The variance for a sample of data, when we are only interested in describing the sample, is represented by the upper case letter S^2, and it too may be calculated with the formula that we have worked with up to this point.

However, if you also recall from Chapter 1, we made a distinction between descriptive statistics that are used to describe a set of data and inferential statistics that are used to infer something about a population parameter by the observation of sample statistics. We often have an interest in doing just that in behavioral science research, or at the very least we are interested in being able to generalize our results to a larger population. It turns out that if you were to actually know the value of a population's variance, and then take a series of samples from the population and compare the sample variances computed for each sample to the actual population variance, you would find that the sample variances tend to underestimate the true size of the population variance. The sample variance is sometimes referred to as a biased estimator of the population variance, and the direction of the bias is to underestimate the true size of the population variance.

We can reduce the bias of the estimate of the population variance when using data from a sample by making a slight adjustment in the formula for the sample variance when we intended it to serve as an estimate of the population variance. Since the direction of the bias is to underestimate the true population variance, we can increase the size of the estimated variance by using the value $n - 1$ in the denominator of the variance formula in place of the usual denominator n. We will then use the lower case letter s^2 to represent a sample variance that is being used as an estimate of the population variance.

It might be useful at this point to review the computational formulas for the three situations for computing variance.

Symbol	Situation	Computational Formula
σ^2	Population Variance	$\sigma^2 = \dfrac{\Sigma X^2 - \dfrac{(\Sigma X)^2}{n}}{n}$
S^2	Sample Variance	$S^2 = \dfrac{\Sigma X^2 - \dfrac{(\Sigma X)^2}{n}}{n}$
s^2	Sample Variance used to estimate the population variance σ^2	$s^2 = \dfrac{\Sigma X^2 - \dfrac{(\Sigma X)^2}{n}}{n-1}$

n – *1 as degrees of freedom*

The use of "$n - 1$" in the denominator of the formula for variance is the result of the concept of **degrees of freedom**. One of the things that we observed when looking at deviation from the mean was that the sum of the deviations from the mean always equaled zero. However, when dealing with sample data from a population there is no guarantee that the sum of the deviations of the individual sample scores from the *population mean* will actually equal zero. (Don't be confused by the fact that the sum of the deviations of the sample scores from the sample mean will equal zero. What we are concerned with here is whether or not the sum of the deviations of the sample scores from the population mean will equal zero.)

Consider this example of a population of $N = 10$ with a mean $\mu = 30$. (You may want to verify that the sum of the deviations of the population scores from the population mean is actually zero.) Suppose I select a sample of $n = 5$ scores from the population as indicated below.

We can calculate a sample mean for the five scores that were selected from the population, and if we calculate the sum of the deviations from the sample mean for the five scores in the sample we will in fact find the sum equal to. But is the sum of the deviations of the sample scores from the population mean equal to zero? No! The sum of the deviations from the sample scores to the population mean is actually –30. How

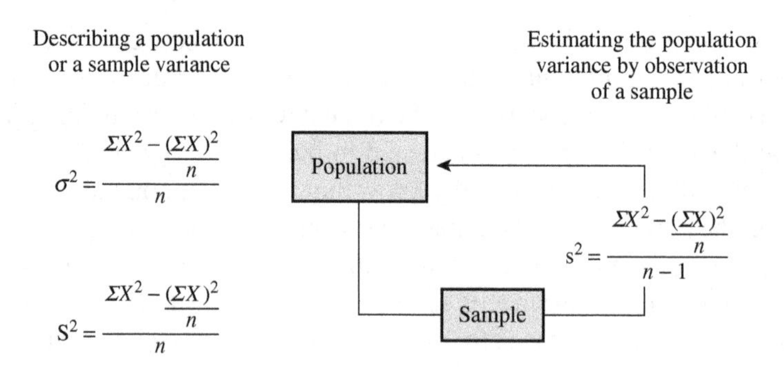

Figure 5.2 Diagram of population and sample with the three formulas

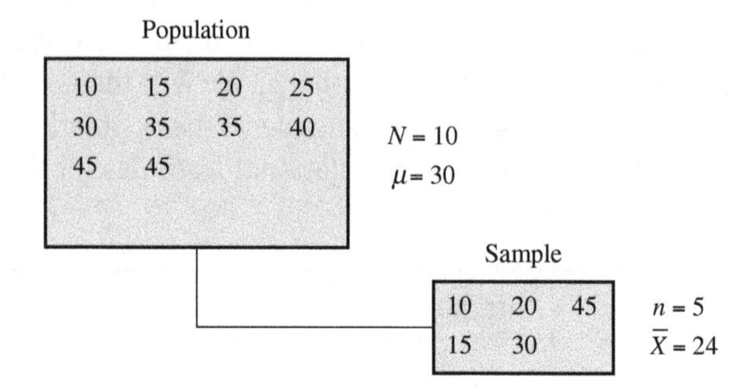

Figure 5.3 Illustration of population data and sample drawn from it

many of the sample scores would I have to change (or control) if I wanted to force the sum of the deviations of the sample scores from the population mean to equal zero? The answer is that *only one* of the sample scores must be controlled. I can always ensure that the sum of the deviations of the sample scores from the population mean will equal zero if I can control one of the sample scores. It does not matter which score I choose to control. I simply must be sure that the value of the score I choose will result in a deviation from the population mean that when added to the other deviations from the population mean will give me a sum of zero. Since I must control only one score, it means that the other scores are free to take on any value. In other words, I have $n - 1$ degrees of freedom. The concept of degrees of freedom will be examined again when we begin our investigation of hypothesis testing in later chapters.

It is easy to be confused by the concept of degrees of freedom, or the difference between describing a sample variance through the use of one formula, and estimating the population variance from sample data using a different formula. Like many of the concepts in statistics, it takes time for reflection and experience before everything falls into place. At this point, the most important thing to keep in mind is what variance tells us about a distribution of scores. Remember, the larger the variance the less likely that the individual scores of a distribution are close to the mean, and the less likely that the mean is a good indicator of what the typical score in a distribution was. *The smaller the variance the more likely that the individual scores of the distribution are close to the mean, and the more likely that the mean is a good indicator of the typical score in a distribution.*

We have examined the logical and computational formulas for calculating the variance when using raw data or a set of individual scores. We will follow a pattern similar to what we did in Chapter 4 on measures of central tendency and also briefly examine the method for computing variance when data are presented in a simple frequency distribution of size $i = 1$, and a frequency distribution when $i > 1$. It is useful to examine these two situations since we do not always control the way data are presented to us. But before moving to these other approaches I want to introduce the concept of **standard deviation**, which is closely related to variance.

Standard deviation

Many more people have heard the term "standard deviation" than the number who actually knows what it means. But what is a "standard deviation?" The **standard deviation** of a set of data is simply the square root of the variance. Just as the variance can be defined as the mean of the squared deviations from the mean for a set of data, the standard deviation can be defined as the square root of the mean of the squared deviations from the mean for a set of data.

You have every reason to be wondering why do we care about the standard deviation when we already know the variance? There are really two reasons. First, recall that we were not able to base a measure of variation in a set of data on the simple deviation from the mean due to the fact that the sum of the simple deviations from the mean was always the same thing—zero. We eliminated that problem by squaring the deviations from the mean, and using the sum of the squared deviations as the basis for our measurement of variance. But in solving one problem we artificially inflated our measure of variance when we squared all of the deviations from the mean. In one sense you can think of the standard deviation as being a measure of variation that is more in line with what we were intending to measure in the first place since by taking the square root of the variance we are "unsquaring" the deviations from the mean.

The information conveyed by the variance and the standard deviation is essentially the same, but the standard deviation is usually a much smaller value. There is an exception of course when the variance is less than 1.00, since the square root of a number greater than zero and less than 1.00 is larger than the original number. For example, the square root of 0.36 is the larger value 0.60, but in most cases the variance is a much larger number, and we will find it much easier to work with the smaller value of a standard deviation.

However, it is the second reason for working with the standard deviation that is much more important. It turns out that by knowing the mean and the standard deviation for certain types of distributions we can also know a great deal about how the individual observations in that distribution are organized. The mean and the standard deviation will tell us a great deal about any distribution that is **normal in form**. We discussed the idea of a distribution being normal in form or having a bell-shaped curve in Chapter 3, and we will be examining the concept of a normal distribution in great detail in the next chapter. Furthermore, the standard deviation, and a related concept of the **standard error** that is based on the standard deviation, will be key terms when we begin our investigation of hypothesis testing in later chapters.

Just as we represented the population variance by the symbol of lower case sigma squared (σ^2), we represent the symbol for a population standard deviation by the lower case sigma (σ). Similarly, a sample standard deviation is represented by the upper case S, and a standard deviation that is being used to infer a population standard deviation is symbolized by a lower case s. We will still have the same problem with standard deviation that we had with variance when we attempt to use sample data to estimate the population standard deviation. We will use the same technique of altering the denominator of the formula by using $n - 1$ in place of the usual denominator n when we wish to estimate the population standard deviation. With this in mind, we can present both the logical formula and the computational formula for the standard deviation.

Logical Formula for Standard Deviation:

$$\sigma \text{ or } S = \sqrt{\frac{\Sigma(X - \bar{X})^2}{n}}$$

Computational formula for Standard Deviation:

$$\sigma \text{ or } S = \sqrt{\frac{\Sigma X^2 - \frac{(\Sigma X)^2}{n}}{n}}$$

When using data from a sample to estimate a population standard deviation we would simply substitute $n - 1$ in the denominator of the formula. For example, the computational formula would become:

Computational formula for Sample Standard Deviation used to estimate the Population Standard Deviation σ:

$$s = \sqrt{\frac{\Sigma X^2 - \frac{(\Sigma X)^2}{n}}{n-1}}$$

Technically, the correction factor of $n - 1$ is sufficient to adjust for the bias when using the sample variance s^2 as an estimate of the population variance σ^2. However, the sample based standard deviation s will still be a biased estimator of the population standard deviation σ even when using the correction factor of $n - 1$. This is especially true when working with a very small sample size. Fortunately, once the sample size moves above an $n = 20$ or so, the bias becomes very slight, and because many research applications in the behavioral sciences involve a large sample size we need not worry in most cases. Those of you with statistical function calculators might take a moment to examine your function keys. Many statistical calculators will automatically compute the mean and standard deviation for a set of data, and some will give you a choice on how you want the standard deviation computed. You may see two keys marked as: σ_n, and σ_{n-1}, providing you with a choice of a descriptive or an inferential computation of the standard deviation.

Computing the standard deviation is relatively simple. Just find the variance, and then take the square root. For example, we previously computed the variance for three simple distributions, and obtained the results of

A		B		C	
X	X^2	X	X^2	X	X^2
10	100	10	100	8	64
8	64	10	100	6	36
6	36	6	36	6	36
4	16	2	4	6	36
2	4	2	4	4	16
$\Sigma X = 30$	$\Sigma X^2 = 220$	$\Sigma X = 30$	$\Sigma X^2 = 244$	$\Sigma X = 30$	$\Sigma X^2 = 188$
$(\Sigma X)^2 = 900$		$(\Sigma X)^2 = 900$		$(\Sigma X)^2 = 900$	
	$S^2 = 8$;	$S^2 = 12.8$;		and	$S^2 = 1.6$.

Standard Deviation Computation		
Distribution A	Distribution B	Distribution C
$\frac{220 - \frac{900}{5}}{5} =$	$\frac{244 - \frac{900}{5}}{5} =$	$\frac{188 - \frac{900}{5}}{5} =$
$\frac{220-180}{5} = \frac{40}{5} =$	$\frac{244-180}{5} = \frac{64}{5} =$	$\frac{188-180}{5} = \frac{8}{5} =$
8	12.8	1.6
$S_A = \sqrt{8} = 2.83$	$S_B = \sqrt{12.8} = 3.58$	$S_C = \sqrt{1.6} = 1.26$

To compute the standard deviation we simply took the square root of each of the observed variances.

Computing Variance and Standard Deviation for Frequency Distributions

As we have seen earlier, we do not always control the way data are presented to us. On occasion we may be presented with data already in the form of a frequency distribution with no access to the original raw scores, and yet we may still want to compute a mean, or a variance and standard deviation. Variance and standard deviation may be computed for frequency distributions by making a simple adjustment in the formula following the same pattern we used when computing the mean for frequency distributions in Chapter 4.

To compute the variance we will need to alter our original computational formula for raw data:

$$S^2 = \frac{\Sigma X^2 - \frac{(\Sigma X)^2}{n}}{n}$$

as follows:

$$S^2 = \frac{\Sigma f X^2 - \frac{(\Sigma f X)^2}{n}}{n}$$

where

fX = each value of X times its frequency
fX^2 = each value of X squared times its frequency

We must simply use the number of frequencies f in each interval to weight the value of ΣX^2 and $(\Sigma X)^2$ in each interval.

Consider the simple frequency distribution below, which we used in Chapter 4 to compute a mean. We can add the appropriate columns to the frequency distribution to obtain the necessary sums.

X	f	fX	fX^2
11	1	11	121
9	5	45	405
8	4	32	256
5	3	15	75
3	3	9	27
	$\Sigma f = 16$	$\Sigma fX = 112$	$\Sigma fX^2 = 884$

Note that the fX^2 column represents the product of $f \times X \times X$, and as such you may obtain it two ways. Consider the top two entries in the fX^2 column of the frequency distribution above. You may first square each X value, and then multiply the resulting sum by the appropriate value of f as follows:

$$X = 11 \quad X^2 = 121 \quad f = 1 \quad fX^2 = 1 \times 121 = 121$$

$$X = 9 \quad X^2 = 81 \quad f = 5 \quad fX^2 = 5 \times 81 = 405$$

and so on, or you might see that you already have the value of fX, and may simply multiply that column by the value of X. Since multiplication is commutative it does not matter what order you perform the computation. That is:

$$fX^2 = f \times X \times X = f \times X^2 = fX \times X$$

So we can also obtain the fX^2 column as follows:

$$X = 1 \quad f = 1 \quad fX = 11 \quad fX^2 = fX \times X = 11 \times 11 = 121$$

$$X = 9 \quad f = 5 \quad fX = 45 \quad fX^2 = fX \times X = 45 \times 9 = 405$$

Use which ever method is easier for you.

To complete the computation for the variance we need only plug the numbers into the formula. Keep in mind that n is equal to the total number of observations in our sample, not the number of intervals in which they happen to be categorized. In this case, $n = 16$, not 5.

$$S^2 = \frac{\Sigma fX^2 - \frac{(\Sigma fX)^2}{n}}{n}$$

$$S^2 = \frac{884 - \frac{12544}{16}}{16}$$

$$S^2 = \frac{884 - 784}{16}$$

$$S^2 = \frac{100}{16} = 6.25$$

So our variance is equal to 6.25.

To compute the standard deviation we would take the square root of the variance.

$$S = \sqrt{6.25} = 2.5$$

Notice that we are assuming that we are interested in describing the observed sample, and are not attempting to estimate a population variance or standard deviation. This is evident from our use of $n = 16$ in the formula instead of $n - 1 = 15$.

We follow the same general procedure when working with a frequency distribution where the interval size is greater than 1. The formula for variance must be adjusted just as it was when working with the simpler frequency distribution. We may write the formula as before as:

$$S^2 = \frac{\Sigma fX'^2 - \frac{(\Sigma fX')^2}{n}}{n}$$

where

X' = the interval midpoint.

We will again use one of the frequency distributions from Chapter 4. Our first step is to find the midpoint of each interval. From that point on we are simply repeating the procedure that we followed for the frequency distribution where $i = 1$. We will need an fX' column, and an fX'^2 column to compute the variance.

X	f	X'	fX'	fX'^2
45–49	6	47	282	13,254
40–44	8	42	336	14,112
35–39	12	37	444	16,428
30–34	10	32	320	10,240
25–29	9	27	243	6,561
20–24	5	22	110	2,420
	$\Sigma f = 50$		$\Sigma fX' = 1{,}735$	$\Sigma fX'^2 = 63{,}015$

$$S^2 = \frac{63015 - \dfrac{(1735)^2}{50}}{50}$$

$$S^2 = \frac{63015 - \dfrac{3{,}010{,}225}{50}}{50}$$

$$S^2 = \frac{63015 - 60204.5}{50}$$

$$S^2 = \frac{2810.5}{50} = 56.21$$

We have computed the variance, and may now compute the standard deviation by simply taking the square root of 56.21.

$$S = \sqrt{56.21} = 7.50$$

Variance as Prediction Error (or Cabo San Lucas Here I Come!)

We have examined the idea of variation in data in several different ways such as the range, the IQR, the SIQR, variance, and standard deviation. Toward that end, we have spent a great deal of time doing a variety of computations. While it is important to be able to take a statistical formula, apply it to a set of data, and generate the correct result, it is much more important to know why we do it. In other words, what does the resulting statistic mean? What does it tell us about a set of data that we did not know before?

Up until now I have stressed the idea that variance is important because it tells us something about the way that individual scores are distributed around their mean. By knowing the size of the variance we know how representative the mean is of the individual scores in the distribution. The variance viewed in those terms is an important piece of information, but in the behavioral sciences we can, and often do, look at variance in another way. We look at variance as a type of prediction error, and try to find ways to reduce the size of prediction error. Or you might think of the process as simply trying to do a better job of predicting the value of some variable that we consider important in our research.

Suppose we have a set of data representing annual income in thousands of dollars for a group of $n = 10$ individuals.

Income (× 1,000)
30
20
50
15
12
75
40
15
22
16
$\Sigma f = 295$

We could calculate the mean income and would find that it is 29.5, or $29,500 per year. By examining the 10 scores in the distribution you can see that there is variation present. That is, not everyone has an income of 29.5; the incomes vary, some are higher and some are lower.

Now suppose I told you that I had the income of each person written on a piece of paper, and that I was going to draw the pieces of paper at random and let you guess what the person's income was. The only restriction is that you have to make the same guess each time. You are free to choose any of the incomes represented, or any other value for that matter as your guess. All 10 incomes will eventually be selected, and I am going to measure how well you guess by comparing your guess to the income that is chosen. Since some of your guesses will be too high and others too low, I am going to square the difference between the actual income and your guess to eliminate any negative numbers. After all 10 incomes have been selected, I will calculate a mean of your squared difference for each guess as an indication of how well you have done.

What income value would you choose as your guess? You want to select a value to guess each time that will give you the smallest amount of squared error possible. (Choose wisely because there might be a prize in this for you if you win. I'm thinking maybe some nice luggage and a trip to Cabo San Lucas, but I haven't made up my mind yet.)

If you examine the distribution you will notice that there are two scores at 15, and you might be tempted to select the value 15 as your guess since your mean squared difference for those two cases would be zero. However, your mean squared difference across all 10 incomes would be 575.9 if you select the value 15 as your guess. (You might want to verify this by subtracting 15 from each of the observed incomes, squaring the difference, and then averaging the 10 results). Is that a winning performance? I doubt it, it seems high.

Selecting the mode of 15 did not seem to be a wise strategy. What if you selected the median income as your guess? If you rearrange the numbers you will find the median to be equal to 21. If 21 becomes your guess, you will not be exactly right on any of the 10 incomes, but your mean squared difference across all 10 incomes will be 437.9, which is much improved over the strategy of selecting 15, but is it the best you could do?

One final strategy might be to select the mean value of 29.5 as your guess. Again, you will not be exactly right on any of the 10 incomes, but how would you do across all 10? By selecting the mean of 29.5 as your guess, your mean squared difference would be 365.65. That is the best we have seen yet, and in fact, there is no other guess that would be any better. Now if you think about how we have measured the accuracy of each guess, you might recognize that the value of 365.65 represents something else we have examined in this chapter. We took each score, subtracted your guess of the mean, squared the difference, and then calculated the mean of the 10 squared deviations. In other words, we computed the *variance*.

Income (× 1,000)	$(X - \bar{X})$	$(X - \bar{X})^2$
30	0.5	0.25
20	−9.5	90.25
50	20.5	420.25
15	−14.5	210.25
12	−17.5	306.25
75	45.5	2070.25
40	10.5	110.25
15	−14.5	210.25
22	−7.5	56.25
16	−13.5	182.25
		$\Sigma(X - \bar{X})^2 = 3656.50$

$$S^2 = \frac{3656.50}{10} = 365.65$$

Since there is no other single value that would serve as a better guess of the individual scores than the mean, we can think of the variance as the maximum amount of prediction error that we would have to accept when trying to predict an individual score. Or to think of it a different way, your best guess of a score in a distribution is the mean, assuming that no additional information is available to you. (For those of you who originally guessed the mean, the prize committee informs me that neither the trip to Cabo nor the luggage is available. You do win our home game allowing you hours of fun guessing anyone's income you please.)

Now let's change the rules a little bit. Suppose before I have you guess the income value, I am willing to give you one piece of additional information. It would be in your best interest to ask for something that might help you better predict income. What sorts of things (variables) are related to income? You can probably think of many things, but certainly education is a key variable that helps determine one's income. Suppose I am willing to tell you if the income I have selected is that of a person who has a college degree or not. Let's also suppose that I am willing to let you provide two different incomes as your guess; one for the college graduates, and one for the noncollege graduates. (I hope you understand that by changing the rules of the game the trip to Cabo San Lucas is definitely out of the question for this round. No, you're not going to get the luggage either!)

As you might suspect, your best strategy for guessing has changed. In the light of this new information, your best strategy is to guess the mean of the college graduates as the selected income when you know I have selected a college graduate, and to guess the mean of the noncollege graduates when you know I have selected a noncollege graduate. Let's look at the income distribution again with the new information added.

College Graduate	Income (× 1,000)	
YES	30	
NO	20	
YES	50	
NO	15	
NO	12	College Graduate's Mean = 43.4
YES	75	Noncollege Graduate's Mean = 15.6
YES	40	
NO	15	
YES	22	
NO	16	

The mean income of the college graduates is 43.4 or $43,400, and the mean income of the noncollege graduates is 15.6, or $15,600. How well can you guess income now if you guess the college graduate mean of 43.4 when you know the individual has a college degree, and guess the noncollege graduate mean of 15.6 when you know the individual does not have a college degree? We will substitute the appropriate mean into the calculation of $(X-\bar{X})$, and $(X-\bar{X})^2$, and then compute what we can think of as a modified variance (symbolized by S'^2).

College Graduate	Income (× 1,000)	$(X-\bar{X})$	$(X-\bar{X})^2$
YES	30	(30 – 43.4) = –13.4	179.56
NO	20	(20 – 15.6) = 4.4	19.36
YES	50	(50 – 43.4) = 6.6	43.56
NO	15	(15 – 15.6) = –0.6	0.36
NO	12	(12 – 15.6) = –3.6	12.96
YES	75	(75 – 43.4) = 31.6	998.56
YES	40	(40 – 43.4) = –3.4	11.56
NO	15	(15 – 15.6) = –0.6	0.36
YES	22	(22 – 43.4) = –21.4	457.96
NO	16	(16 – 15.6) = 0.4	0.16
			$\Sigma(X-\bar{X})^2 = 1724.40$

$$S'^2 = \frac{1724.40}{10} = 172.44$$

When we use the mean income of the college graduates to guess a college graduate's income, and the mean income of the noncollege graduates to guess the mean income of the noncollege graduates, and then find the mean squared deviation, we arrive at a modified measure of variance; one that uses the specific group mean in place of the overall group mean. In this case using educational level to help predict income results in a modified variance of 172.44, which we can compare with our previous variance of 365.65. By using the appropriate mean income for each group we have been able to reduce the amount of variance by 193.21 points (365.65 – 172.44), or we can express that difference as a percentage of the original variance and say that we have reduced the variance by 52.8%.

Using educational level to help predict income has explained over half (52.8%) of the variance in income. Not everyone in our small sample of 10 individuals has the same income, variation is present. By knowing what the individual's educational level is we are able to explain or account for over half of the variation in income. This idea of being able to explain or **reduce variance** in one variable by knowing the value of a second variable is one of the more important concepts in statistical analysis in the behavioral sciences. We will deal with this concept again when looking at the interpretation of correlation between two variables in Chapter 9, and in assessing the quality of a linear regression analysis in Chapter 10.

Computer Applications

1. Select several variables from the GSS data set, and generate descriptive statistics.
2. Be sure to click on the "Options" button and request the variance and range in addition to the default statistics of mean, standard deviation, minimum, and maximum.
3. Enter data from one of the short examples from the text and compare the results to SPSS. Is SPSS using the variance formula with "n" as the denominator, or "$n - 1$?"

How to do it

Open the GSS data set, and then click on "Analyze," "Descriptive Statistics," and then "Descriptives." Highlight the desired variables and select them by clicking on the direction arrow. Click on "Options" to request additional statistics such as the variance and range. Click on "OK" to run the procedure.

Clear the GSS data set by clicking on "File," "New," and then "Data." Use the new empty Data Editor Screen to input data from one of the simple examples from the text. Run the descriptive statistics procedure with "variance" requested, and determine the formula used.

Summary of Key Points

Measures of variation are a group of statistics that indicate how a set of scores is distributed. Some measures of variation indicate the variation from the bottom or lower end of the distribution to the top or the upper end of the distribution, while other measures of variation indicate the variation of each score from a central point such as the mean. The former techniques typically measure the range of the distribution, while the latter techniques typically measure deviation from the mean.

Range—The range is the distance from the smallest score in a distribution to the largest score. It is one of the simplest measures of variation.

Interquartile Range—The interquartile range is the distance from the 75th percentile to the 25th percentile in a distribution.

Semi-Interquartile Range—The semi-interquartile range is the interquartile range divided by 2.

Deviation from the Mean—Deviation from the mean is a measure of each score's distance from the mean. The sum of the deviations from the mean is always zero.

Variance—The variance is the average of the squared deviations from the mean for a set of scores. We square the deviations to keep the positive and negative deviations from cancelling each other out. The smaller the variance the more closely the scores of a distribution are to their mean.

Standard Deviation—The standard deviation is the square root of the variance.

Degrees of Freedom—The number of values in a sample that are free to take on any value and still represent an unbiased estimate of a population parameter.

Normal in Form—A distribution of scores whose shape approximates that of a normal or bell-shaped curve.

Reduce Variance—The ability to reduce the amount of error when predicting a variable by making use of information obtained from a second variable. Reducing or explaining variance is an important concept that is central to several statistical procedures.

Questions and Problems for Review

1. Compute the range for the following sets of data:
   A. 24 26 28 30 38 49 55 56 67 75
   B. 0.50 1.25 3.50 4.55 8.95 10.50
2. Under what circumstances would it be wise to compute the IQR or the SIQR?
3. Examine the frequency distribution below. Do you think it would be better to report the range for these data, or the IQR? Why?

X	f	Cf
95–99	3	135
90–94	3	132
85–89	20	129
80–84	25	109
75–79	29	84
70–74	24	55
65–69	20	31
60–64	4	11
55–59	5	7
50–54	2	2
	$\Sigma f = 135$	

4. Compute the range, IQR, and SIQR for the data in Problem 3 above.
5. Examine the summary information presented below. What can you conclude about the income distribution for each occupational category? For which groups does the mean seem to be a better measure of central tendency? For which groups is the mean less indicative of overall group income?

	Income	
Occupation	Mean ($)	Standard Deviation ($)
Accountant	35,500	8,500
Attorney	72,200	23,100
Engineer	57,800	12,100
Evangelist	66,600	57,800
Physician	108,000	8,200
Psychic	23,700	4,250

6. Compute the variance and standard deviation for the two sections of the statistics class illustrated in Figure 5.1.

7. Compute the variance and standard deviation for the simple frequency distribution below.

X	f
20	2
18	4
15	7
12	4
10	3
	$\Sigma f = 20$

8. What does it mean to be able to explain variance?

9. Scores for the verbal section of the SAT are presented below for a group of $n = 10$ students.

A. Compute the variance for the entire group.

B. Compute the mean score of the females, and the mean score of the males.

Gender	Verbal SAT
Female	500
Female	650
Female	485
Female	720
Male	450
Male	395
Male	700
Male	630
Male	450
Male	585

10. How much of the variance in the SAT scores in problem 9 can be explained by gender? (Hint: you will need to use the mean score of the females when predicting a female score, and the mean score of the males when predicting a male score to compute a modified variance).

Part III The Bridge to Inferential Statistics

Part III consists of Chapters 6–8. The chapters in Part III include statistical concepts that are important in their own right, but they also serve to link us to the final section of the text dealing with inferential statistics. In descriptive statistics, we are trying to "describe" a population or a sample. In inferential statistics, we want to take information from a sample and make generalizations (inferences) about the population from which the sample was drawn. Often times our ability to legitimately make inferences from a sample depends on the existence of a normal distribution in the population from which the sample was drawn. Chapter 6 discusses the concept of a normal distribution. Any time we make a decision based on statistical analysis we have a chance (or probability of being wrong), Chapter 7 discusses several types of probability. Chapter 8 presents the logic of hypothesis testing; the chief mechanism we have in statistics for testing our ideas about the causal relationship between two variables.

CHAPTER 6

The Normal Distribution

Key Concepts

- The Normal Distribution
- A Normal Distribution
- *Z* Scores
- The Sampling Distribution
- Standard Error
- Statistical Significance
- Confidence Intervals
- The t distribution

Introduction

This chapter continues a logical theme that has tied several of the previous chapters together. In the first three chapters, we have examined the ideas of research in the behavioral sciences, selecting a sample, collecting raw data, and reducing the data into a more easily interpreted form. In Chapter 4, we examined several statistics that give us a sense of where the central point is in the distribution of a measured variable. In Chapter 5, we examined several statistics that give us a sense of how much variation there is in the distribution of a measured variable. It turns out that many of the variables that we examine in the behavioral sciences have a similar distribution. The variables do not have the same mean, nor do they have the same variance or standard deviation, but they are similar in the way in which the observed values are distributed about their respective means. We often find that the variables are distributed in a manner conforming to the shape of a normal distribution. In this chapter, we will examine the characteristics of **The normal distribution**, as well as the concept of **A normal distribution**. We will make extensive use of the standard deviation, and you will see that the standard deviation can be used as a measure of the distance between the mean and an observed value for a variable. That distance based on the standard deviation is called a ***Z* score**. *Z* scores will enable us to make meaningful comparisons between different distributions, by allowing us to convert a raw score from one distribution to a raw score on another distribution. Finally, we will examine the concept of a **sampling distribution** that was introduced in Chapter 2, and which sets the foundation for the concepts of **statistical significance** and hypothesis testing in Chapter 8.

We will focus on three different types of normal distributions in this chapter. The first type of normal distribution we will examine is **The** normal distribution. I have emphasized the word "The" because there is only one. The normal distribution is a theoretical distribution consisting of *Z* scores. *Z* scores are points along the normal distribution whose distance from the mean of the distribution is measured in units of standard deviation. For example, suppose we had a distribution of raw scores on income of a group of $n = 250$ individuals, and find that the mean income is $\overline{X} = \$55{,}650$; and, the standard deviation $s = \$5{,}000$. A particular member of the sample might have an income of $65,650, which we can think of as being $10,000 above the mean. The *Z* scores that make up the normal distribution can also be thought of as being some distance from the mean, but instead of the distance being measured in points, a *Z* score's distance from the mean is measured in units of standard deviations. In the case of our sample of $n = 250$ individuals with a mean income of $\overline{X} = \$55{,}650$; and, a standard deviation is equal to $s = \$5{,}000$, our individual in the sample with an income of $65,650 can be thought of as being $10,000 above the mean when his or her distance from the mean is measured in dollars, or as being two standard deviations above the mean since one standard deviation is equivalent to $5,000. All of the values (*Z* scores) in the normal distribution are expressed as distances from the mean in units of standard deviation.

The second type of normal distribution we will examine is **A** normal distribution that is probably more familiar to you. **A** normal distribution is the distribution of a measured variable of the type that we have seen in the previous chapters. A particular variable of interest is measured and the resulting raw scores constitute the distribution. The variable will have some mean, $\overline{X}$; and, some standard deviation, *s*, and the raw scores in the distribution will happen to be distributed in a fashion similar to **The** normal distribution. There are countless numbers of "normal distributions." The third type of normal distribution we will examine is **A** sampling distribution, which, as you should recall from Chapter 2, is a distribution of all possible samples of some size drawn from a population. The sampling distribution is not a distribution of raw scores or *Z* scores, but is a distribution of all possible sample means computed for samples of a given size. Again, "A" is emphasized because there are countless numbers of sampling distributions that are "normal."

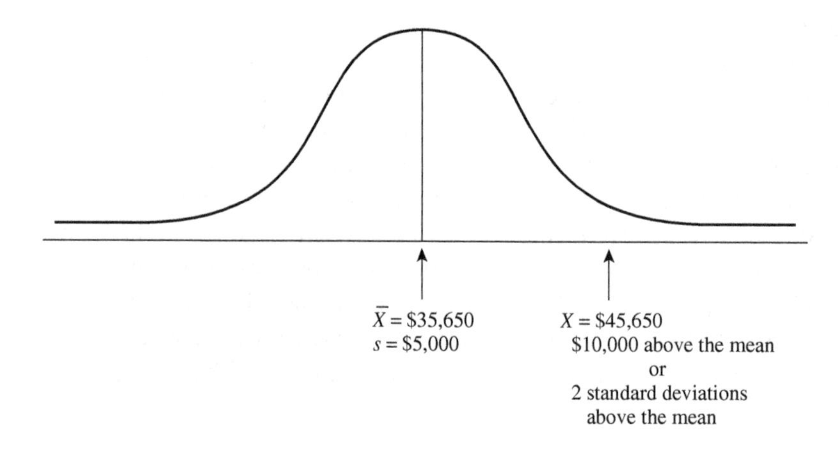

Figure 6.1 A normal distribution for the variable "income" with the mean, standard deviation, and an observed score identified

The Normal Distribution

The normal distribution, sometimes referred to as the standard normal distribution and the *Z* distribution, was first described in the 1600s, and was later extensively studied by a German mathematician named Karl Friedrich Gauss. Gauss' work with the normal distribution was so significant that it is also sometimes referred to as the Gaussian normal distribution. **The** normal distribution is a theoretical distribution with a mean μ (mu) arbitrarily set equal to zero, and a standard deviation σ (sigma) arbitrarily set equal to one. The normal distribution is symmetrical about the mean with half of the distribution above the mean, and half of the distribution below the mean. The shape of the normal distribution is often described as a "bell-shaped" curve. The distribution itself is made up of normal scores or *Z* scores, that is to say that the values of the distribution are expressed as distances from the mean in units of standard deviation. Since the standard deviation of the normal distribution is equal to 1.00, and the mean is equal to 0.00, a *Z* score of 1.00 is one standard deviation above the mean, and a *Z* score of –1.00 is one standard deviation below the mean.

In the normal distribution, the mean μ is in the center, and approximately 34% of the distribution will be from the mean to one standard deviation above the mean with another 34% of the distribution from the mean to one standard deviation below the mean. Or we can say that plus and minus one standard deviation around the mean of the normal distribution will contain approximately 68% of the total distribution. Plus and minus two standard deviations around the mean will contain approximately 95% of the total distribution, and plus and minus three standard deviations around the mean will contain approximately 99% of the total distribution. Figure 6.2 presents the normal distribution with the mean, and the percent of the distribution associated with various standard deviations around the mean indicated. Notice the location of one standard deviation from the mean on the normal curve in Figure 6.1. One standard deviation away from the mean is at the inflection point of the curve. The inflection point is the point where the slope of the curve changes from the steep decline beginning at the top of the curve above the mean to the point where the curve begins to fall away in a more gradual fashion toward the tail or extreme end of the distribution.

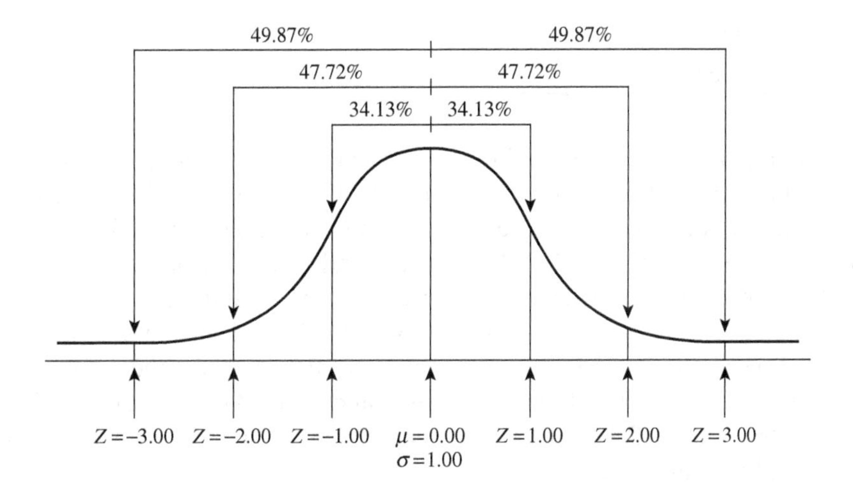

Figure 6.2 The normal distribution showing the percent of the distribution between ±1, ±2, and ±3 *Z* scores

The *Z* Table, Areas Under the Normal Curve

For reasons that will become clear later, it is often desirable to know the proportion of all scores in the normal distribution between a given *Z* score and some other point. For example, we might want to know the proportion of scores from a *Z* score of 1.00 and the mean at 0.00, or we might want to know the proportion of all scores above a *Z* score of 1.50, and so on. Table 1 in the appendix can be used to obtain this information. The table is arranged in the following manner. Column A contains the *Z* score values to two decimal places. Only the *Z* scores of the positive half of the distribution (the half above the mean) are listed, but the negative half of the distribution is identical to the positive half given the symmetrical nature of the normal distribution. Column B contains the area under the normal curve expressed as a proportion from the *Z* score to the mean, and column C contains the area under the normal curve expressed as a proportion from the *Z* score to the tail of the distribution. Because half of the normal distribution is above the mean and half of the normal distribution is below the mean, the values in columns B and C will always equal .5000 or 50% of the distribution. The total area under the normal curve is 1.00 or 100% of the distribution.

Columns B and C simply indicate how the half of the normal distribution above the mean is divided at any particular *Z* score listed in column A. I indicated earlier that approximately 34% (or .34 when expressed as a proportion) of the normal distribution is between the mean and one standard deviation above the mean. One standard deviation above the mean is equivalent to a *Z* score of 1.00, because the standard deviation of the normal distribution σ is arbitrarily set equal to 1.00. We can use Table 1 to see exactly how much of the normal distribution can be found between the mean and $Z = 1.00$. A portion of Table 1 from the appendix is reproduced below.

	Z	
A	B	C
.97	.3340	.1660
.98	.3365	.1635
.99	.3389	.1611
1.00	**.3413**	.1587
1.01	.3438	.1562
1.02	.3461	.1539

To find the proportion of the normal curve between the mean and $Z = 1.00$, we will search down the values in column A until we find the *Z* score 1.00. Column B contains the proportion of the curve between the mean and the *Z* score indicated in column A, so we simply need to find the value in column B associated with the *Z* score of 1.00. The value, as indicated in bold, is .3413, which represents the proportion of the entire normal distribution between the mean and a $Z = 1.00$. If we express the area of the curve as a percentage, we would say that 34.13% of the distribution is between the mean and a $Z = 1.00$. The proportions or areas under the curve are the same for the negative side of the normal curve, so between a $Z = -1.00$ and the mean we would find .3413 or 34.13% of the normal curve as well. Between a $Z = -1.00$ to the mean and continuing on to a $Z = 1.00$, we would find a total of .6826 or 68.26% of the total normal curve.

Column C contains the area under the normal curve or proportion of the distribution from the *Z* score to the tail of the curve. In the case of a $Z = 1.00$, there is .1587 or 15.87% of the distribution beyond

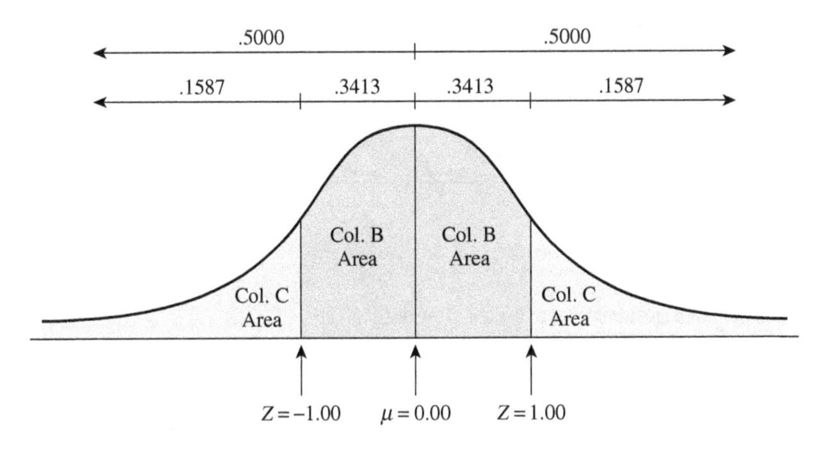

Figure 6.3 The normal distribution with areas from a $Z = 1.00$ to the mean, and to the tail of the distribution indicated

a Z score of 1.00. From the mean to a $Z = 1.00$ we have .3413 of the distribution, and from the Z score of 1.00 to the tail we have .1587 of the distribution. If we add the two values together, $.3413 + .1587 = .5000$, we have the entire half of the distribution that is above the mean as illustrated in Figure 6.3.

There are several common problems concerning areas or proportions under the normal curve that can be solved easily by use of Table 1 in the appendix. Strategies for solving these types of problems are outlined below, but it is not until we begin looking at the other two types of distributions in this chapter (A normal distribution, and the sampling distribution) that you will really understand why we would want to solve these types of problems.

Problem Type	**Strategy for Solving**
Find the proportion of the normal curve from the mean to a given Z score	Locate the Z score in Column A The proportion of the curve from the mean to the Z score is in Column B

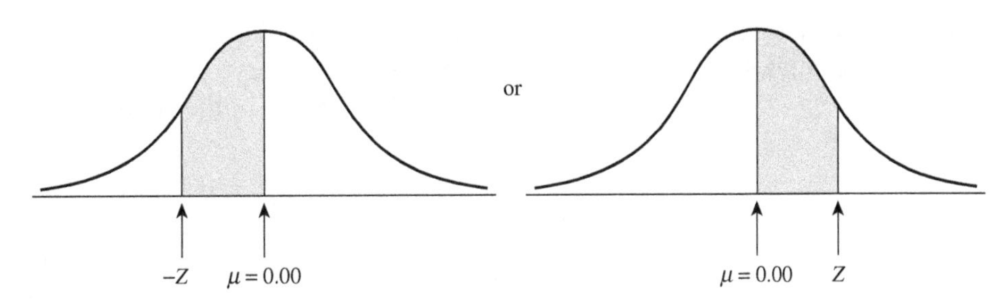

Figure 6.4 The normal distribution with areas from $\pm Z$ to the mean

Problem Type	**Strategy for Solving**
Find the proportion of the normal curve from a given Z score to the tail of the curve on the same side of the normal distribution	Locate the Z score in Column A The proportion of the curve from the Z score to the tail is in Column C

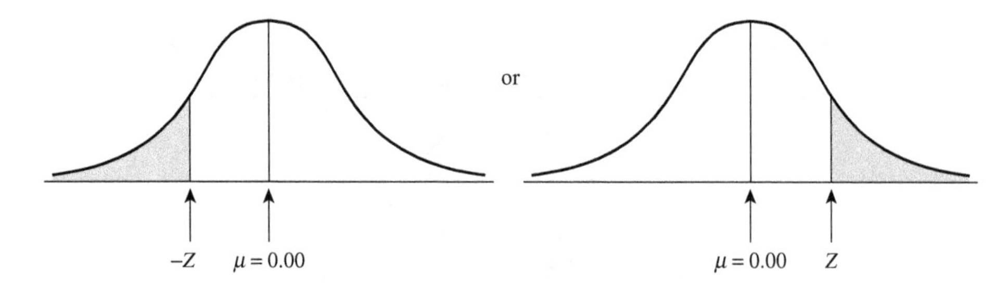

Figure 6.5 The normal distribution with areas from $\pm Z$ to the tail of the distribution

Problem Type	**Strategy for Solving**
Find the proportion of the normal curve below a given Z score on the positive side of the mean	Locate the Z score in Column A Add the proportion of the curve in Column B to .5000 to account for the half of the curve below the mean

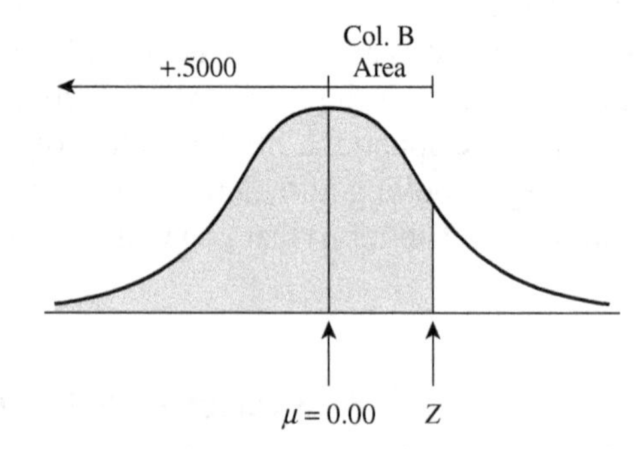

Figure 6.6 The normal distribution with the total area below a positive Z score indicated

Problem Type	**Strategy for Solving**
Find the proportion of the normal curve above a given Z score on the negative side of the mean	Locate the Z score in Column A Add the proportion of the curve in Column B to .5000 to account for the half of the curve above the mean

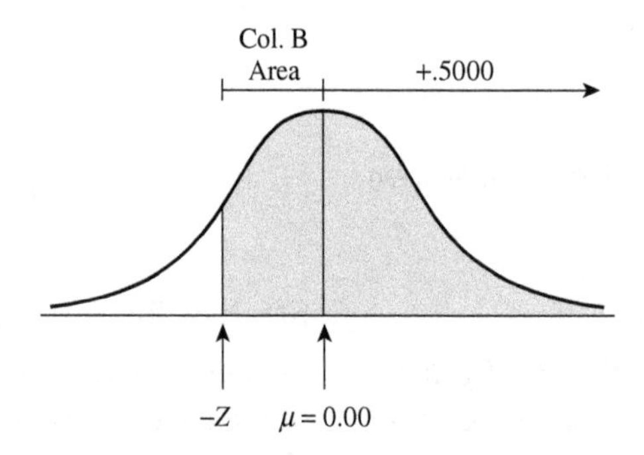

Figure 6.7 The normal distribution with the total area above a negative Z score indicated

Problem Type

Find the proportion of the normal curve between two Z scores on opposite sides of the mean (one positive Z and one negative Z)

Strategy for Solving

Locate the first Z score in Column A and find the area of the curve associated with it in Column B. Locate the second Z score in Column A and find area of the curve associated with it in Column B. Add the two Column B values together

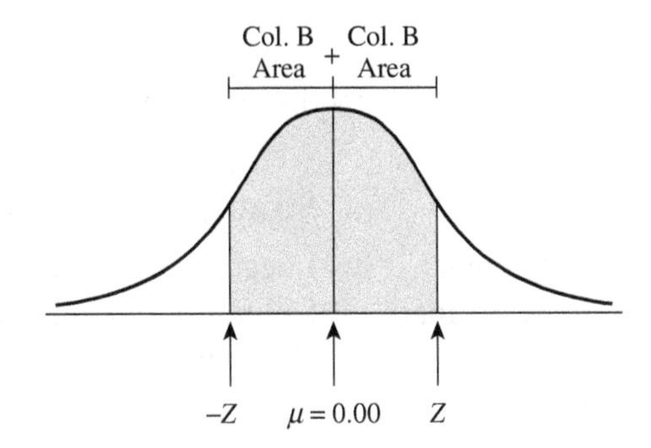

Figure 6.8 The normal distribution with the total area between a negative and a positive Z score indicated

Problem Type

Find the proportion of the normal curve between two Z scores on the same side of the mean

Strategy for Solving

Locate the larger Z score in Column A and find the area of the curve associated with it in Column B. Locate the smaller Z score in Column A and find area of the curve associated with it in Column B. Subtract the Column B value of the smaller Z score from the Column B value of the larger Z score

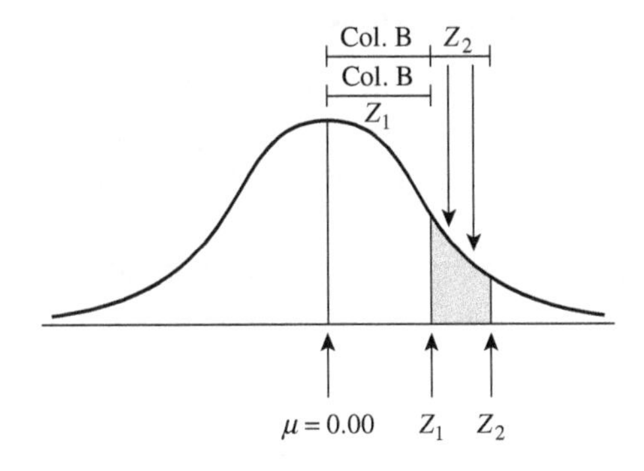

Figure 6.9 The normal distribution with the area between two positive Z scores indicated

Problem Type

Find the Z score that is associated with a given proportion of the normal curve between it and the mean

Strategy for Solving

Locate the proportional area of the curve in Column B, and then find the associated Z score in Column A

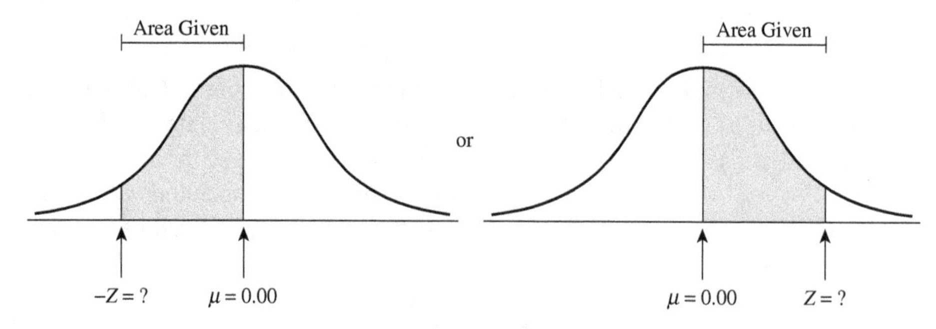

Figure 6.10 The normal distribution finding a *Z* score associated with a given area between the mean and the *Z* score

Problem Type

Find the *Z* score that is associated with a given proportion of the curve in the tail of the normal distribution

Strategy for Solving

Locate the proportional area of the curve in Column C, and then find the associated *Z* score in Column A

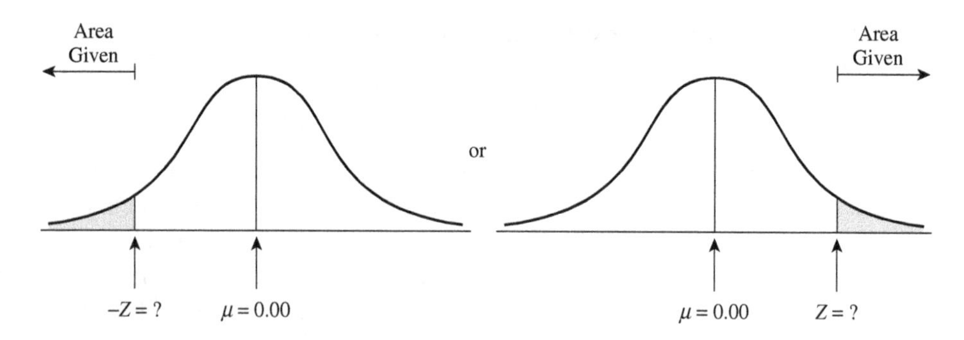

Figure 6.11 The normal distribution finding a *Z* score associated with a given area in the tail of the distribution

Using the *Z* Table to Solve Some Common Problems

Using the strategies outlined above as a guide, we can solve some common problems involving the normal curve. When solving problems on your own, you will usually find it useful to draw the normal curve, label it with the information provided, and shade the area or areas that you are trying to identify.

Find the proportion under the normal curve below a *Z* score of 1.25.

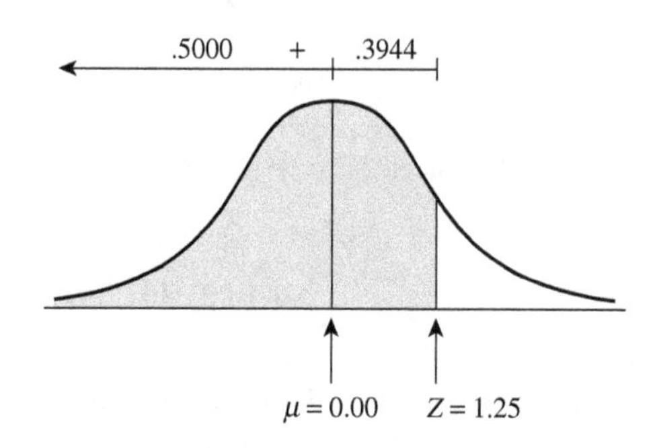

Figure 6.12 graphic illustration of the problem

Solution: the area from the $Z = 1.25$ to the mean is provided in Column B of Table 1 in the appendix, and to that value we must add .5000 representing the half of the normal curve below the mean. The total area below a $Z = 1.25$ is $.3944 + .5000 = .8944$. We may also think of the area under the normal curve as a probability. For example, if we were to select a Z score at random from the normal distribution we would have a .8944 probability (or 89.44% chance) of selecting one that was equal to or less than $Z = 1.25$.

Find the area between the Z scores $Z = -2.25$ and $Z = 1.75$.

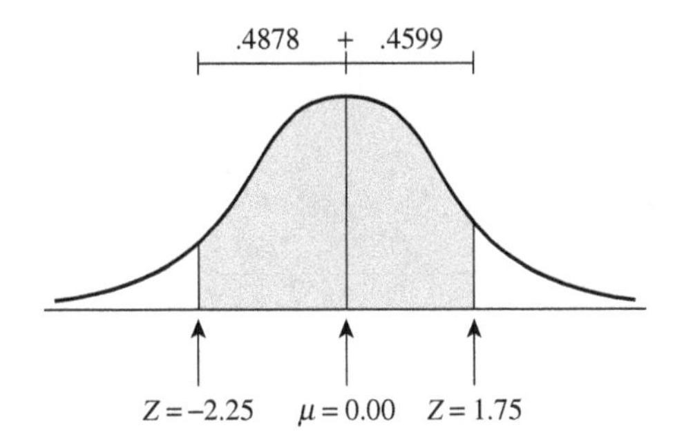

Figure 6.13 Graphic Illustration of the Problem

Solution: Simply add the two Column B areas for the two Z scores. The total area between $Z = -2.25$ and $Z = 1.75$ is $.4878 + .4599 = .9477$.

Find the area above a $Z = 1.96$.

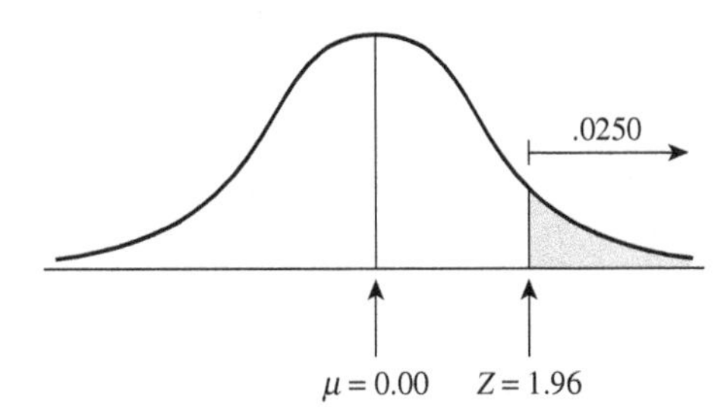

Figure 6.14 Graphic Illustration of the Problem

Solution: The area above $Z = 1.96$ can be found directly in Column C. The total area above $Z = 1.96$ is .0250.

Find the area between the Z scores, $Z = 1.30$ and $Z = 2.48$.

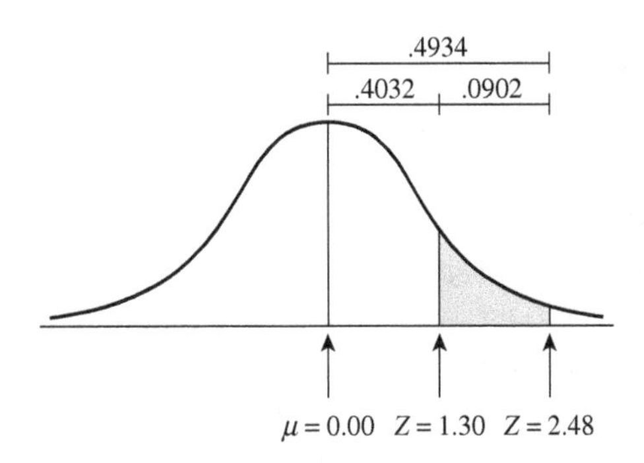

Figure 6.15 Graphic Illustration of the Problem

Solution: The area in Column B for the Z score 2.48 is .4934, which contains not only the area for which we are looking, but also a portion of the normal curve that we do not want. The portion that we do not want is the area from the mean to the Z score, $Z = 1.30$, but the size of that area is easily found in Column B associated with the Z score, $Z = 1.30$. The Column B area associated with $Z = 1.30$ is .4032. The area under the normal curve between $Z = 1.30$ and $Z = 2.48$ can be found by subtracting .4032 from the larger area of .4934. The total area under the normal curve between $Z = 1.30$ and $Z = 2.48$ is: $.4934 - .4032 = .0902$.

Find the positive Z score that has .0500 beyond it in the tail of the curve.

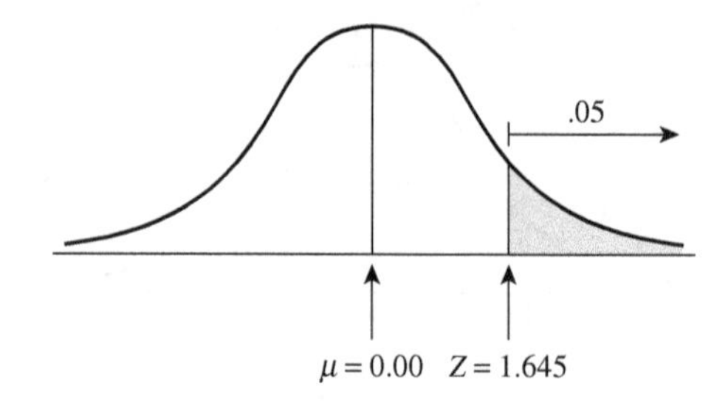

Figure 6.16 Graphic Illustration of the Problem

Solution: Find the area .0500 in Column C of Table 1, and then find the associated Z score in Column A. The Z score on the positive side of the mean that has .0500 beyond it is $Z = 1.645$. Note that the Z score with .0500 of the normal curve beyond it is reported to three decimal places, as is the Z score with .0100 of the normal curve beyond it. (These might be special Z scores for some reason.)

Find the negative Z score that has .025 of the curve below it.

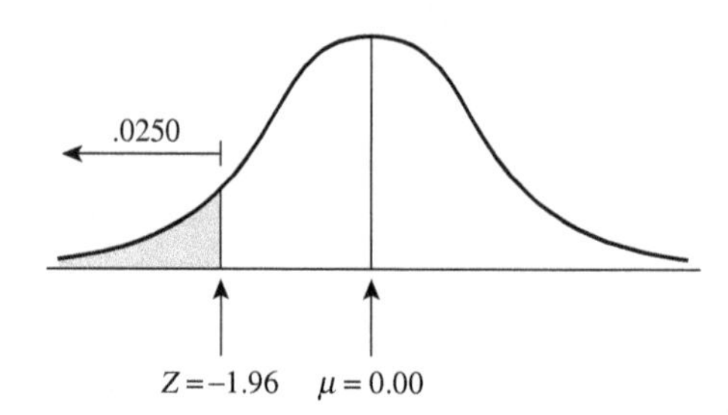

Figure 6.17 Graphic Illustration of the Problem

Solution: Find the area .025 in Column C of Table 1, and then find the associated Z score in Column A. The Z score on the negative side of the curve with .025 of the curve below it is $Z = -1.96$.

A Normal Distribution

There is only one **The** normal distribution with its mean of $\mu = 0$, and its standard deviation of $\sigma = 1.00$, but many variables that we encounter in the behavioral sciences are distributed in a similar or normal fashion. When a measured variable shares the characteristics of The normal distribution, we speak of the variable as being **normally distributed**, or as being an example of **A** normal distribution. There are countless examples of variables that may be normally distributed, and which share the important characteristics of The normal distribution. Variables that are normally distributed will have a mean and a standard deviation, which may take on any value, but the mean will always be in the center of the distribution; the observed values of the variable will be distributed symmetrically around the mean in the characteristic "bell-shaped curve"; half of the distribution will be above the mean, and half of the distribution will be below the mean; and, the observed values of the variable will be distributed around the mean in the same proportions as The normal distribution. That is, approximately 68% of A normal distribution will be found plus and minus one standard deviation on either side of the mean, approximately 95% of the distribution will be found plus or minus two standard deviations on either side of the mean, and, approximately 99% of the distribution will be found plus or minus three standard deviations on either side of the mean.

The same proportional distribution of A normal distribution with respect to its mean and standard deviation will be found no matter what the observed values of the mean and standard deviation are. For example, in a normally distributed variable with a mean, $\bar{X} = 30$, and a standard deviation, $s = 5$, approximately 68% of the distribution will be found between the values 25 and 35 (plus and minus one standard deviation from the mean of 30). In a normally distributed variable with a mean, $\bar{X} = 120$, and a standard deviation, $s = 20$, approximately 68% of the distribution will be found between the values 100, and 140 (plus and minus one standard deviation from the mean of 120). Also recall that 99% of a normal distribution will be found between plus and minus three standard deviations around the mean, so a quick way to estimate the size of the standard deviation for a variable that is normally distributed is to divide the range (which represents 100% of the distribution) by 6 (to account for the three standard deviations above the mean and the three standard deviations below the mean).

Predicting the Distribution of Scores in A Normal Distribution

When a variable is normally distributed, we can accurately predict the proportion of scores between any two points by using what we know about The normal distribution, and Table 1 provided in the appendix. For example, suppose we know that children's attention spans are normally distributed with a mean, $\bar{X} = 2.4$ minutes, and a standard deviation, $s = .8$ minutes. What proportion of children should we expect to have an attention span of 3.6 minutes or longer? At first glance, it may seem that not enough information is provided to answer the question because we do not have a table of proportions under the normal curve for a distribution with this particular mean and standard deviation. We do know that 3.6 minutes is a distance of 1.20 minutes above the mean, and we also know that the standard deviation is equal to .8 minutes. All that we have to do is convert the distance of 1.20 minutes above the mean into a distance based on units of standard deviation. A distance of 1.20 minutes above the mean is equal to 1.50 standard deviations above the mean ($1.20/.8 = 1.50$) and when distance from the mean is expressed in units of standard deviation it becomes a Z score. And we just happen to have a table in the appendix that allows us to find areas under the normal curve when the distance is expressed as a Z score.

All normal distributions are distributed in the same fashion regardless of the observed values of the specific mean and standard deviation. We can convert any raw score's distance from its mean in any normal distribution to a Z score with a simple mathematical transformation. We simply subtract the mean for any normal distribution from the raw score of interest, and then divide by the standard deviation. The formula to convert any raw score into a Z score is as follows:

Formula to Convert any Raw Score to a *Z* Score

$$Z = \frac{X - \bar{X}}{s}$$

where

X = the raw score of interest
$\bar{X}$ = the mean of the distribution
s = the standard deviation of the distribution.

Once a raw score has been converted to a Z score we can turn to the table of areas under the normal curve in the appendix and solve problems similar to those earlier when we were working with The normal distribution. For example, in our distribution of children's attention spans, what proportion of children would be expected to have an attention span between 1.4 minutes and 3.8 minutes? Just as before, you will find it helpful to draw and label the normal distribution with which you are working. The mean of any normal distribution will be in the center, and one standard deviation from the mean will coincide with the inflection point of the curve. Convert the raw scores to Z scores, and then apply the techniques you used earlier when we were working directly with the normal distribution.

To find the area under the normal curve between 1.4 and 3.8 minutes, we will first need to convert both values to Z scores.

$$Z = \frac{1.4 - 2.4}{.8} = \frac{-1.0}{.8} = -1.25$$

The raw score of 1.4 minutes converts to a Z score of -1.25. Note that a raw score does not need to be a negative value to result in a negative Z score. A negative Z score simply means that you are below the mean on the normal distribution. Consequently, any raw score below the mean (in this case, any raw score below 2.4 minutes) will convert to a negative Z score. We will need to convert the raw score 3.8 minutes to a Z score as well.

$$Z = \frac{3.8 - 2.4}{.8} = \frac{1.4}{.8} = 1.75$$

The area under the normal curve between the Z scores -1.25 to 1.75 will be the same as the area between the raw scores of 1.4 to 3.8 minutes. To find the area we simply turn to Table 1, and since we have Z scores on opposite sides of the mean we can just add the two areas found in column B together.

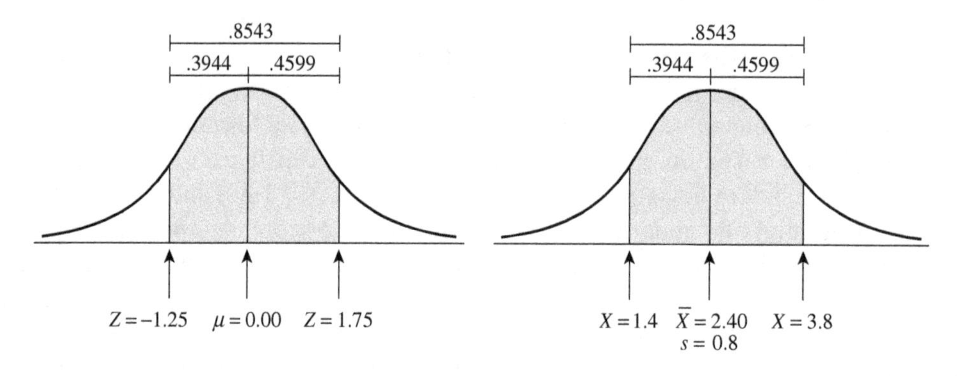

Figure 6.18 Illustration showing the area between the two Z scores is the same as the are between the two raw scores

Just as a total of .8543 or 85.43% of the normal distribution can be found between the Z scores -1.25 and 1.75, a total of .8543 or 85.43% of the children will have an attention span between 1.4 and 3.8 minutes.

What proportion of children would we expect to have an attention span between .8 and 2.0 minutes?

In this problem, we are interested in the proportion of the distribution between two raw scores on the same side of the mean. We will need to convert both raw scores to Z scores, and then subtract the area from the mean to the first Z score from the larger area consisting of the distance from the mean to the second Z score. Note that both of the raw scores are below the mean, so the Z scores corresponding to them will be negative. That does not mean that the areas under the curve are negative, it simply means that we are dealing with two raw scores that are below the mean of $\bar{X} = 2.4$.

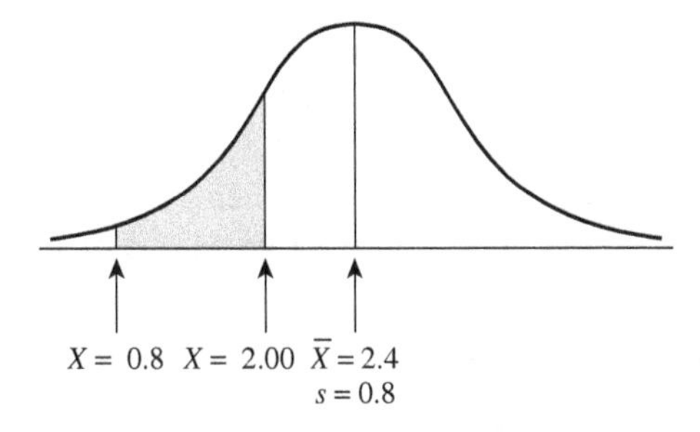

Figure 6.19 Illustration showing the desired area between the two raw scores

Converting the raw scores to Z scores yields

$$Z = \frac{.8 - 2.4}{.8} = \frac{-1.6}{.8} = -2.00$$

$$Z = \frac{2.0 - 2.4}{.8} = \frac{-.4}{.8} = -.50$$

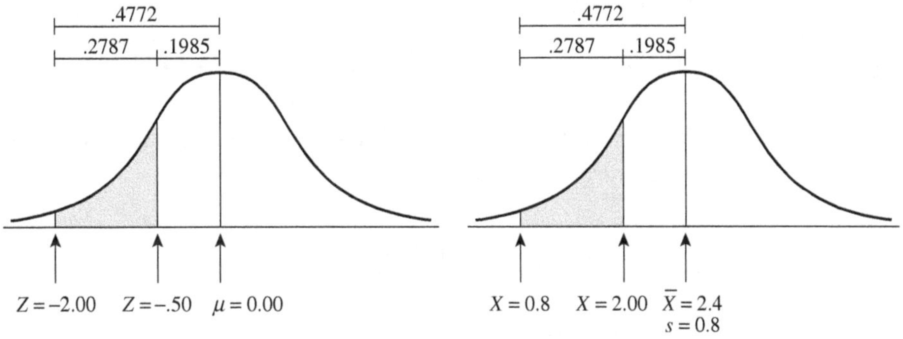

.4772
.2787 .1985
Z = −2.00 Z = −.50 μ = 0.00

.4772
.2787 .1985
X = 0.8 X = 2.00 X̄ = 2.4
s = 0.8

Figure 6.20 Illustration showing the area between the two *Z* scores is the same as the area between the two raw scores

The proportion of the normal distribution between the mean and $Z = -.50$ is .1985, which represents the area that we want to subtract from the larger area from the mean to the Z score −2.00. The proportion of the normal distribution from the mean to $Z = -2.00$ is .4772, so the proportion under the normal distribution between the Z scores −.50 to −2.00 is equal to .4772 − .1985 = .2787. Similarly, .2787 or 27.87% of the children in our normal distribution of attention spans will have an attention span between .8 and 2.0 minutes.

Comparing Raw Scores from One Normal Distribution to Raw Scores of Another Normal Distribution

One useful application of the fact that we can convert any raw score from a normal distribution to a Z score is that it also allows us to compare raw scores from any normal distribution to raw scores on any other normal distribution. A typical application might be where we have data on intelligence scores (IQ) for two samples, but a different IQ test has been used for each of the two samples. As long as both IQ tests are normally distributed, we can convert all of the scores to a single standard by converting them to Z scores. We can then compare raw scores from two different distributions by using the normal distribution as a single standard. This is why *the* normal distribution is sometimes referred to as the standard normal distribution, and Z scores are sometimes referred to as standard scores. *The* normal distribution can serve as a common standard for the countless numbers of variables that are normally distributed, and the Z scores become the common standard by which we can compare them.

For example, we might have IQ scores for some individuals based on test "A" that is normally distributed with a mean $\bar{X} = 100$, and a standard deviation of $s = 15$. We might have IQ scores for other individuals based on test "B" that is normally distributed with a mean $\bar{X} = 80$, and a standard deviation of $s = 10$. Given information on the form of the IQ test each individual took, and each individual's raw score, we can then make direct comparisons across the two exams by converting all scores to *Z* scores.

Raw scores from Two Different IQ Tests

IQ Test	Raw Score
A	110
A	115
A	90
B	80
B	100
B	95

At this point, all we can do is directly compare individuals within each test group. For example, the individual who scored 115 on test "A" has the highest score of the three individuals who took exam "A," and the individual who scored 100 on test "B" has the highest score of the three individuals who took exam "B." To compare all size individuals, we will need to put them on a single scale that can be used as a common standard—the standard normal distribution. We can easily convert all six scores to *Z* scores, and then compare all six individuals directly as follows:

Raw scores from Two Different IQ Tests Converted to Z scores

IQ Test	Raw Score	*Z* Score	Rank
A	110	$\frac{110-100}{15} = \frac{10}{15} = 0.67$	4
A	115	$\frac{115-100}{15} = \frac{15}{15} = 1.00$	3
A	90	$\frac{90-100}{15} = \frac{-10}{15} = -0.67$	6
B	80	$\frac{80-80}{10} = \frac{0}{10} = 0.00$	5
B	100	$\frac{100-80}{10} = \frac{20}{10} = 2.00$	1
B	95	$\frac{95-80}{10} = \frac{15}{10} = 1.50$	2

By converting each of the raw scores to *Z* scores, we have now placed each of the six raw scores on the same standard (*the* normal distribution), and we can now compare them directly.

Converting a Raw Score from A Normal Distribution to a Raw Score on Another Normal Distribution

In other cases, we might want to convert the raw scores from one distribution to raw scores on the other distribution. Converting raw scores from one normal distribution to raw scores on another normal distribution is particularly likely when the majority of the sample is on one scale and only a few members of the total sample have scores on the other scale. For example, we can convert the raw scores of the three individuals who took test "B" to raw scores on test "A" with a simple algebraic transformation of the Z score formula.

In most cases when we work with the Z score formula, we have information on the raw score (X), the mean of the distribution ($\overline{X}$), and the standard deviation of the distribution (s), and we then use the formula to solve for the Z score. When we want to convert a raw score from one normal distribution to a raw score on another normal distribution we are making use of the fact that equivalent raw scores from two different normal distributions will equal the same Z score. Once we have the Z score associated with a particular raw score from one distribution, we can use the Z score along with the mean ($\overline{X}$), and the standard deviation (s) from the other normal distribution to solve for the raw score (X) on the second distribution. We simply transform the Z score formula from its usual form to solve for "X" as indicated below.

$$X = \overline{X} + (Z \times s)$$

We found that the raw scores on test "B" of $X = 80$, $X = 100$, and $X = 95$ were equivalent to the Z scores of $Z = 0.00$, $Z = 2.00$, and $Z = 1.50$, respectively. Using this information, along with the mean and standard deviation of test "A" ($\overline{X} = 100$; $s = 15$), we can convert the Z scores associated with the raw scores on exam "B" to raw scores on test "A." Before actually computing the raw scores for test "A" we might want to examine the Z scores associated with the raw scores from test "B" to estimate what the raw scores on test "A" will be. For example, the raw score of $X = 80$ on test "B" converts to a Z score of 0.00, indicating that it is no distance from the mean. Therefore, the raw score on test "A" corresponding to the raw score of $X = 80$ on test "B" should also be no distance from the mean of test "A." As a result, the raw score should equal the mean of $\overline{X} = 100$. By formula, we obtain the same result when we convert the Z score of 0.00 to a raw score on test "A."

$$\begin{aligned} X &= \overline{X} + (Z \times s) \\ X &= 100 + (0.00 \times 15) \\ X &= 100 + 0.00 \\ X &= 100 \end{aligned}$$

Similarly, the raw score of $X = 100$ on test "B" converted to a Z score of 2.00, indicating that it was two standard deviations above the mean. The corresponding raw score on test "A" should be two standard deviations above the test "A" mean, which is 130 (the mean of 100 plus two standard deviations of 15 points each). By formula we obtain the following result:

$$\begin{aligned} X &= \overline{X} + (Z \times s) \\ X &= 100 + (2.00 \times 15) \\ X &= 100 + 30 \\ X &= 130 \end{aligned}$$

Finally, we will convert the raw score of 95 on test "B" which corresponds to a Z score of 1.50 to a raw score on test "A." The raw score on test "A" should be one and one-half standard deviations above the mean, or 122.5 (the mean of 100 plus one and one-half standard deviations of 15 points each). By formula we obtain the following result:

$$\begin{aligned} X &= \overline{X} + (Z \times s) \\ X &= 100 + (1.50 \times 15) \\ X &= 100 + 22.5 \\ X &= 122.5 \end{aligned}$$

Z Scores Provide an Easy Way to Compute Percentile Rank

Recall that a percentile represents the percentage of a distribution equal to or lower than a given score. For example, the 60th percentile is the score in a distribution with 60% of all other scores in the distribution equal to or below it. Once a raw score in a normal distribution has been converted to a *Z* score, you can easily find the raw score's percentile rank by using the table of areas under the normal curve in the appendix. If the corresponding *Z* score is negative, below the mean, then column C in the table will indicate the proportion of the distribution at or below the *Z* score. The proportion of the distribution can be converted to a percentage, or percentile rank, by multiplying the proportion by 100. For example, the raw score $X = 90$ on test "A" in the distribution of IQ scores converted to a *Z* score of $-.67$. The area at or below a *Z* score of $-.67$ found in column C of the Table 1 in the appendix is .2514 or 25.14% of the distribution. Therefore, the raw score $X = 90$ on test "A" represents approximately the 25th percentile.

If a raw score converts to a positive *Z* score, both the raw score and the *Z* score will be above the mean, then column B in Table 1 will provide the proportion of the curve from the *Z* score to the mean, and to this area we will add .5000 to account for the 50% of the distribution below the mean. For example, the raw score $X = 100$ on test "B" from above converts to a *Z* score of 2.00. The proportion of the normal distribution from a *Z* score of 2.00 back to the mean can be found in column B, and is .4772 or 47.72%. The percentile rank can be computed by adding the remaining 50% of the distribution below the mean, which yields a total of .9772 or 97.72%. Therefore, the raw score $X = 100$ represents approximately the 98th percentile, meaning that 98% of all individuals taking test "B" would score 100 or less.

Types of Normal Distributions

A Normal Distribution is a distribution of raw scores (*X*s) with some mean $\bar{X}$, and some standard deviation s.

The Normal Distribution is a theoretical distribution of *Z* scores (scores whose distance from the mean is expressed in units of standard deviation) with a mean $\mu = 0$, and a standard deviation $\sigma = 1.00$.

A Sampling Distribution is a distribution of sample means ($\bar{X}$s) of a particular sample size *n*, with a mean $\mu_{\bar{X}}$ equal to the population mean μ from which the samples are drawn, and a standard deviation (called the **standard error,** $\sigma_{\bar{X}}$), equal to the standard deviation of the original population divided by the square root of the sample size.

A Sampling Distribution

The final example of a normal distribution we will examine in this chapter is the sampling distribution. The sampling distribution is unlike the other two types of normal distributions we have examined thus far. We have seen that a normal distribution is a distribution of raw scores with some mean, $\bar{X}$, and some standard deviation, s. There is, quite literally, an infinite number of possible normal distributions that only differ by the size of the observed mean or standard deviation. Other than the possible difference in the value of the mean and/or standard deviation, all of the normal distributions are similar. The mean will always be in the center, and the raw scores that make up the distribution will be distributed around the mean just as scores are distributed around the normal distribution. We have also spent a great deal of time examining the normal distribution, which is a theoretical distribution of *Z* scores with a mean, $\mu = 0$, and a standard deviation, $\sigma = 1.00$.

The sampling distribution, as you should recall from Chapter 2, is not a distribution of raw scores, and it is not a distribution of *Z* scores. The sampling distribution is the distribution of all possible sample means based on samples of size "*n*" taken from a population. So the sampling distribution is made up of sample means based on a given sample size, "*n*." One of the most important characteristics of the sampling distribution is that for any given population with a mean μ, and a standard deviation σ, *as the sample size "n" becomes large, the sampling distribution of all possible sample means approaches a normal distribution with the mean of the sampling distribution equal to the original population mean μ, and a standard deviation equal to the original population standard deviation σ divided by the square root of the sample size.* This tendency for the sampling distribution to become normal in form as the sample size becomes large is referred to as the central limit theorem, and it holds true no matter what the shape of the original population distribution from which the samples are drawn. Keep in mind that the mean of the sampling distribution is a mean or average of all of the separate sample means contained in it, so the mean of the sampling distribution is a "mean of means." We usually symbolize the mean of the sampling distribution as $\mu_{\bar{X}}$, and we refer to the standard deviation of the sampling distribution as the **standard error** ($\sigma_{\bar{X}}$).

A given population can have any number of sampling distributions associated with it depending on the size of the samples that are drawn from it, but they will all be similar in several respects. First, all of the sampling distributions will be normal in form provided they are based on a "large" sample size. There is no hard and fast definition of "large" with respect to the size of the sample upon which a sampling distribution is based, but generally a sample size greater than or equal to 30 is sufficient. Second, the mean of the sampling distribution, $\mu_{\bar{X}}$, will always be in the center of the sampling distribution, and it will always equal the population mean μ from which the samples were drawn. Third, because the sampling distribution is normal in form, half of all the possible samples of size "*n*" will be equal to or greater than the mean, and half of all of the possible samples of size "*n*" will be equal to or less than the mean. Finally, the distribution of the sample means in the sampling distribution will follow the same pattern as any other normal distribution. That is, approximately 68% of all of the sample means in the sampling distribution will fall between plus and minus one standard error (the name for the standard deviation of the sampling distribution) on either side of the mean; approximately 95% of all of the sample means will fall between plus and minus two standard errors on either side of the mean; and, approximately 99% of all of the sample means will fall between plus and minus three standard errors on either side of the mean.

The difference among all of the possible sampling distributions associated with a given population is found in the size of the standard error. Recall that the standard error is always equal to the original population's standard deviation divided by the square root of the sample size. As a result, as the sample size increases the size of the standard error will decrease. One important consequence of the effect of the sample size on the standard error is that a larger sample size results in all of the possible samples being closer to the mean (see Figure 6.21).

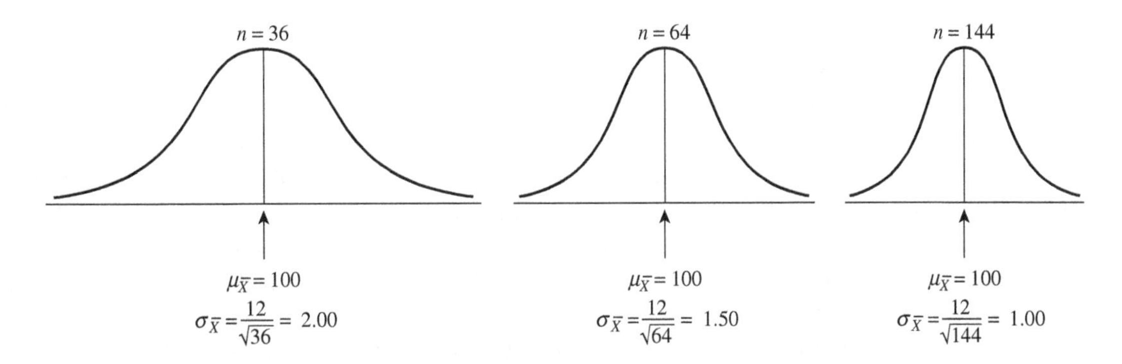

Figure 6.21 Illustration of the sampling distributions when $n = 36$; $n = 64$; and, $n = 144$

As you can see in Figure 6.21 in each case the mean remains $\mu_{\bar{X}} = 100$, but as the sample size increases the standard error $\sigma_{\bar{X}}$ is reduced.

With a sample size of $n = 36$, the standard error of the sampling distribution is equal to 2.00 (the population standard deviation of 12 divided by the square root of the sample size $n = 36$). The sampling distribution is normal in form and centered on the population mean of 100, so we can expect approximately 68% of all of the samples in the first sampling distribution to fall within the range of plus and minus one standard error (this is the case because any normal distribution will have approximately 68% of the distribution within plus or minus one standard deviation around the mean). In this first sampling distribution, approximately 68% of all of the possible samples will fall within the range of 98–102 (the mean of 100 plus and minus one standard error). Approximately 99% of all of the possible samples would fall within the range of 94–106 (the mean of 100 plus and minus three standard errors).

In the second sampling distribution based on samples of size $n = 64$, the standard error falls to a value of 1.50 (the population standard deviation of 12 divided by the square root of the sample size $n = 64$). In the second sampling distribution based on a slightly larger sample size, we would still expect to find approximately 68% of all of the possible samples within plus and minus one standard error around the mean of 100, but in this case the range of plus and minus one standard error is 98.5–101.5. Approximately 99% of all of the possible samples to fall within the range of plus and minus three standard errors, but in this case that range becomes 95.5–104.5 (the mean of 100 plus and minus three standard errors).

In the final sampling distribution based on samples of size $n = 144$, the standard error falls to a value of 1.00 (the population standard deviation of 12 divided by the square root of the sample size $n = 144$). We will still expect to find approximately 68% of all of the possible samples within plus and minus one standard error around the mean of 100, but in this final case the range of plus and minus one standard error is 99–101. Approximately 99% of all of the possible samples fall within the range of plus and minus three standard errors, but in this case that range is even smaller, and becomes 97–103 (the mean of 100 plus and minus three standard errors).

Range in Which We Find Approximately 99% of All Possible Samples from Three Sampling Distributions

Sample Size:	36	64	144
Mean of Sampling Distribution:	100	100	100
Standard Error:	2.00	1.5	1.00
Range of 99% of All Possible Samples:	94–106	95.5–104.5	97–103

With So Many Possible Samples How do We Know We Have a "Good" One?

In most cases when we do research in the behavioral sciences, we base our conclusions on the observation of a single sample drawn from a much larger population. In the case of inferential statistics, we then want to generalize the results found among the sample to the larger population from which it was drawn. When we first examined the concept of the sampling distribution in Chapter 2, you saw how many unique samples it is possible to select from even a small population. For example, we found that with a population size of $N = 40$ we would be able to select a total of 658,008 unique samples of size $n = 5$! You can imagine how many unique samples could be drawn from a population the size of the United States consisting of over 300 million people.

Each of the possible samples of size "n" for a given population can be thought of as existing in a sampling distribution. From what we have seen in this chapter, we know that the mean of the sampling distribution is equal to the population mean from which the samples are drawn, and we know that the sampling distribution is normal in form with half of the possible samples above the mean and half of the possible samples below the mean, and that the sampling distribution has a standard error equal to the population standard deviation divided by the square root of the sample size. All of that we know. What we do not know when we do research is, "which of the possible samples in the sampling distribution did we actually select and observe?" We know what the sampling distribution looks like, and we know that our sample came from some part of the sampling distribution, but which part? Which sample did we select? Some of the samples we can think of as being "good" samples, that is, the sample mean is very close to the center of the sampling distribution, and as a result is very close to the true population mean. But some of the samples are toward the tail of the sampling distribution, which puts them farther away from the center of the sampling distribution, and farther away from the true population mean.

It may seem like we have two problems, if we are interested in generalizing our results based on the observation of the sample to the total population. The first problem deals with the question of whether we can legitimately generalize our results to the total population if we have one of the "bad" samples from either extreme end of the sampling distribution? The second problem is that we can never know which particular sample we actually have. We never know if it is one of the "good" samples close to the center of the sampling distribution, or one of the "bad" samples from either of the extreme ends of the sampling distribution. Fortunately, we do not have to worry about either problem; it does not matter which sample we have selected because as long as a probability sampling technique is used and the sample size is reasonably large, then even the "bad" samples in the sampling distribution are going to turn out to be very good. Consider the following illustration.

Suppose we take a random sample of size $n = 1{,}600$ from the U.S. population to estimate the mean age of the total population. A sample size of $n = 1{,}600$ is typical of many national surveys. Suppose we know that the true mean age of the U.S. population is $\mu = 32$ years, with a standard deviation $\sigma = 16$ years. Our particular sample will have a sample mean that we can compute, but we can never know if we have one of the "good" samples from close to the center of the sampling distribution, or one of the "bad" samples from one of the extremes of the sampling distribution. At this point, we have to make a value judgment. Let's say that a "good" sample is one that is no more than 2 years away from the true population mean, that is, any sample with a mean age between 30 and 34 years will be considered "good."

Now let's take a look at what the sampling distribution would look like for all possible samples of size $n = 1{,}600$ drawn from the U.S. population (see Figure 6.22). We know that the sampling distribution will be normal in form (remember that this trait holds true no matter how the population itself is distributed), the sampling distribution mean will equal the mean of the population from which the samples were drawn (in this case the mean of the sampling distribution will equal 32 years), and, the sampling distribution will have a standard error equal to the standard deviation of the population divided by the square root of the sample size (in this case the standard error will equal 16 divided by 40, or .4 years). Because the sampling distribution is just like any other type of normal distribution, we also know that approximately 68% of the distribution will be found between plus and minus one standard error around the mean, approximately 95% of the distribution will be found between plus and minus two standard errors around the mean, and approximately 99% of the distribution will be found between plus and minus three standard errors around the mean. The sampling distribution for all samples of size $n = 1{,}600$ is presented below with the mean, and the age ranges associated with plus and minus one, two, and three standard errors.

Notice that 68.26% of all of the possible samples that we could possibly select will have a mean age between 31.6 and 32.4 years (plus and minus one standard error on either side of the mean), and 95.44% of all of the possible samples that we could possibly select will have a mean age between 31.2 and 32.8 years (plus and minus two standard errors on either side of the mean). Finally, 99.74% of all of the possible sample means that we could possibly select will have a mean age between 30.8 and 33.2 years (plus and minus three standard errors on either side of the mean). Note that plus and minus three standard

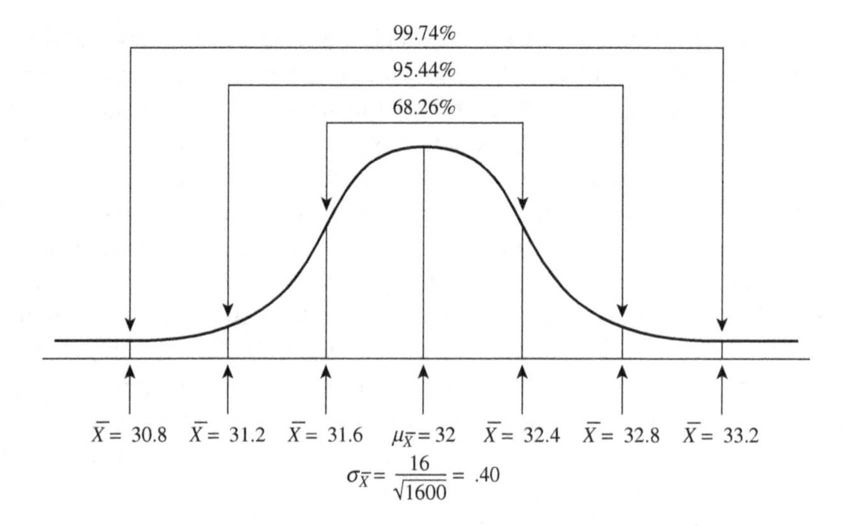

Figure 6.22 Illustration of the sampling distribution for $n = 1{,}600$ showing percentage of the distribution ± 1, 2, and 3 standard errors from the mean

errors on either side of the mean is a range of only plus or minus 1.2 years, which is well within our original limit of plus or minus 2 years to qualify for a "good" sample. In fact, for a sample of size $n = 1{,}600$ to have a mean that is 2 years away from the true population mean, it would have to be five standard errors away from the center of the sampling distribution! For all practical purposes, you can consider obtaining a sample mean that is five standard deviations away from the center of the sampling distribution as being impossible. As long as a probability sampling plan is used in combination with a reasonably large sample size, then any sample from the sampling distribution will be a good one.

Areas Under the Sampling Distribution

The sampling distribution is just another example of a normal distribution, so we can solve the same types of problems for the sampling distribution as we did with examples of other normal distributions. The only difference is that instead of converting a raw score "X" from a normal distribution to a Z score on the normal distribution, we will be converting a sample mean "$\bar{X}$" to a Z score, and then placing the Z score on the normal distribution. The areas under the normal curve provided in Table 1 in the appendix will be distributed in the same fashion in the sampling distribution. To convert a sample mean from a sampling distribution to a Z score, we simply alter the Z score formula slightly.

Formula to Convert any Sample Mean to a *Z* Score

$$Z = \frac{\bar{X} - \mu_{\bar{X}}}{\sigma_{\bar{X}}}$$

where

$\bar{X}$ = the sample mean of interest
$\mu_{\bar{X}}$ = the mean of the sampling distribution
$\sigma_{\bar{X}}$ = the standard error of the distribution $(\sigma / \sqrt{n})$.

Once a sample mean from the sampling distribution has been converted to a Z score and the resulting area under the normal curve has been identified, we can consider the size of the area under the normal curve as a probability. Consider the following example. A population is known to have a mean, $\mu = 90$, with a standard deviation $\sigma = 10$. The sampling distribution for all samples of size $n = 64$ will appear normal

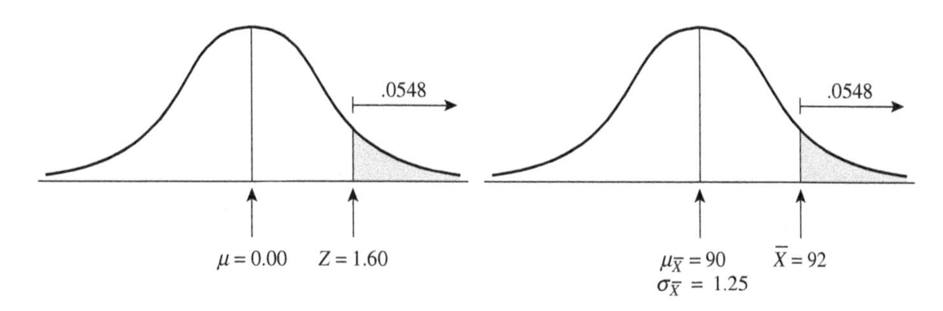

Figure 6.23 Illustrations showing the area under the normal curve and the sampling distribution

in form with a mean, $\mu_{\bar{X}} = 90$, and a standard error, $\sigma_{\bar{X}} = 1.25$ (10 divided by 8). We might be interested in knowing the size of the area of the sampling distribution equal to or above a sample mean of $\bar{X} = 92$.

Solving problems associated with areas under the curve of a sampling distribution follows the same general guidelines that we used earlier with a normal distribution. We first convert the sample mean to a *Z* score, and then use the table provided in the appendix to find the appropriate area under the normal curve. In this case, we find the *Z* score as follows:

$$Z = \frac{92 - 90}{1.25} = \frac{2}{1.25} = 1.60$$

The area beyond a $Z = 1.60$ in the normal distribution is found in column C of Table 1 in the appendix, and is equal to .0548, so the corresponding area under the sampling distribution beyond a mean of $\bar{X} = 92$ will also equal .0548 or 5.48%. We can also think of the area beyond a particular sample mean in the sampling distribution as a probability. For example, in this case, knowing that only 5.48% of all of the possible sample means in the sampling distribution are equal to or greater than 92 means that our probability of selecting a sample of $n = 64$ individuals from the population whose mean is 92 or greater is only 5.48%.

In a similar fashion, we might want to know the area under the sampling distribution between the sample means of $\bar{X} = 89$ and $\bar{X} = 91$ (a distance of 1 point on either side of the true population mean of 90). We would begin by converting both sample means to *Z* scores, and then finding the appropriate areas associated with them on the normal distribution.

$$Z = \frac{89 - 90}{1.25} = \frac{-1}{1.25} = -.80$$

$$Z = \frac{91 - 90}{1.25} = \frac{1}{1.25} = .80$$

The sample mean $\bar{X} = 91$ converts to a *Z* score of .80, and the sample mean of $\bar{X} = 89$ converts to a *Z* score of −.80. We are interested in the total area between the two *Z* scores, and because they are on opposite sides of the mean of the normal distribution we will simply need to add the two column B areas associated with the two *Z* scores. The area between the mean and the *Z* score of .80 on the normal distribution is .2881, and there is an additional area of .2881 between the mean and the *Z* score of −.80. The total area between the two *Z* scores −.80 and .80 is .5762, so there will be a total area of .5762 between the corresponding sample means $\bar{X} = 89$ and $\bar{X} = 91$ on the sampling distribution (see Figure 6.24).

Once again we can think of an area under the sampling distribution as a probability. In this case, the area between the two sample means of $\bar{X} = 89$ and $\bar{X} = 91$ is equal to the area .5762 or 57.62%, which indicates that 57.62% of all of the possible samples of size $n = 64$ in the sampling distribution will have a sample mean greater than or equal to 89 and less than or equal to 91. As a result, we have a 57.62%

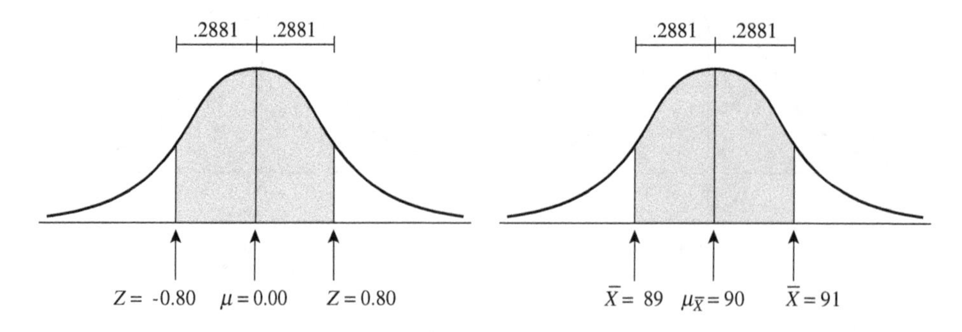

Figure 6.24 Illustration showing the area between the two sample means, and the equivalent *Z* scores

chance of selecting a sample mean from the sampling distribution within that range from 89 to 91. Specifying a range in the sampling distribution and associating a probability with it is referred to as a **confidence interval**.

Point Estimation and Interval Estimation

This chapter begins Part III of the text which I have called the "bridge to inferential statistics" in which we make inferences about unknown population parameters based on our direct observation of sample statistics. The normal distribution and the sampling distribution are two important concepts on which inferential statistics is based, and you have seen a brief illustration of one type of inferential statistics with the example of a confidence interval. Very often we wish to estimate the value of an unknown population parameter, and we use our observation and measurement of a sample drawn from the population to do so. There are two types of estimation procedures: point estimation and interval estimation.

Point estimation is by far the simpler procedure, and consists solely of using the observed value of a sample statistic as the estimate of an unknown population parameter. For example, if we want to estimate the value of a population mean (μ), and we know that the mean ($\bar{X}$) of a sample taken from the population is equal to a particular value, then in the absence of any other information our best guess for the population mean is the observed value of the sample mean. For example, if we observe a sample mean to be $\bar{X} = 80$, then our **point estimate** for μ would be 80. The same logic applies to estimating a population standard deviation (σ). Our best point estimate of the population standard deviation would be the observed value of the sample standard deviation (being sure to use $n - 1$ as the denominator in the computation of the sample standard deviation).

Interval Estimation also provides an estimate of an unknown population parameter based on the observed value of a sample, but the estimate consists of a range of values with an associated probability. Interval estimation is most often used to create a **confidence interval** for a population mean, and our degree of confidence is equal to the probability that the specified range of values contains the population mean we are trying to estimate. The confidence interval is created by adding a range both above and below our observed value of the sample mean. The size of the range varies directly with our degree of confidence, so a high degree of confidence requires a larger range on either side of the sample mean than does a lower degree of confidence. The confidence interval is often commonly referred to as the "margin for error."

The size of the range applied to either side of the observed sample mean can be easily computed, given our knowledge of the sampling distribution, and the fact that it represents another type of normal distribution. For example, if we want to be 68.26% sure that our confidence interval contains the population mean, we would need to have a range equal to one standard error on either side of the sample mean (because plus and minus one standard error on either side of the mean of the sampling distribution contains 68.26% of all possible samples, just as plus or minus one standard deviation on either side of the mean contains 68.26% of any normal distribution). If we want to be 95.44% sure that our confidence interval contains the population mean, then we would need a range equal to two standard errors on either side of the sample mean (because plus and minus two standard deviations contains 95.44% of any normal distribution).

The one problem is that we seldom want to compute a 68.26% or a 95.44% confidence interval. Because we have an affinity for round numbers, we are more likely to want a 90% confidence interval, or a 95% confidence interval. The solution is simple; all we must do is equate the size of the desired confidence interval to the number of standard errors it represents in the sampling distribution. We actually determine the number of standard errors needed to equal the desired confidence interval by working with half the size of the desired confidence interval because half will be on one side of the mean and half will be on the other side of the mean. For example, if we want to compute a 90% confidence interval, we would need to find how many standard errors it takes to have 45% of the sampling distribution on one side of the mean, and 45% of the sampling distribution on the other side of the mean. Expressed as a proportion, 45% is equal to .4500, so all we must do is find the number of standard errors it takes to have .4500 of the sampling distribution from the mean to that point. Again we will turn to Table 1 in the appendix containing areas under the normal curve. By looking in column B, containing areas from the mean to a given Z score, we can find the Z score that has .4500 of the normal curve from the mean to that Z score. In this case, you will find that the appropriate Z score is $Z = 1.645$, meaning that plus or minus 1.645 standard deviations on either side of the mean will contain the center .90, or 90% of the normal curve. Similarly, for any sampling distribution, if we move 1.645 standard errors on either side of the mean, we will have the center 90% of that particular sampling distribution. Consequently, we can be 90% confident that the resulting range of plus and minus 1.645 standard errors on either side of the mean contains the true population mean, μ. The formula for a confidence interval can be written as follows:

Formula for Computing a Confidence Interval

$$\text{CI} = \overline{X} \pm \text{Range}$$

where

$\overline{X}$ = the observed sample mean

Range = the number of standard errors required to equal the desired confidence interval. The range in turn is equal to the following

$$\text{Range} = \sigma_{\overline{x}} \times Z \text{ associated with } \tfrac{1}{2} \text{ desired interval.}$$

A Computational Example of a Confidence Interval

A sample of $n = 64$ individuals is observed to have a mean, $\overline{X} = 75$, with a standard deviation, $s = 10$. The 90% confidence interval is computed as follows. First, we will need to estimate the standard error of the sampling distribution ($s_{\overline{x}}$).

$$s_{\overline{x}} = \frac{s}{\sqrt{n}} = \frac{10}{8} = 1.25$$

(Note that we use $s_{\overline{x}}$ as the symbol for the estimated standard error of the sampling distribution if we must estimate the population standard deviation (σ) by the value of the sample standard deviation (s)).

Second, we will need to find the number of standard errors we must go on either side of the mean to have 45% (or .4500) of the sampling distribution. From Table 1 in the appendix, we will find the Z score (the number of standard deviations) that provides us with .4500 of the normal curve from the mean to it. We move down column B until we find the value .4500, and then note the Z score associated with it. The Z score is $Z = 1.645$, so that means we need to establish a range on either side of our observed sample mean of 1.645 standard errors to define the middle 90% of the sampling distribution.

One standard error is equal to 1.25, so 1.645 standard errors is equal to a range of $(1.645 \times 1.25) = 2.056$. Therefore, our 90% confidence interval for the population mean, μ, is plus and minus 2.056 on either side of the observed sample mean. We can compute that range as $75 - 2.056 = 72.944$; and, $75 + 2.056 = 77.056$. Confidence intervals are conventionally written as follows:

$$90\% \text{ CI} = (72.944 \leq \mu \leq 77.056)$$

This confidence interval would be read as, "The 90% confidence interval is equal to mu greater than or equal to 72.944, and less than or equal to 77.056." Again, we can be 90% sure that the population mean, μ, is within this range, because 90% of all possible samples of size $n = 64$ taken from that population would be in that range. Note that even though we have computed a confidence interval, it is still based on two point estimates. We have made a point estimate that the population mean is equal to 75 (the value of the sample mean), and we have made a point estimate that the population standard deviation is equal to 10 (the value of the sample standard deviation).

Computing the 95% Confidence Interval

To compute the 95% confidence interval for this same population, we only need to adjust the size of the range on either side of the mean. The observed value of the sample mean, and the standard error will remain the same. We need to find the Z score that has half of the desired interval size between the mean of the normal distribution and it. One half of the desired interval size is .4750, and we want to find that value in column B of Table 1. The Z score associated with .4750 in column B is $Z = 1.96$, indicating that we need a range equal to 1.96 standard errors on either size of the sample mean to have a total of 95% of the sampling distribution. One standard error is equal to 1.25, so 1.96 standard errors is $(1.25 \times 1.96) = 2.45$. The 95% confidence interval can then be written as:

$$95\% \text{ CI} = (72.55 \leq \mu \leq 77.45)$$

Note that the 95% confidence interval is slightly larger than the 90% confidence interval. The greater our degree of confidence, the larger the associated confidence interval must be.

The Sampling Distribution Is the Foundation for Two Important Statistical Concepts

The last two types of problems we have just examined with the sampling distribution—finding the probability of selecting a sample mean beyond a given point in the sampling distribution, and finding the probability associated with a range in the sampling distribution—represent two very important statistical concepts that we will explore in depth in upcoming chapters. The entire concept of statistical significance, in which we try to determine if two observed values are significantly different, is based on the relative position of the observed values in the sampling distribution. We will encounter the concept of statistical significance and hypothesis testing in a number of different ways in upcoming chapters, but let me give you one quick example of how it works.

Suppose we know that a population has a mean $\mu = 50$, with a standard deviation $\sigma = 12$. We are presented with a sample of $n = 36$ individuals whose observed mean is $\bar{X} = 52$. We will be trying to determine if the observed sample mean $\bar{X}$ is significantly different from the population mean μ. When we say that an observed sample mean is significantly different from a population mean, we are concluding that the difference between the two values is so large that it is unlikely that the sample was selected from the given population. When we say that an observed sample mean is not statistically different from a population mean, we are concluding that the difference between the two values is small enough that the sample could easily come from the given population. We will base our decision on how likely it is that the observed sample comes from a given population on the location of the observed sample mean in the population's sampling distribution. In our example, we know what the sampling distribution will look like. It will be normal in form, it will have a mean equal to the population mean (in this case $\mu_{\bar{X}} = 50$), and it will have a standard error equal to the population standard deviation divided by the square root of the sample size (in this case the standard error is 12 divided by 6 or $\sigma_{\bar{X}} = 2.00$). The sampling distribution for all samples of size $n = 36$ for the population is presented below, along with the sample mean of $\bar{X} = 52$ which is in question.

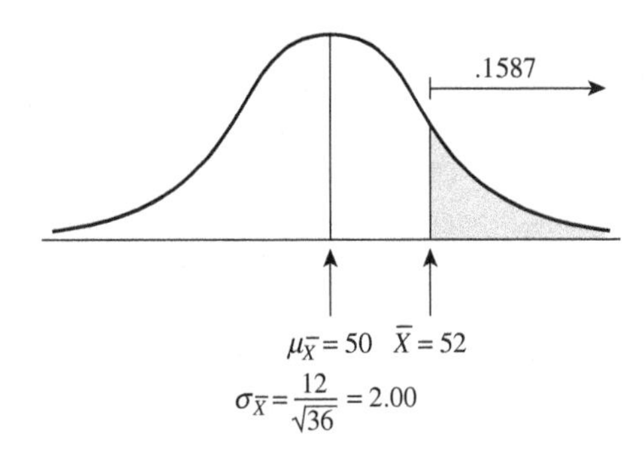

Figure 6.25 Illustration of the area beyond $\bar{X} = 52$ in the tail of the curve.

If we convert the sample mean $\bar{X} = 52$ to a Z score, we find that it is $Z = 1.00$. Using Table 1 in the appendix, we know that .1587 or 15.87% of the normal distribution is above a $Z = 1.00$. Similarly, there will be 15.87% of the sampling distribution above the sample mean $\bar{X} = 52$. A sample mean of $\bar{X} = 52$ could easily come from this sampling distribution. If we were to conclude that any sample mean equal to or larger than 52 did not come from the sampling distribution, then we could be wrong almost 16% of the time (15.87% to be exact). When we begin testing hypotheses concerning the likelihood of a sample mean coming from a given population in Chapter 11, we will define an acceptable risk of being wrong, such as 5%. This will mean that if an observed sample mean is in the extreme 5% of a sampling distribution, we will conclude that it is not likely that the observed sample came from the given population. It is unlikely that we would select a sample that "bad" given what we know about a sampling distribution. Of course we could be wrong, but we only have a 5% chance of being wrong. If an observed sample is not in the extreme 5% of the sampling distribution, then we will conclude that it is likely that the sample came from the given population. We will be examining all of these ideas in greater detail in later chapters, but it is useful to get a beginning sense of the logical basis of statistical significance at this point while examining the nature of the sampling distribution.

Sometimes the Sampling Distribution Is Not Normal: The t Distribution

I indicated earlier that the sampling distribution would be the final type of normal distribution that we would examine in this chapter, and it was. However, the sampling distribution is not the final distribution that we will examine, that distinction belongs to **the t distribution**. The characteristics of the sampling distribution described in the central limit theorem are based on the assumptions that the population standard deviation (σ) is known and serves as the basis for our computation of the standard error ($s_{\bar{X}}$), and that a "large" sample has been selected from the population. I indicated earlier that in this case we can take "large" to mean a sample size of approximately 30 or larger. Unfortunately, when σ is unknown and must be estimated by the sample standard deviation (s) to estimate the standard error of the sampling distribution ($s_{\bar{x}}$), and when the sample size is small, the sampling distribution is not exactly normal in form. The sampling distribution based on $s_{\bar{x}}$ and small sample sizes still has the characteristic bell-shaped curve, and the mean of the sampling distribution will still equal the population mean from which the samples are drawn, but the distribution of all of the sample means in the sampling distribution does not follow the same pattern as that of a true normal distribution. For example, plus and minus one standard error on either side of the mean in a sampling distribution when σ is estimated and a small sample size is used will not quite contain 68.26% of all of the possible sample means the way a true normal distribution would. Further, the smaller the sample size, the greater the sampling distribution will deviate from the pattern of a normal distribution.

The tendency for sampling distributions based on an estimate of the standard error and small sample sizes to deviate from normality was the subject of extensive research by a statistician named W. S. Gossett. Gossett found that in such cases one needed to work with a series of sampling distributions, each of which is adjusted depending on the sample size. For example, the smaller the sample size, the farther one must go beyond plus and minus one standard error on either side of the mean to capture the middle 68.26% of the distribution. Conversely, the larger the sample size, the closer the sampling distribution will approximate a normal distribution. By the time the sample size reaches approximately $n = 30$, the sampling distribution is close enough to a normal distribution that the difference from normality becomes inconsequential for most applications.

Gossett considered his research to be quite important, and wanted to publish the results so that the information could be shared by the scientific community. There was just one slight problem, his research was based on data from the Irish brewing company where he worked, and the company had a policy discouraging the disclosure of research based on their data. Gossett chose to publish the results anyway, but used the pseudonym of Student. He chose to call his distribution the "t" distribution just as the normal distribution is sometimes referred to as the "Z" distribution.

Table 2 in the appendix contains values of Student's t distribution. The table is organized quite differently than Table 1 containing areas under the normal curve. The difference is due to the fact that the t tables, as they are often called, are used primarily for hypothesis testing. As you will see in later chapters, when testing statistical hypotheses we are most interested in the *Z* score or, in the case of research based on small samples, the t score that cuts off the extreme 1%, 5%, or 10% of the sampling distribution. Consequently, Table 2 in the appendix is organized differently than Table 1 which provides areas under the normal curve associated with each *Z* score. Table 2 provides the t values that cut off the extreme 1% and 5% of the sampling distribution for selected sample sizes. We will be working with values from Tables 1 and 2 in the process of testing hypotheses regarding sample and population means in Chapter 11, and the organization and logic behind Table 2 will be explained in detail at that time. But if you take a moment to examine Table 2, you will see that it contains two major sections—one for a two-tail test, and one for a one-tail test. Again, these concepts are best explained in a later chapter, but for now take a look at the one-tail section of the table under the heading of "level of significance $\alpha = .05$." The values in this column are t scores that cut off the extreme 5% of the tail of the sampling distribution. We saw earlier in this chapter that the *Z* score that cuts off the extreme 5% of the tail of the sampling distribution was $Z = 1.645$. The numbers in the column headed "df" are degrees of freedom which equals the sample size minus one ($df = n - 1$). You may recall the discussion of degrees of freedom from Chapter 5. Notice that as the sample size increases (as measured by the degrees of freedom value) the t score in the column becomes smaller and smaller. Note also that at around a sample size of $n = 30$, the t scores in the table are very close to the value of the *Z* score (1.645) that cuts off the extreme 5% of the sampling distribution, and that at the very end of the table indicated by the infinity symbol the t value equals exactly the *Z* value which cuts off the extreme 5% of the sampling distribution.

Computer Applications

1. Load the GSS data set and run descriptive statistics on the variable "AGE."
2. Use the values for the mean and standard deviation to create a new variable called "AGEZ" computed as the *Z* score for the value of "AGE."
3. Create *Z* scores for the variable "AGE" by using the "Save Standardized Values as Variables" option under the "Analyze," "Descriptive Statistics" "Descriptives" procedure, and compare the results to your computed variable.

How to do it

Load the GSS data set and use the "Analyze" "Descriptive Statistics" "Descriptives" option to obtain the mean and standard deviation for the variable "AGE." Note the values for the mean and standard deviation, and then use them to create a new variable called "AGEZ" using the "Transform" "Compute" option. Enter the numeric expression to compute a *Z* score for age.

Next use the "Analyze," "Descriptive Statistics" "Descriptives" procedure again for the variable AGE, but this time select the "Save Standardized Values as Variables" box (located under the list of variables). SPSS will automatically compute a new variable called "ZAGE" consisting of the normal scores for the variable AGE." Compare the results of the "AGEZ" variable you created with the "ZAGE" variable created by SPSS. The results should be the same with the exception of minor rounding errors.

Summary of Key Points

This chapter dealing with the normal distribution is a pivotal chapter in the text. In one sense, the chapter continues a theme that we have followed for the first five chapters. In the first three chapters we have examined the research process, looked at methods of sampling, and examined ways to reduce raw data to a more easily interpretable form. In Chapter 4, we examined a series of statistics that give us a sense of where the central point is in a distribution. In Chapter 5, we examined some statistics that give us a sense of the variation that is present in a distribution. Finally, in this chapter, we have examined the normal distribution in various forms. Chapter 6 is a link to the previous chapters in that many of the variables that we examine in the behavioral sciences turn out to be normally distributed. But Chapter 6 is also important because the normal distribution and, in particular, the sampling distribution are key components of the process of hypothesis testing and the concept of statistical significance that we will encounter in upcoming chapters. A thorough understanding of Chapter 6 is necessary if you are to really understand these very important upcoming topics.

The Normal Distribution—A theoretical distribution appearing as a bell-shaped curve with a mean equal to 0.00, and a standard deviation equal to 1.00.
***Z* Scores**—Scores whose distance from the mean is expressed in units of standard deviation.
A Normal Distribution—Any distribution with some mean ($\bar{X}$), and some standard deviation (s), and whose raw scores are distributed around their mean in the same proportion as the normal distribution.
The Sampling Distribution—The distribution of all sample means based on a sample of size "*n*" selected from a population with a mean (μ), and a standard deviation (σ). When σ is known and the sample size is large, the sampling distribution will be normal in form with a mean ($\mu_{\bar{x}}$) equal to the population mean from which the samples are drawn, and a standard error ($\sigma_{\bar{x}}$) equal to the population standard deviation divided by the square root of the sample size.
Standard Error—The standard deviation of a sampling distribution. The standard error ($\sigma_{\bar{x}}$) is equal to the population standard deviation (σ) divided by the square root of the sample size.

Point Estimate—The estimation of a population parameter by using the observed value of a sample statistic.

Confidence Interval—The estimation of a population parameter by creating a range of values on either side of a sample statistic. The size of the range is determined by our degree of confidence that it contains the population parameter being estimated.

Statistical Significance—In the case of testing statistical hypotheses about sample means, statistical significance is the point at which the difference between two means is sufficient to suggest that the samples came from different populations. Statistical significance is examined in greater detail in Chapter 8.

The t Distribution—A set of sampling distributions based on small sample sizes ($n < 30$) when the population standard deviation (σ) is unknown and must be estimated by the sample standard deviation (s). The t distribution shares many of the characteristics of a sampling distribution based on a larger sample size, but the t distribution is not exactly normal in form. As the sample size increases, the t distribution becomes more similar to the normal distribution.

Questions and Problems for Review

1. Explain the difference between the normal distribution and a normal distribution.
2. What is the difference between a Z score and a raw score?
3. Indicate the proportion of the normal curve associated with the following Z scores:
   A. Between $Z = -1.35$ and the mean
   B. Below a $Z = -1.68$
   C. Above a $Z = 2.28$
   D. Below a $Z = 1.96$
4. Find the Z scores that define the following percentiles in the normal distribution (if the exact value is not in the table, choose the Z score that is closest to the given percentile):
   A. The 75th percentile
   B. The 50th percentile
   C. The 99th percentile
   D. The 10th percentile
5. Convert the following raw scores from a normal distribution with a mean $\overline{X} = 72$, and a standard deviation $s = 8$ to Z scores.
   A. $X = 80$
   B. $X = 64$
   C. $X = 75$
   D. $X = 112$
6. What is unusual about the raw score $X = 112$ from problem 5 above? (Hint: what percentage of the normal distribution is within plus and minus 3 standard deviations?)
7. Indicate the proportion of the normal distribution associated with the following Z scores:
   A. Between $Z = -1.10$ and $Z = 1.00$
   B. Between $Z = 1.00$ and $Z = 1.45$
   C. Between $Z = -2.00$ and $Z = -1.10$
   D. Above $Z = 2.33$
8. Indicate the Z scores associated with the following areas in the normal curve (report the closest Z score if you cannot find the exact area in the table):
   A. .10 in the tail on the positive side of the mean
   B. .05 in the tail on the negative side of the mean
   C. .025 in each tail of the curve
   D. .01 in the tail on the positive side of the mean

9. A set of examination scores is normally distributed with a mean $\overline{X} = 200$, and a standard deviation $s = 50$. Indicate the proportion of scores:

A. Between the mean and $X = 275$
B. Between $X = 180$ and $X = 280$
C. Between $X = 250$ and $X = 300$
D. Below $X = 240$

10. What raw score in the distribution in problem 9 would be equal to the 80th percentile?

11. A student earned a score of $X = 90$ on an exam with a mean $\overline{X} = 75$, and a standard deviation $s = 10$. What raw score would the student be expected to earn on a similar exam with a mean $\overline{X} = 50$, and a standard deviation $s = 12$?

12. What is the relationship between a standard deviation and a standard error?

13. A population is known to have a mean $\mu = 80$, and a standard deviation $\sigma = 16$. Describe the sampling distribution of all samples of size $n = 64$. Draw the sampling distribution, and label it with the mean, and standard error.

14. In the sampling distribution from problem 13, what proportion of the sample means would be the following:

A. $\overline{X} = 83$ or larger
B. $\overline{X} = 84$ or less
C. $\overline{X} = 75$ or less
D. Between $\overline{X} = 75$ and $\overline{X} = 78$

15. In a sampling distribution with a mean of $\mu_{\bar{x}} = 30$ and a standard error of $\sigma_{\bar{x}} = 1.5$, what is the probability of selecting a sample mean from the distribution between the values of $\overline{X} = 28.5$ and $\overline{X} = 31.5$?

16. Given a sample of $n = 36$ with a mean, $\overline{X} = 50$, and a standard deviation, $s = 12$, compute the 90% and 95% confidence intervals.

17. What effect does increasing the sample size to $n = 100$ have on the confidence intervals in problem 16?

CHAPTER 7

Probability

Key Concepts

Simple Probability
Events
Sample Space
Independent Events
Related Events
Compound Probability
Binomial Probability

Introduction

All of the chapters up to this point in the text have had a great deal in common. For example, the previous chapter dealing with the normal distribution began by tracing the logical theme that extends from the first chapter through to Chapter 6. You may find that this chapter on probability appears to have little in common with any of the previous chapters, and in one sense that is true. We did interpret areas under the normal curve as probabilities in a few of the examples from Chapter 6, but other than that, much of what we will examine in this chapter will not seem to be directly related to the previous chapters.

Like Chapter 6 on the normal distribution, Chapter 7 dealing with various issues in probability is a transitional chapter. The normal distribution, and in particular the sampling distribution, along with probability serves as the foundation for two extremely important statistical concepts: hypothesis testing and statistical inference (inferential statistics). We begin our examination of hypothesis testing in the next chapter, and begin working with inferential statistical procedures in Chapter 9 on measures of correlation. We will continue to deal with a combination of both descriptive and inferential statistical procedures, and hypothesis testing throughout the remainder of the text. As you may recall from Chapter 1, descriptive statistics refers to procedures that are used to describe a population or a sample, and inferential statistics are procedures that are used to infer something about unknown population parameters based on the observation of a sample. As you will see, inferential statistics is based to a large degree on probability theory.

Origins of Probability Theory

Much of the study of probability theory began in the 17th century (the 1600s) when a French nobleman approached one of the leading mathematicians of the time, Blaise Pascal, for help in understanding the odds relating to gambling on dice games. Pascal began a collaboration on the subject with another leading mathematician, Pierre de Fermat, and the origins of probability theory began from their work.

You may or may not be familiar with many games of chance, but most people have at least a vague understanding of the casino game of roulette. The roulette wheel consists of 36 small slots that are numbered from 1 to 36, and that are alternately colored either red or black. There is also either one additional slot numbered zero and colored green, or two additional slots numbered zero and double zero respectively both of which are colored green. The game is played by having the croupier spin a small ball around the perimeter of the spinning roulette wheel. Gamblers may place a bet by putting one or more chips on one of the numbers from 1 to 36 representing the slots on the wheel, or by putting one or more chips on the color red or black. If the small ball comes to rest in the green slot, the house (the casino) automatically wins. If the small ball comes to rest in the slot corresponding to the number you selected, then you win your bet, usually with a payoff of 35 chips for each one you wagered. If the ball comes to rest on any red slot when you have bet red, then you will also win, usually with a payoff of two chips for each one you have wagered. Probability theory can be used to estimate the odds of a player winning on any given spin of the wheel. For example, if you place your wager on a particular number, then your odds of winning are 1 out of 37, because there are 36 slots numbered 1 to 36 plus the green slot marked 0. (Of course your odds of winning would be 1 out of 38 if there are 2 green slots on the wheel.)

The point of describing the roulette wheel is simply this. You can think of the wheel itself as a "population" of known possible outcomes. We may not know which particular outcome will be selected when the ball comes to rest, but at least we know the identity of all possible outcomes. You can think of each spin of the wheel as taking a "sample" from the population. Probability theory allows us to calculate the odds of a particular outcome (sample) from the population of possibilities because we know the exact composition of the population. Now think about the case of inferential statistics. With inferential statistics we have an unknown population, and a known sample that has been selected from it. The sample is known because we observe or measure the members of the sample, and we then generalize those results back to the unknown population. In statistical terms, we infer something about unknown population parameters based on our knowledge of known sample statistics. In a sense, inferential statistics represents the reverse process of computing probabilities based on probability theory.

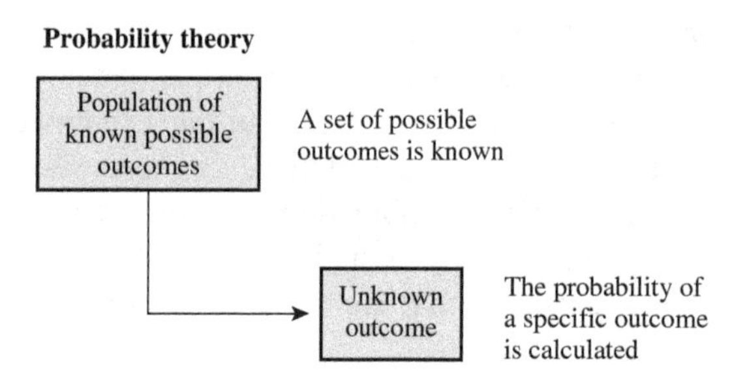

Figure 7.1 The Basis for Probability Theory

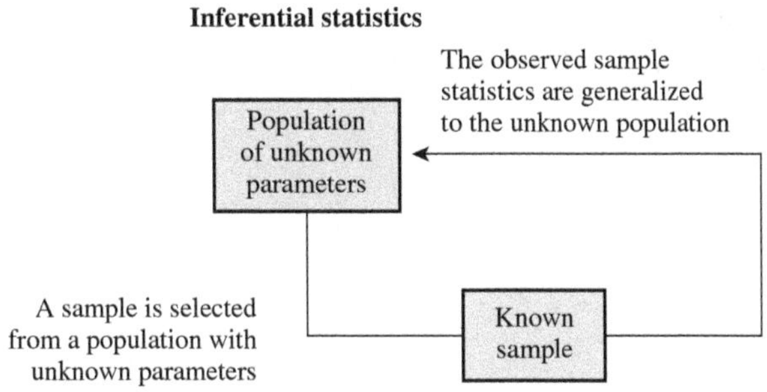

Inferential statistics

The observed sample statistics are generalized to the unknown population

Population of unknown parameters

A sample is selected from a population with unknown parameters

Known sample

Figure 7.2 The Basis for Inferential Statistics

Probability

In this chapter we will examine three different types of probability: simple probability, compound probability, and **binomial probability**. Simple probability is not very difficult to understand, but it derives its name from the fact that simple probability deals with a single set of possible outcomes. Compound probability involves the consideration of two sets of possible outcomes at the same time. Binomial probability involves the analysis of binomial experiments in which the probability of a specific outcome is calculated for an event that is repeated for a specific number of trials. It is termed "binomial" because only two outcomes are possible on each trial: success, or failure. An example of a binomial probability problem is to compute the probability of flipping a coin 5 times, and having the coin turn up heads on 3 of the 5 flips of the coin. We will examine each of these types of probability in turn after examining some basic terms and concepts relating to probability.

Basic concepts

Probability theory is based on the fact that a particular outcome results from a known population of all possible outcomes. In probability theory, the set of all possible outcomes for a probability experiment is called the **sample space**. The sample space is usually represented by an uppercase "S," and the possible outcomes are listed in braces. For example, if we consider flipping a coin as a probability experiment, then the sample space consists of the two possibilities of: heads or tails. Symbolically we would represent this by:

$$S = \{\text{Heads, Tails}\}.$$

Another probability experiment might consist of drawing a card from a standard deck, and if we were interested only in the suit of the card (whether it is a heart, club, diamond, or spade), then the sample space would be represented as follows:

$$S = \{\text{Heart, Club, Diamond, Spade}\}.$$

An **event** is defined as an outcome from a probability experiment to which we can assign a probability. Events are usually designated by uppercase letters. For example, we might define the event "E" as flipping a coin and having it turn up as "heads." Symbolically we would represent the probability, represented by an uppercase "P," of a given event, "E," as:

$$P(E) =$$

Which would be read as, "the probability of event E equals." Often times the probability of an event can easily be computed by examining the sample space. For example, in the case of event "E" (flipping a coin) we know that the sample space consists of only two possibilities: heads or tails, and if we assume that the coin is fair, then each outcome is equally likely. Therefore the probability of event "E" is .50, or written symbolically:

$$P(E) = .50$$

Note that probabilities are usually expressed as proportions.

Simple probability

Calculations of **simple probability** are based on a single set of outcomes, and are governed by six basic rules. In the simplest case, such as the examples of sample spaces above where all of the possible outcomes are equally likely, the calculation of simple probabilities are often intuitive. Let's consider some examples before looking at the rules for simple probability. We saw earlier that the sample space for a deck of cards (if we are just interested in the suit of the card selected) is

$$S = \{\text{Heart, Club, Diamond, Spade}\}.$$

If we define event "K" as selecting a heart on a single draw from the deck, and we can see that hearts represent one of four suits, all of which are equally represented in the deck, then the probability of satisfying event "K" (drawing a heart) must be 1 out of 4, or .25. In symbolic notation we would write the probability of "K" as:

$$P(K) = .25$$

Suppose we have another sample space based on a box of 10 colored balls. The box contains 4 red balls, 3 green balls, and 3 blue balls. The sample space may be represented as:

$$S = \{\text{Red, Green, Blue}\},$$

but unlike the other sample spaces we have examined, the probabilities of each of the three possible outcomes are not equal. The box contains more red balls than any other color, so the probability of selecting red must be greater. What would the probability of selecting a red ball be? If we define event "F" as selecting a red ball from the box on a single selection, and we know that there are 4 red balls among the total of 10 balls in the box, then the probability of satisfying event "F" must be 4 out of 10, or .40.

$$P(F) = .40$$

Similarly, the probability of selecting a green ball must be 3 out of 10 or, .30, and the probability of selecting a blue ball must be 3 out of 10, or .30.

If we define event "T" as selecting an orange ball from the box, what would be the probability of satisfying event "T"? There are no orange balls in the box, so the probability of satisfying event "T" must be 0 out of 10, or 0.00.

If we define event "C" as selecting a red or a blue ball from the box on a single draw, what is the probability of satisfying event "C"? Note that we are only going to select one ball from the box, but our probability event that we are calling "C" will be satisfied if the selected ball is either a red one or a blue one. The box contains 4 red balls and 3 blue balls, so there is a total of 7 balls that would satisfy event "C." The probability of satisfying event "C" must be 7 out of 10, or .70.

$$P(C) = .70$$

Along these same lines, what if we define event "J" as selecting either a red, or a blue, or a green ball from the box on a single draw. All 10 of the balls in the box satisfy the condition of being either red, or green, or blue, so the probability of satisfying event "J" must be 10 out of 10, or 1.00.

$$P(J) = 1.00$$

One final probability situation that we may encounter with simple probability involves conducting a probability experiment based on successive trials. For example, if we define event "L" as selecting a red ball from the box on two successive draws with replacement, then what is the probability of satisfying event "L"? This probability experiment is a little different than the others with which we have dealt thus far. We are still working with a single set of outcomes (the balls in the box), but our probability experiment requires us to do two things in succession. In this case we will draw a ball from the box, note its color to see if it is red or not, put the ball back in the box, then draw a second ball from the box, and note its color to see if it is red or not. Putting the first ball back in the box is what is meant by "drawing with replacement." Unless the first ball selected from the box is replaced, the odds of selecting a red ball on the second draw will change.

The calculation of the probability of satisfying event "L," selecting a red ball on two successive draws from the box with replacement is not as intuitive as the other probability experiments, but it is still a relatively easy calculation. To compute the probability of two successive outcomes that are independent of each other, you simply multiply the probability of the first outcome by the probability of the second outcome. In our example the two outcomes do meet the condition of being independent of each other. We insure independence when we replace the first ball drawn from the box, otherwise the result of the first outcome would affect the probability of the second outcome. We know from the composition of the box that the probability of selecting a red ball on one draw is .40, so the probability of selecting 2 red balls on successive draws is $.40 \times .40$, or .16. In symbolic terms:

$$P(L) = .40 \times .40 = .16.$$

We can use the same strategy to compute other similar events. For example, we can define event "M" as selecting a red ball, and then a green ball, and then a blue ball on three successive draws with replacement. The probability of satisfying event "M" would be found by multiplying the probability of selecting a red ball (.40) times the probability of selecting a green ball (.30) times the probability of selecting a blue ball (.30) or:

$$P(M) = .40 \times .30 \times .30 = 0.036$$

At this point, with the possible exception of the last example, we have intuitively arrived at the six basic rules for simple probability.

Rules for simple probability

1. The probability for any event E is $(0 \leq P(E) \leq 1)$. A probability cannot be less than 0 (at zero the event will never occur) or greater than 1 (at 1.00 the event will always occur).
2. The sum of the probabilities associated with an event must equal 1.00. If six things are possible, then their collective probability will equal 1.00 indicating that one of the six things must occur. For example, the probability of selecting a red, or a green, or a blue ball on one draw from the box was seen to be $.40 + .30 + .30 = 1.00$. If you draw a ball from the box it must be either red, green, or blue, since those are the only color of balls in the box. It is therefore a certainty that one of those colors must be selected.
3. If a group of outcomes is equally likely, then the probability of any single outcome is

$$\frac{1}{\text{the number of possible outcomes}}$$

For example, the probability of selecting a heart from a standard deck is $\frac{1}{4} = .25$

4. If several events share the same probability, then the probability of that outcome is equal to:

$$\frac{\text{number of events with shared probability}}{\text{the number of possible outcomes}}$$

For this example think about the box with 10 colored balls in it. Four of the balls are red, three of the balls are green, and three of the balls are blue. The probability of selecting a red ball is equal to: $\frac{4}{10}$

5. The probability that one of a group of alternative events will occur is equal to the sum of the probabilities of the separate events. Under the assumptions of this rule we are still performing a single action, but there are several outcomes that satisfy the event. This rule is often referred to as the addition rule for mutually exclusive events.

   For example, the probability of selecting a red ball or a blue ball on a single draw from the box is equal to the sum of the probability of each of those outcomes separately, or $.40 + .30 = .70$. This rule holds when the individual outcomes that satisfy the probability experiment are mutually exclusive. That is, the individual outcomes all belong to one and only one category.

6. The probability that a combination of **independent events** will occur over separate trials is equal to the product of the probabilities for each event. Under the assumptions of this rule we are performing several independent actions in succession. This rule is often referred to as the multiplication rule. (Note that the multiplication rule does not apply to **related events** in which one outcome affects the probability of a later event.)

For example, the probability of selecting three green balls, with replacement, in succession from the box is equal to the probability of selecting a green ball on the first draw times the probability of selecting a green ball on the second draw times the probability of selecting a green ball on the third draw, or: $.30 \times .30 \times .30 = .027$

Applying the Rules for Simple Probability, or I Have Two Boxes of Soxes (with apologies to Dr. Suess)

I have 2 boxes in which I keep my socks (I keep my socks in boxes for this particular illustration in the book in case you are wondering). In box "A" I have 10 blue socks, and 10 black socks for a total of 20 socks. In box "B" I have 6 blue socks, and 4 black socks for a total of 10 socks. Event "K" is defined as selecting a black sock from box "A." What is the probability of satisfying event "K?"

Solution: Box "A" contains 20 socks, and 10 of them are black. The probability of satisfying "K," P(K) must equal 10 divided by 20, or .50 (rule 4).

Event "T" is defined as selecting a blue sock 2 times in a row (with replacement) from box "B." What is the probability of satisfying event "T?"

Solution: The probability of selecting a single blue sock from box "B" is 6 divided by 10, or .60. The probability of selecting 2 blue socks in succession is $.60 \times .60$ or .36, therefore P(T) = .36.

How about a harder question? If I select a single sock from each box, what is the probability that I will end up with a matching pair? This question requires the application of several of the rules for simple probability.

Solution: To find a solution to this problem we should first identify all of the possible things that can happen if we select a single sock from each box, then compute the probability of each of the possible outcomes, and then identify the outcomes that satisfy the condition of being a matching pair.

To identify all of the possible outcomes resulting from selecting a single sock from each box, we will let "black" represent a black sock being selected, and "blue" represent a blue sock being selected. There are 4 possible outcomes when selecting a single sock from each box. They are represented below with the sock from box "A" listed first, and the sock from box "B" listed second.

Possible outcomes

(blue, blue)   (blue, black)   (black, blue)   (black, black)

From the composition of the 2 boxes, we know that the probability of selecting a blue sock from box "A" is .50, the probability of selecting a black sock from box "A" is also .50, the probability of selecting a blue sock from box "B" is .60, and the probability of selecting a black sock from box "B" is .40 (rule 4 in application).

Each of the 4 possible outcomes requires us to do 2 things in succession, selecting a sock from box "A," and then selecting a sock from box "B." Because we have 2 separate boxes, the selecting of a sock from each box represents independent events. Therefore, by application of rule 6, the multiplication rule, we can compute the probability of each of the 4 possible outcomes by multiplying the probabilities of the 2successive actions that make up each outcome. The 4 possible outcomes are represented below with their associated probabilities, and each of the possible outcomes that satisfies our original condition of being a matching pair is represented in bold.

Probability of each possible outcome

(blue, blue)	(blue, black)	(black, blue)	**(black, black)**
.50 × .60	.50 × .40	.50 × .60	**.50 × .40**
.30	.20	.30	**.20**

Out of the 4 possible outcomes resulting from the selection of a single sock from each box, 2 outcomes satisfy the condition of being a matching pair. The first outcome consisting of 2 blue socks has a probability of .30, and the final outcome consisting of 2 black socks has a probability of .20. Because the color of the matching pair of socks does not matter, we can satisfy the criterion of a matching pair by the first outcome or the last outcome. Therefore, the overall probability of selecting a single sock from each box and having a matching pair is .30 + .20 = .50 (rule 5, the addition rule in application).

Also note that the sum of the probabilities of each of the 4 possible outcomes (.30 + .20 +.30 + .20) is equal to 1.00, the sum of the probabilities of all possible outcomes must equal 1.00 (rule 2 in application).

Compound probability

Compound probability involves the consideration of two sets of possible outcomes at the same time. We will use generalized outcomes to demonstrate compound probability. A set of generalized outcomes is presented below, along with the probability of each specific outcome.

Outcome	Probability
O_1	.20
O_2	.10
O_3	.40
O_4	.10
O_5	.20

The set of generalized outcomes consists of five possible outcomes called: O_1 through O_5, and each outcome has a specific probability. Note that the sum of the probabilities for all five outcomes is 1.00. If you want to think of a more concrete example of where the individual probabilities come from, then you can think of a box with 100 balls in it. Twenty of the balls are marked "O_1," 10 of the balls are marked "O_2," 40 of the balls are marked "O_3," 10 of the balls are marked "O_4," and 20 of the balls are marked "O_5." The individual probabilities result from the application of rule 4 for simple probability. For example, the probability of selecting a ball marked "O_3" on a single draw from the box would be 40 divided by 100, or .40. The other individual probabilities can be computed in a similar manner.

The generalized outcomes can be used to calculate simple probabilities as well as compound probabilities. In fact, compound probabilities are based on the consideration of two simple probabilities at the same time. For example, suppose we define event "E" as outcomes (O_1, or O_2, or O_4). That is, (using the language of the concrete example), on a single draw from the box, event "E" is satisfied if the ball selected is marked O_1, or O_2, or O_4. We can compute the probability of event "E," P(E) by the application of rule 5 for simple probability. The probability of event "E" will be the sum of the individual probabilities that satisfy it, or .20 + .10 + .10 = .40.

We can also define event "F" as outcomes (O_2, or O_3, or O_4). The probability of event "F" can be computed just as event "E". P(F) = .10 + .40 + .10 = .60. Compound probability involves the consideration of two sets of outcomes at the same time. In this case, we would be concerned with the outcomes of event "E" and event "F."

Three types of compound events

We will examine three types of compound events where two sets of outcomes (outcome associated with event "E," and outcomes associated with event "F") are considered.

1. P(E or F) The probability of E or F is satisfied if any outcome satisfying event "E" occurs, or if any outcome satisfying event "F" occurs, or if any outcome common to both "E" or "F" occurs.
2. P(E and F) The probability of E and F is satisfied only when one or more outcomes common to both event "E" and event "F" occur.
3. P(E|F) (read as the probability of E given F) The probability of E given F is the probability of satisfying event "E" given the fact that event "F" has already been satisfied. This final example of compound probability involves conditional probability. By knowing that event "F" has been satisfied, we know that only those outcomes that satisfy event "F" are possibilities, and outcomes which do not satisfy event "F" are no longer possible.

To illustrate the computation of compound probability we will use the events "E" and "F" as defined above. We began with a set of generalized probabilities that are presented again below.

Outcome	Probability	Simple Events E and F
O_1	.20	E = (O_1 or O_2 or O_4) P(E) = .40
O_2	.10	F = (O_2 or O_3 or O_4) P (F) = .60
O_3	.40	
O_4	.10	
O_5	.20	

Based on these probabilities we have defined two events called event "E" and event "F." Event "E" is defined as outcomes (O_1, or O_2, or O_4), and has a probability of P(E) = .40. Event "F" is defined as outcomes (O_2, or O_3, or O_4), and has a probability of P(F) = .60. Note that the probability of event "E" and event "F" as we have just computed them are still based on simple probability following the rules that we examined earlier. It is only when we are dealing with the outcomes that satisfy events "E" and "F" simultaneously that we begin working with compound probability. Keep in mind when working with probabilities for compound events that we are still dealing with the occurrence of a single outcome, or in other words, we are doing one thing such as drawing a card from a deck, or selecting a ball from a box. The event is "compound" because we are considering the probability that both probability events will be satisfied by the one outcome which occurs.

P(E or F) The first type of compound event is P(E or F), read as the probability of "E" or "F." The key to computing the probability of this and the other compound events correctly is to forget what you think should be the solution, and to focus on the actual definition of what satisfies the various compound events. The compound event (E or F) is satisfied if any of the individual outcomes that satisfy event "E" occur, or if any of the individual outcomes that satisfy event "F" occur, or if any of the individual outcomes that are common to both "E" and "F" occur. To compute the probability of (E or F) simply identify all of the outcomes that satisfy event "E," all of the outcomes that satisfy event "F," and all of the outcomes common to both, but do not count any outcome twice (for example if a particular outcome satisfies both "E" and "F," just include it in the list of all outcomes satisfying (E or F) a single time).

Event "E" is satisfied by outcomes (O_1, or O_2, or O_4), and event "F" is satisfied by outcomes (O_2, or O_3, or O_4). The probability of event (E or F) would therefore be satisfied by outcomes (O_1, or O_2, or O_4, or O_3), which has an overall probability of $.20 + .10 + .10 + .40 = .80$, so the P(E or F) = .80. The solution is found by identifying all of the outcomes that satisfy the individual events "E," and "F" without including any outcome twice, and then computing the simple probability of the identified outcomes by adding their individual probabilities.

P(E and F) The compound event P(E and F), read as the probability of E and F, is satisfied when a single outcome occurs that will satisfy both event "E" and event "F." For a single outcome to satisfy both event "E" and event "F," it must be one of the outcomes that is included in the set of outcomes common to both events. If there are no outcomes common to both event "E" and event "F," then there is a zero probability of satisfying the event (E and F). The solution to P(E and F) is to identify all of the outcomes (if any) that satisfy both event "E" and event "F," and then to compute the simple probability of that set of common outcomes.

Event "E" is satisfied by outcomes (O_1, or O_2, or O_4), and event "F" is satisfied by outcomes (O_2, or O_3, or O_4). In this case, the outcomes O_2 and O_4 are common to both sets. Therefore, the probability of satisfying (E and F) on a single occurrence is equal to the probability of (O_2, or O_4), and the simple probability of (O_2, or O_4) is equal to $.10 + .10 = .20$, so P(E and F) = .20.

P(E|F) The final compound event is P(E|F), read as the probability of E given F. In this application of compound probability we are interested in computing the probability of satisfying event "E," given the fact that event "F" has already been satisfied. Or stated in another way, we do not know which individual outcome has occurred, but we do know that it must be one of the ones that satisfies event "F." The question becomes, given the knowledge that an outcome has occurred which satisfies event "F," what is the probability that the particular outcome is also one of the ones that will satisfy event "E?" As was the case with calculating P(E and F), the P(E|F) will equal zero unless there is at least one outcome common to both event "E," and event "F." In fact, the solution to P(E|F) is based on the probability of (E and F), and on the fact that we know that certain outcomes are no longer possible (those that are not capable of satisfying event "F").

The solution to P(E|F) is equal to:

$$P(E \mid F) = \frac{P(E \text{ and } F)}{P(F)}$$

The probability of P(E|F) is equal to the probability of the compound event (E and F), as we have just calculated it, divided by the probability of the event that is given, in this case P(F). We know that the probability of P(E and F) = .20, and that the probability of the simple event P(F) = .60, so:

$$P(E \mid F) = \frac{P(E \text{ and } F)}{P(F)} = \frac{.20}{.60} = .33$$

Examination of the probabilities of the individual outcomes that satisfy event "E" and event "F" helps explain why the probability of P (E|F) = .33. We know that event "F" is satisfied by outcomes (O_2, or O_3, or O_4), and we know that one of those outcomes must have occurred. We also know that event "E" is satisfied by outcomes (O_1, or O_2, or O_4). Event "E" can be satisfied if the outcome that has satisfied event "F" is (O_2, or O_4). The remaining outcome, (O_3), is capable of satisfying event "F," but not event "E." You might be tempted to think that the probability of P(E|F) should be .67, because two of the three possible outcomes that can satisfy event "F" can also satisfy event "E." The probability of P(E|F) would be .67, if the three outcomes that can satisfy event "F" were equally likely, but in this case they are not. Outcome O_2 has a probability of .10, outcome O_4 has a probability of .10, but outcome O_3 (that will satisfy event "F" but not event "E") has a probability of .40. Overall, event "F" has a .60 probability of being satisfied, and the outcome that can satisfy event "F" and event "E" have a combined probability of .20 (.10 + .10 for outcomes O_2 and O_4). Therefore, the probability of satisfying event "E" given the fact that event "F" has been satisfied will be equal to .20 divided by .60, or .33.

We can also compute the probability of event "F" given the fact that event "E" has been satisfied. In effect, we are turning the problem around. In the case of P(F|E), we do not know which outcome has occurred, but we do know that it is one of the ones that will satisfy event "E" (O_1, or O_2, or O_4). The logic behind the solution to P(F|E) is the same as before, we will divide the probability of the compound event (E and F) by the probability of the event that is given, in this case event "E." So the probability of P(F|E) is

$$P(F|E) = \frac{P(E \text{ and } F)}{P(E)} = \frac{.20}{.40} = .50$$

Applying compound probability

Let's look at an application of compound probability using a concrete example. Suppose we have a total of $N = 50$ individuals who are located in four different rooms. Room "A" contains 5 males and 5 females; room "B" contains 10 males; room "C" contains 10 females; and, room "D" contains 5 males and 15 females.

Event "E" is defined as selecting an individual from room "A." Event "F" is defined as selecting a male from any room. Event "E" and event "F" are simple probabilities that can be computed as we have done previously. For example, the probability of event "E" is .20, because there is a 10 individuals in

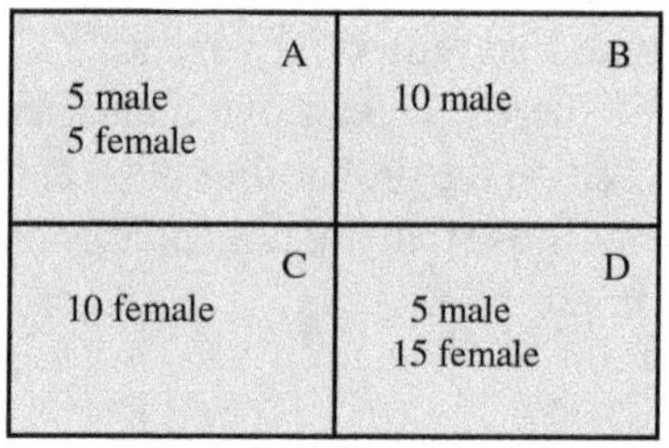

A	B
5 male 5 female	10 male
C	**D**
10 female	5 male 15 female

Figure 7.3 Fifty Individuals in Four Rooms

room "A" of the 50 individuals present, so P(E) = .20. The probability of event "F" is .40, because there are 20 males out of the 50 individuals, so P(F) = .40. Once we have computed the probability of the simple events of "E," and "F," we can move on to the compound events of: P(E or F); P(E and F); and, P(E|F).

The probability of event (E or F) is satisfied if any person from room "A" is selected, or if any male is selected. The P(E or F) can be computed by beginning with the total number of males among the 50 individuals, and then adding the number of females who are in room "A." We do not want to add the number of males who are in room "A" because they are already included in the count of all males, and we do not want to count any individual twice. There is 20 males in the group, and an additional 5 females in room "A," so the probability of (E or F) is 25 out of the 50 individuals, P(E or F) = .50.

The probability of event (E and F) is satisfied if the individual selected is both a male, and in room "A." There is 5 individuals who meet the criteria of being male, and being in room "A," so the probability of (E and F) is 5 out of the 50 individuals, P(E and F) = .10.

The probability of event "E" given event "F" has been satisfied, P(E|F), is the probability of selecting a person from room "A" given the fact that the selected individual is a male (event "F"). We know that there is 20 males among the 50 individuals, and we know that 5 of the males are in room "A," so logically the probability of selecting someone from room "A" given the fact that the selected individual is a male should be 5 out of 20, or .25. By formula, we obtain the same result:

$$P(E|F) = \frac{P(E \text{ and } F)}{P(E)} = \frac{.10}{.40} = .25$$

Again, we can turn this type of compound event around, and compute the probability of selecting a male given the fact that a person from room "A" has been selected, or stated symbolically, P(F|E). Logically, there is 10 individuals in room "A," and half of them are male, so the probability of selecting a male given the fact that the individual selected is from room "A" should be 5 out of 10, or .50. By formula we obtain the same result:

$$P(F|E) = \frac{P(E \text{ and } F)}{P(E)} = \frac{.10}{.20} = .50$$

Binomial probability

Binomial probability is so named because the outcome of a binomial experiment can have only two possible outcomes: success, or failure. The probability associated with a binomial experiment is based on the **binomial probability** distribution, which is the distribution of the number of successes in a fixed number of repeated, independent trials. An example of a binomial experiment is to flip a coin 5 times hoping to have exactly 3 heads, or to draw a ball from a box 4 times hoping to have exactly 2 red balls and 2 "not red" balls. Binomial experiments have the following properties.

Properties of a binomial experiment

1. There are a fixed number of trials (n).
2. Each trial is the same, and has only two possible outcomes: success (p), or failure (q).
3. The probability of success (p), and the probability of failure (q) are the same for each trial. (Note: q = 1 − p).
4. Each trial is independent—one outcome will not influence the next.
5. It is the total number of successes (r) that are of interest; not the order of occurrence. (For example, if we are flipping a coin $n = 5$ times, and are hoping to observe $r = 3$ heads, we do not care about the order in which the 3 heads are obtained.)

Binomial experiments involve a specified number of repeated trials, and the corresponding probability refers to achieving a certain number of successes. For example, let's go back to the box of colored balls that we used in one of the simple probability examples. As you should recall, the box contained 4 red balls, 3 green balls, and 3 blue balls for a total of 10 balls. A typical binomial experiment would begin by defining the selection of one color of ball as success, and either of the two remaining colors as failure. Let's arbitrarily define selecting a red ball as success, and either of the other two colors as failure. The binomial experiment involves a specific number of trials during which we hope to select a certain number of red balls. In this case we will select a ball from the box (with replacement) a total of 4 times. What is the probability of selecting exactly 2 red balls?

There is a rather simple formula for computing binomial probabilities. The formula is as follows:

The Formula for a Binomial Experiment

$$\text{Probability} = \text{C}\binom{n}{r} \times \text{p}^r \times \text{q}^{n-r}$$

where

$\text{C}\binom{n}{r}$ = the number of combinations of "n" things taken "r" at a time (we worked with the combination notation previously in Chapter 2)

n = number of trials

r = number of successes

p = probability of success on any single trial

q = probability of failure on any single trial.

There, what can be simpler than that!

Actually the formula is rather simple, but to appreciate that fact we first need to compute some binomial probabilities the hard way. In our first binomial experiment, we are going to select a ball, with replacement, a total of 4 times, and we want to know the probability of selecting exactly 2 red balls and 2 balls that are not red (in this case either green or blue). One way we can compute the probability of this experiment is to identify every possible way that we can select 4 balls and have 2 of them be red and 2 of them not be red. It turns out that there are 6 ways that we can select exactly 2 red balls out of 4 trials. The 6 outcomes that satisfy the condition of the binomial experiment are listed below. Each line represents one trial consisting of 4 draws from the box. The selection of a red ball is symbolized by "**R**," and the selection of a ball that is not red is symbolized by "E" (representing something else).

Outcomes satisfying 2 red balls from 4 draws:

Trial				
Outcome	1	2	3	4
1	**R**	**R**	E	E
2	**R**	E	**R**	E
3	**R**	E	E	**R**
4	E	**R**	**R**	E
5	E	**R**	E	**R**
6	E	E	**R**	**R**

Each of these 6 outcomes represents a simple probability that we can compute. For example, the first outcome would require us to select a red ball on each of the first 2 trials, and then another color ball on each of the final 2 trials. In other words, we are simply doing 4 things in succession, and rule 6 for simple probability (the multiplication rule) provides us with an easy way to compute this particular outcome. The probability of selecting a red ball on any single trial is .60 (because there are 6 red balls among the10 balls in the box). The probability of selecting a "not red" ball is .40 (because there are 4 other colored balls among the 10 balls in the box). The probability of the first outcome satisfying the conditions of our binomial experiment must be:

$$\textbf{R} \quad \textbf{R} \quad \text{E} \quad \text{E} = .60 \times .60 \times .40 \times .40 = 0.0576$$

In fact, the probability of each of the 6 possible outcomes satisfying the conditions of our binomial experiment is the same. If you examine each outcome, you will see that there are always 2 red balls, each of which always has a probability of .60, and 2 "not red" balls each of which always has a probability of .40. The only difference from one outcome to the next is when the 2 red balls are selected, and because multiplication is commutative it does not matter in what order we multiply the values: .60, .60, .40, and .40; the product will always come out to be 0.0576, as illustrated below.

Trial					
Outcome	1	2	3	4	Probability of Each Outcome
1	**R**	**R**	E	E	$.60 \times .60 \times .40 \times .40 = 0.0576$
2	**R**	E	**R**	E	$.60 \times .40 \times .60 \times .40 = 0.0576$
3	**R**	E	E	**R**	$.60 \times .40 \times .40 \times .60 = 0.0576$
4	E	**R**	**R**	E	$.40 \times .60 \times .60 \times .40 = 0.0576$
5	E	**R**	E	**R**	$.40 \times .60 \times .40 \times .60 = 0.0576$
6	E	E	**R**	**R**	$.40 \times .40 \times .60 \times .60 = 0.0576$
					Total Probability = 0.3456

Finally, because we do not care which of the 6 outcomes satisfying the conditions of the probability experiment occurs, the final probability of selecting 4 balls with replacement and having exactly two of them be red is obtained by adding the probability of each of the outcomes satisfying the conditions of the experiment (or multiplying 0.0576 by 6). So the overall probability of selecting exactly 2 red balls on 4 draws is 0.3456.

Consider a second binomial experiment. In this case we are going to flip a coin 5 times, and want to observe exactly 3 heads. Again we can compute the probability the hard way by identifying each of the ways in which 5 flips of the coin can result in 3 heads. It turns out that there are 10 ways it can happen. Each of the 10 ways is listed below along with the probability of each. Note that the probability of success (heads) and failure (tails) is the same in this experiment. Both success and failure each have a probability of .50.

Trial						
Outcome	1	2	3	4	5	Probability of Each Outcome
1	H	H	H	T	T	$.50 \times .50 \times .50 \times .50 \times .50 = 0.03125$
2	H	H	T	H	T	$.50 \times .50 \times .50 \times .50 \times .50 = 0.03125$
3	H	H	T	T	H	$.50 \times .50 \times .50 \times .50 \times .50 = 0.03125$
4	H	T	H	H	T	$.50 \times .50 \times .50 \times .50 \times .50 = 0.03125$
5	H	T	H	T	H	$.50 \times .50 \times .50 \times .50 \times .50 = 0.03125$
6	H	T	T	H	H	$.50 \times .50 \times .50 \times .50 \times .50 = 0.03125$
7	T	H	H	H	T	$.50 \times .50 \times .50 \times .50 \times .50 = 0.03125$
8	T	H	H	T	H	$.50 \times .50 \times .50 \times .50 \times .50 = 0.03125$
9	T	H	T	H	H	$.50 \times .50 \times .50 \times .50 \times .50 = 0.03125$
10	T	T	H	H	H	$.50 \times .50 \times .50 \times .50 \times .50 = 0.03125$
						Total Probability = 0.3125

Each of the 10 ways that we can flip a coin 5 times and observe exactly 3 heads has a probability of 0.03125. Because, there are 10 ways, the overall probability of flipping a coin 5 times and observing exactly 3 heads is 0.3125.

Now that we have computed the probability of two binomial experiments by identifying all of the possible ways the experiment may be satisfied, and summing the probability associated with each particular outcome to achieve the total probability, let's take another look at the formula for calculating binomial probability.

The Formula for a Binomial Experiment

$$\text{Probability} = \text{C}\binom{n}{r} \times \text{p}^{\text{r}} \times \text{q}^{n-r}$$

where

$\text{C}\binom{n}{r}$ = the number of combinations of "*n*" things taken "*r*" at a time

n = number of trials
r = number of successes
p = probability of success on any single trial
q = probability of failure on any single trial.

The formula has two main components. The first component, consisting of the combination problem of "*n*" things taken "*r*" at a time, should be familiar to you. We used that combination method to calculate the number of unique samples of some size that we could draw from a larger population in Chapter 2 on sampling. In the binomial probability formula, the combination component is going to tell us how many outcomes there are which satisfy a particular binomial experiment. For example, we were able to identify 6 ways that we can achieve 2 red balls and 2 "not red" balls when selecting a ball on 4 trials. We later were able to identify 10 ways that we could achieve 3 heads and 2 tails when flipping a coin for 5 trials. The same results can be obtained with the formula, as we will see below, but first we should probably review the use of the combination formula. Recall that the formula tells us the number of unique combinations of "*n*" things taken "*r*" at a time. In the case of a binomial experiment, "*n*" is the number of trials, and "*r*" is the number of successes we are expecting. The combination formula is expanded as follows:

$$\text{C}\binom{n}{r} = \frac{n!}{r! \times (n - r)!}$$

Also recall that the factorial of a number is the product of that number times itself minus 1 successively until we reach 1. For example, $4! = 4 \times 3 \times 2 \times 1 = 24$, and $3! = 3 \times 2 \times 1 = 6$. By definition $0! = 1$.

In our first binomial experiment we were drawing a ball from a box 4 times with replacement and wanting to observe exactly 2 red balls, so "*n*" = 4, and "*r*" = 2. We found that there was a total of 6 ways this could occur. Solving by formula we have:

$$\text{C}\binom{4}{2} = \frac{4}{2! \times (4 - 2)!} = \frac{4 \times 3 \times 2 \times 1}{2 \times 1 \times 2 \times 1} = \frac{24}{4} = 6$$

Or we can cancel terms and achieve the same result.

$$\text{C}\binom{4}{2} = \frac{4!}{2! \times (4-2)!} = \frac{\overset{2}{\cancel{4}} \times 3 \times \cancel{2} \times \cancel{1}}{\cancel{2} \times 1 \times \cancel{2} \times \cancel{1}} = 6$$

In our second binomial experiment we were going to flip a coin 5 times and wanted to observe 3 heads, so "n" = 5 and "r" = 3. We were able to identify10 ways that this could occur. By formula we obtain:

$$C\binom{5}{3}=\frac{5!}{3!\times(5-3)!}=\frac{5\times4\times3\times2\times1}{3\times2\times1\times2\times1}=\frac{120}{12}=10$$

Or, again by cancelling terms, we can achieve the same result.

$$C\binom{5}{3}=\frac{5!}{3!\times(5-3)!}=\frac{5\times\overset{2}{\cancel{4}}\times\cancel{3}\times\cancel{2}\times\cancel{1}}{\cancel{3}\times\cancel{2}\times\cancel{1}\times\cancel{2}\times1}=10$$

So the combination component of the binomial probability formula tells us how many outcomes are associated with a particular binomial experiment. The other component of the formula:

$$p^r\times q^{n-r}$$

tells us the probability of any single outcome. This component of the formula requires us to raise "p," the probability of success on any single trial, to the "r" power (the number of successes we want to achieve), and to raise "q," the probability of failure on any single trial, to the "$n-r$" power (the number of nonsuccesses we expect out of "n" trials).

In the case of our first binomial experiment, we had

$n = 4$ trials
$r = 2$ successes (a red ball)
$p = .60$ (the probability of a red ball on any single trial)
$q = .40$ (the probability of a not red ball on any single trial).

By solving the combination problem earlier, we already know that there are 6 outcomes that satisfy the conditions of the experiment. We can now compute the probability of any single outcome with the formula as follows:

$$p^r\times q^{n-r}=.60^2\times.40^2=.36\times.16=0.0576$$

So the overall probability is equal to $6\times0.0576=0.3456$.

The binomial probability formula is easy to use once you see the purpose of each component.

$$\text{Probability}=\underbrace{C\binom{n}{r}}_{\uparrow}\times\underbrace{p^r\times q^{n-r}}_{\uparrow}$$

Tells us the number of outcomes — Tells us the probability of any single outcome

Using the formula to solve the coin example provides the following solution:

$n = 5$ flips
$r = 3$ heads
$p = .50$ (heads on any single flip)
$q = .50$ (not heads on any single flip)

$$\text{Probability} = C\binom{5}{3} \times .50^3 \times .50^{5-3}$$

$$= 10 \times .50^3 \times .50^2$$
$$= 10 \times 0.125 \times 0.25$$
$$= 10 \times 0.03125 = 0.3125$$

It is useful to recall some of the basic rules for exponentials and factorials, when using the formula for computing binomial probability. For example, for any number X raised to a power: X to the first power is equal to X; and, X raised to the zero power is equal to 1.00. The value of 1! is equal to 1.00, and by definition the value of 0! is equal to 1.00.

Exponentials	Factorials
$X^1 = X$	$1! = 1$
$X^0 = 1$	$0! = 1$

Let's use the formula for computing binomial probability to compute the entire sample space for drawing a red ball from the box on 4 trials. We have already computed the probability of selecting 2 red balls. What other results can occur on 4 independent trials? In addition to selecting exactly 2 red, we can also obtain the following results: 4 red, 3 red, 1 red, or 0 red. Each of these other results can be thought of as a separate binomial experiment, and together with the result we have already computed for selecting exactly 2 red will comprise the entire sample space of what happens with respect to selecting a red ball when we draw a ball from the box 4 times. That is, out of 4 independent trials, all 4 balls will be red, or 3 will be red, or 2 will be red, or 1 will be red, or 0 will be red. Using the formula to solve each of the remaining results yields:

$n = 4$ draws
$r = 4$ red balls
$\text{p} = .60$ (red on any single draw)
$\text{q} = .40$ (not red on any single draw)

$$\text{Probability} = C\binom{4}{4} \times .60^4 \times .40^{4-4}$$

$$= 1 \times .60^4 \times .40^0$$
$$= 1 \times 0.1296 \times 1 = 0.1296$$

$n = 4$ draws
$r = 3$ red balls
$\text{p} = .60$ (red on any single draw)
$\text{q} = .40$ (not red on any single draw)

$$\text{Probability} = C\binom{4}{3} \times .60^3 \times .40^{4-3}$$

$$= 4 \times .60^3 \times .40^1$$
$$= 4 \times 0.126 \times .40 = 0.3456$$

$n = 4$ draws
$r = 1$ red ball

p = .60 (red on any single draw)
q = .40 (not red on any single draw)

$$\text{Probability} = C\begin{pmatrix} 4 \\ 1 \end{pmatrix} \times .60^{1} \times .40^{4-1}$$

$$= 4 \times .60^{1} \times .40^{3}$$

$$= 4 \times 0.60 \times .064 = 0.1536$$

and finally,

$n = 4$ draws
$r = 0$ red balls
p = .60 (red on any single draw)
q = .40 (not red on any single draw)

$$\text{Probability} = C\begin{pmatrix} 4 \\ 0 \end{pmatrix} \times .60^{0} \times .40^{4-0}$$

$$= 1 \times .60^{0} \times .40^{4}$$

$$= 1 \times 1 \times .0256 \times 0.0256$$

Putting the probability from all of our binomial experiments together provides us with the entire sample space.

Outcome of 4 Draws	Probability
4 of 4 red	0.1296
3 of 4 red	0.3456
2 of 4 red	0.3456
1 of 4 red	0.1536
0 of 4 red	0.0256
Sum of the sample space =	1.0000

Binomial probability distribution has practical applications (or why are most presidential polls accurate to plus or minus 3%?)

The binomial probability distribution has some practical applications, one of which may be very familiar to you. One of the reasons that the binomial probability distribution has practical applications is that, like the sampling distribution, when "*n*" is large, the binomial distribution becomes normal in form with a mean $\bar{X} = n \times \text{p}$, and a standard error,

$$S_{\bar{X}} = \sqrt{\frac{p \times q}{n}}$$

A typical presidential poll is usually based on a sample of approximately $n = 1{,}000$, and might have a proportion of .55 favoring the leading candidate. In this case:

$n = 1{,}000$
p = .55 (the proportion favoring the leading candidate)
q = .45 (the proportion not favoring the leading candidate)

So, 55% favor the leading candidate. Invariably, the poll is said to be accurate to plus or minus 3%, but where does the plus or minus 3% come from? The plus or minus 3% is actually a confidence interval based on a normally distributed sampling distribution. We saw in the previous chapter that approximately 95% of the sampling distribution is within plus or minus 2 standard errors (to be precise, 95% of the sampling distribution will be found plus or minus 1.96 standard errors around the mean). The standard error is equal to:

$$S_{\bar{x}} = \sqrt{\frac{p \times q}{n}}$$

or in this case:

$$S_{\bar{x}} = \sqrt{\frac{.55 \times .45}{1,000}} = .0157 \text{ or } 1.57\%$$

If one standard error is 1.57%, then 1.96 standard errors will equal: $1.96 \times 1.57\% = 3.08\%$, so we are 95% sure that the leading candidate is favored by 55% of those surveyed, plus or minus 3%.

The Link Between Probability, Hypothesis Testing, and Statistical Inference

The study of Probability is important for its own sake, but it is also an important topic in statistics due to its link to hypothesis testing and statistical inference. Probability theory, along with the fact that the sampling distribution of the mean is normal in form when "n" is large, serve as the foundation for much of inferential statistics. Think about the question we raised in the last chapter where we considered the fact that we usually take a single sample from a population when we do research. The question becomes, "with so many possible samples that can be drawn from a population, how do we know that the one sample we select is a good one, that is, a sample that is reasonably representative of the population?" We were able to demonstrate that even the extreme samples are still very close to the true population mean when the sample size is reasonably large. We were also able to create a confidence interval consisting of an area of the sampling distribution that we interpreted as a probability. For example, we can be 68.26% sure that the sample we draw is within plus or minus one standard error of the true population mean; we can be 95.44% sure that the sample we draw is within plus or minus two standard errors of the true population mean; and, we can be 99.74% sure that the sample we draw is within plus or minus three standard errors of the true population mean.

Hypothesis testing is also based to a large extent on probability. We will be examining the logic of hypothesis testing in much more detail in the next chapter, and at times it can seem very complicated, but hypothesis testing is really based on a set of simple ideas. In the case of a sample mean, we know what the sampling distribution must look like, and the farther out into the tail of the sampling distribution that a sample mean is found, the less likely that the sample came from that particular population. It is more likely that it came from some other population. We will always have a probability of being wrong, but we can make that probability as large or small as we want within conventional standards. The next chapter presents the logic of hypothesis testing in detail, and also provides an overview of the types of hypotheses that we will encounter in the remainder of the text.

Summary of Key Points

Probability can be one of the more interesting topics in statistics, and as you have seen in this chapter, probability has important links to many other topics in statistical analysis. In this chapter, we have examined the basic rules that apply to simple probability, which deals with a single set of outcomes. We have examined compound probability, which deals with two separate sets of outcomes, and includes the events: P(E or F) the probability of "E" or "F;" P(E and F) the probability of "E" and "F;" and, P(E|F) the probability of "E" given "F." Finally, we examined binomial probability involving a certain number of successes over a fixed number of repeated, independent trials, and saw that one major application of the binomial probability distribution was in determining the confidence interval for proportions such as those resulting from political polls.

Simple Probability—Probabilities based on a single set of outcomes. Simple probability is based on six rules.
Event—An outcome from a probability experiment to which we can assign a probability.
Sample Space—The set of all possible outcomes for a probability experiment.
Independent Events—Events whose outcome is not affected by the outcome of a prior event. Independence of events is often achieved by sampling with replacement.
Related Events—Events whose outcome is related to the probability of some other event.
Compound Probability—Probability based on satisfying two sets of outcomes with a single event. Compound events include: P(E or F); P(E and F); and, P(E|F).
Binomial Probability—Probability based on binomial experiments involving a fixed number of independent trials each of which will result in one of only two possible outcomes, success or failure.

Questions and Problems for Review

1. A box contains 10 green balls, 10 blue balls, 20 orange balls, 20 red balls, and 40 yellow balls.
   a. What is the probability of selecting a green ball?
   b. What is the probability of selecting a yellow ball?
   c. What is the probability of selecting an orange ball on two successive draws (with replacement)?
   d. What is the probability of selecting a ball that is not yellow?
2. Given the box of balls in problem 1, selecting an orange ball is defined as success, and any other color is defined as failure. If a ball is selected 6 times (with replacement) what is the probability of selecting exactly 4 orange balls over the 6 independent trials?
3. Given the generalized outcomes below:

Outcome	Probability
O_1	.20
O_2	.50
O_3	.20
O_4	.10

   Event "E" is defined as (O_2 or O_4), and event "F" is defined as (O_1 or O_2). What is the probability of:
   a. P(E)
   b. P(F)
   c. P(E or F)
   d. P(E and F)
   e. P(E|F)
   f. P(F|E)
4. The leading candidate in a presidential election race has 60% of the vote based on a poll of $n = 1{,}200$ likely voters. Compute the 95% margin of error for the poll (remember that 95% of the sampling distribution will be found within plus and minus 1.96 standard errors).

5. What happens to the margin of error in the presidential poll from question 4 if the sample size is reduced to $n = 500$?

6. Professor Seuss keeps his socks in two boxes. In box "A" he has 20 red socks and 30 blue socks. In box "B" he has 30 red socks and 20 black socks. If he selects one sock from each box, what is the probability that he will have a matching pair?

7. A population is known to have a mean $\mu = 50$, with a standard deviation $\sigma = 8$. A sample of $n = 100$ individuals is found to have a mean $\overline{X} = 49$.

a. Is it likely that the sample of $n = 100$ individuals was drawn from the population with the mean $\mu = 50$? Defend your answer.

b. What proportion of all possible samples of size $n = 100$ would have a mean of $\overline{X} = 49$ or less?

8. How many unique samples of size $n = 5$ can be drawn from a population of $N = 12$ individuals?

9. Create the entire sample space for observing a "heads," with associated probabilities, for flipping a coin 4 times. That is, what is the probability of observing exactly: 4 heads, 3 heads, 2 heads, 1 heads, and 0 heads? Which outcome is most likely to occur if a fair coin is flipped 4 times.

10. Given the generalized outcomes below:

Outcome	Probability
C1	.10
C2	.30
C3	.20
C4	.20
C5	.20

Event "A" is defined as (C1 or C3 or C5)
Event "B" is defined as (C2 or C3 or C4)
Compute the following probabilities:

a. P(A)
b. P(B)
c. P(A and B)
d. P(A or B)
e. P(B|A)
f. P(A|B)

11. Compute the 95% margin for error for the following election polls:

a. Leading candidate has 80% of the vote, sample size $n = 1{,}200$.
b. Leading candidate has 65% of the vote, sample size $n = 1{,}200$
c. Leading candidate has 50% of the vote, sample size $n = 1{,}200$.
d. What happens to the size of the margin for error as the leading candidate's share moves toward 50%?

CHAPTER 8

Hypothesis Testing

Key Concepts

Hypothesis	Statistical Significance	Critical Region
Research Hypothesis	Alpha Level	Type I Error
Null Hypothesis	Level of Significance	Type II Error
One-Tail Test	Test Statistic	Inferential Statistics
Two-Tail Test	Critical Value	

Introduction

Hypothesis testing is the primary mechanism for making decisions based on statistical analysis. You will find that much of the logic of hypothesis testing is based on our understanding of probability theory, and the nature of the sampling distribution. The material in this chapter can be quite challenging because it requires you to integrate several important concepts from previous chapters, and it is also important because it sets the foundation for the material on inferential statistics where we will make inferences about population parameters based on observed sample statistics. In addition to the ideas of probability theory and the sampling distribution, we will also be discussing the nature of hypotheses, types of relationships between variables, strategies for uncovering evidence in support of a hypothesis, and methods to test the adequacy of evidence for supporting a hypothesis. Before we begin to examine these important issues, it will be useful to review the nature of the sampling distribution along with some related issues.

The Sampling Distribution

As you should recall from Chapter 2 on sampling and Chapter 6 on the normal curve, the sampling distribution of the mean is the set of all possible sample means of a given size "n," drawn from a population. From the Central Limit Theorem, we know that μ when the sample size is sufficiently large, the sampling distribution of the mean has the important characteristics of being normal in form (it is an example of a normal distribution), with a mean "$\mu_{\bar{X}}$" equal to the population mean "μ" from which the samples are drawn, and a standard deviation (called the standard error, "$\sigma_{\bar{X}}$") equal to the population's standard deviation "σ" divided by the square root of the sample size "n."

For example, suppose we have a population with a mean income of μ = \$35,000, with a standard deviation of σ = \$5,000. The sampling distribution of all possible means for samples of size "n" = 100 would take the form of a normal distribution with a mean $\mu_{\bar{X}}$ = \$35,000, and a standard error ($\sigma_{\bar{X}}$) equal to the population standard deviation σ = \$5,000 divided by the square root of the sample size of $n = 100$ ($\sigma_{\bar{X}}$ = \$5,000/10 = \$500). The sampling distribution in this case is shown in Figure 8.1.

Further suppose that we are presented with a sample of $n = 100$ individuals who are known to have a mean income of $\bar{X}$ = \$35,400. We do not know if the sample of $n = 100$ individuals came from the population described above or from some other population, but it is our task to decide if it did or not. Of course, we can never be sure if the sample in question came from the population with a mean μ = \$35,000 or not, but we can use our knowledge of the nature of the sampling distribution, the normal curve, and probability theory to come to a logical decision.

If the sample with a mean of $\bar{X}$ = \$35,400 came from the population above, then the sample mean must reside in that population's sampling distribution along with all of the other possible sample means computed for all samples of size $n = 100$. In fact, we can place the sample mean in question on the sampling distribution, and then reach a decision as to how likely it is that it came from the population with a mean of μ = \$35,000. For example, the closer the sample mean in question is to the center of the sampling distribution (the closer it is to the true population mean), then the more likely it is that the sample mean came from that population's sampling distribution. By the same logic, the farther away the sample mean in question is from the center of the sampling distribution (the closer it is to one or the other extreme regions in the tail of the sampling distribution), the less likely it is that the sample mean came from that population's sampling distribution, and the more likely that the sample came from some other population (a population with a higher or lower mean income).

The sample mean in question appears reasonably close to the population mean, but we might want to quantify our decision, or state it in terms of a specific probability. We can do that by converting the distance between the population mean and the sample mean we are examining into a Z score. Converting the distance between the two means to a Z score allows us to return to Table 1 in the appendix containing

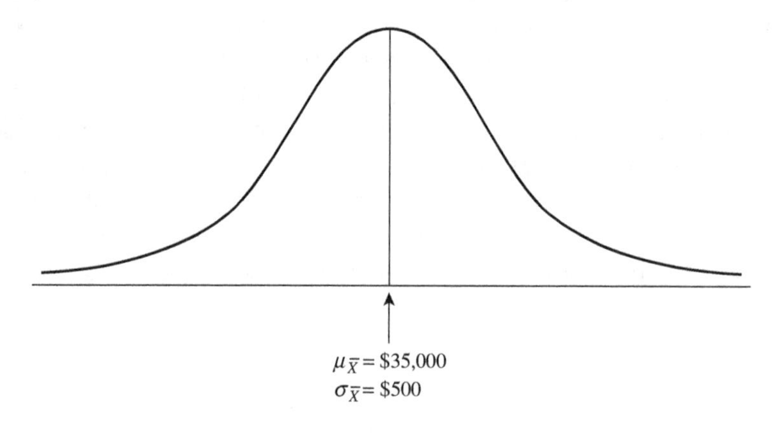

Figure 8.1 The Sampling Distribution for all Samples of $n = 100$

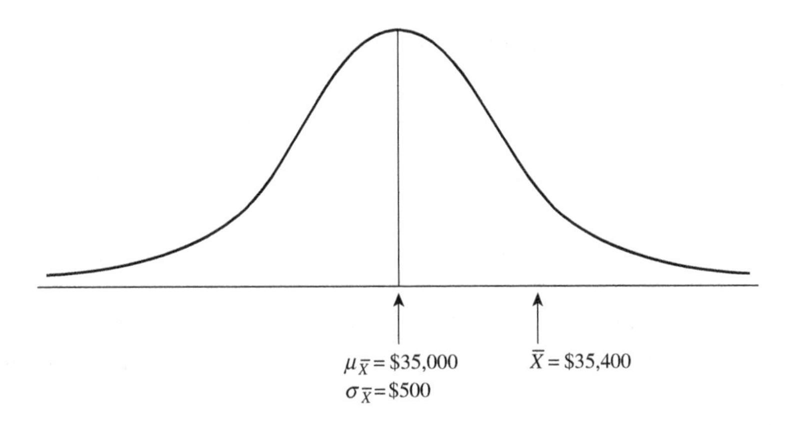

Figure 8.2 The Location of $\overline{X} = \$35{,}400$ on the Sampling Distribution

the area under the normal curve associated with a given Z score. As we saw in Chapter 6 dealing with the normal curve, the area under the normal curve associated with a particular Z score can be thought of as a probability. We also saw that the difference between a sample mean and the mean of the sampling distribution can be converted to a Z score by the following formula.

Formula to Convert the Difference between a Sample Mean and a Population Mean to a *Z* Score

$$Z = \frac{\overline{X} - \mu}{\sigma_{\overline{X}}}$$

where

$\overline{X}$ = the sample mean of interest
μ = the population mean (also equal to the mean of the sampling distribution)
$\sigma_{\overline{X}}$ = the standard error of the distribution ($\sigma/\sqrt{n}$)

In this case the computation would be as follows:

$$Z = \frac{35{,}400 - 35{,}000}{500}$$
$$= \frac{400}{500} = 0.80$$

The resulting value of $Z = .80$ indicates that the sample mean in question is only .80 standard errors away from the population mean. If we were to conclude that the sample mean of $\overline{X}$ = $35,400 did not come from the population with a mean μ = $35,000 described above, we would also logically be excluding any sample mean equal to or larger than $35,400 as coming from the population. By turning to Table 1 in the appendix, which provides areas under the normal curve associated with various Z scores, we can identify the proportion of the normal curve at or beyond a $Z = .80$. That proportion will correspond to the area in the sampling distribution equal to or beyond a sample mean of $35,400.

As you may recall from Chapter 6, when working with Table 1 in the appendix, column "A," of the table contains the Z scores, column "B," contains the proportion of the normal curve from the mean to the Z score, and column "C," contains the proportion of the normal curve at or beyond the Z score out toward the tail of the distribution. In this case, we see that the value in column "C," is .2119, indicating that 21.19% of the normal curve is equal to or beyond a Z score of .80. Similarly, 21.19% of all of the possible sample means that legitimately comprise the sampling distribution above would be equal to or greater than the sample mean of $\overline{X} = \$35{,}400$.

If we were to conclude that the sample mean of \$35,400 did not come from the population in question, we would have a 21.19% chance of being wrong! Most of us would not be comfortable with being wrong over 21% of the time when making an important decision, so we would conclude that the sample mean in question probably did come from the population.

Let's consider a second sample mean of $n = 100$, who are observed to have a mean income of \$34,000, and again we are asked to decide if this sample comes from the above population. We can pursue the same strategy of placing the sample mean in question on the population's sampling distribution, and then converting the distance between the population mean and the sample mean to a Z score to reach a decision.

Converting the distance between the population mean and the sample mean to a Z score is accomplished by the following computation.

$$Z = \frac{34{,}000 - 35{,}000}{500}$$
$$= \frac{-1{,}000}{500} = -2.00$$

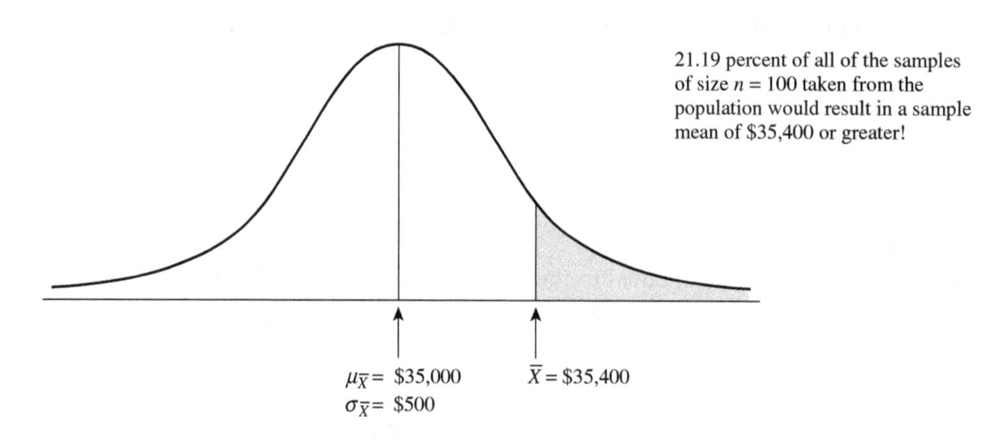

Figure 8.3 The Percent of the Sampling Distribution Beyond a Sample Mean of $\overline{X} = \$35{,}400$

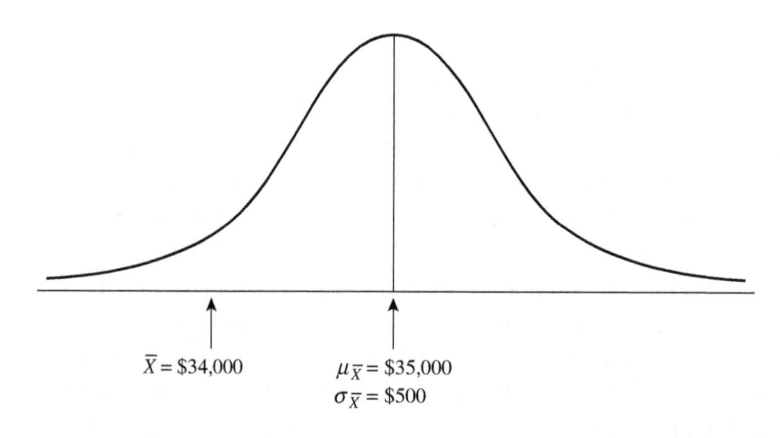

Figure 8.4 The Location of $\overline{X} = \$34{,}000$ on the Sampling Distribution

The area of the normal curve below a Z score equal to –2.00 can be found in column C of Table 1, and that value is .0228. The area in the sampling distribution equal to or below a sample mean of $\bar{X}$ = \$34,000 will also equal to .0228. Again, we can interpret the area in the tail of the sampling distribution beyond the sample mean as our risk of being wrong if we were to conclude that the sample mean in question did not come from the above population. In this case, there is still a chance that a sample of n = 100 individuals with a sample mean of $\bar{X}$ = \$34,000 came from the population above, but the chance is rather small.

Only .0228, or 2.28% of all of the possible samples of size n = 100 drawn from the population would have an observed mean of $\bar{X} \geq$ \$34,000, so if we conclude that the sample did not come from the population above, we only have a 2.28% chance of being wrong. In this case, we would conclude that the sample mean of $\bar{X}$ = \$34,000 probably did not come from the population above, it is more likely to have come from some other population (a population with a true mean income much closer to μ = \$34,000). We still have a risk of being wrong, but remember that we must always accept some risk of being wrong. Clearly, a risk of being wrong only 2.28% of the time is certainly a more reasonable level of risk to accept than the level of a 21.19% chance of being wrong that we would have had to accept if we had concluded that the sample mean of $\bar{X}$ = \$35,400 had not come from the population above.

We just saw that we would probably say that a sample of n = 100 individuals with a mean income of $\bar{X}$ = \$34,000 is not likely to have come from the population with a true mean income of μ = \$35,000. What would we conclude if we were to consider a sample of n = 100 individuals with a mean income of $\bar{X}$ = \$36,000? You might notice that a sample mean of $\bar{X}$ = \$36,000 is just as far away from the population mean of μ = \$35,000 as was the previous example of a sample mean of $\bar{X}$ = \$34,000. The difference is simply that \$36,000 represents a point in the sampling distribution \$1,000 above the true population mean, and \$34,000 represents a point in the sampling distribution \$1,000 below the true population mean. As a result, if we convert the distance between the population mean and the sample mean of $\bar{X}$ = \$36,000 to a Z score we will find that it is the same size as the Z score associated with the sample mean $\bar{X}$ = \$34,000, both sample means are two standard errors away from the true population mean. The only difference would be the sign of the resulting Z score. A negative Z score is associated with any sample mean below the population mean, and a positive Z score is associated with any sample mean above the population mean. Just as a Z score of $Z = -2.00$ has only 2.28% of the sampling distribution below it, a Z score of $Z = 2.00$ has only 2.28% of the sampling distribution above it. So we would conclude that a sample of n = 100 individuals with a mean income of \$36,000 is no more likely to have come from the population above any more than the sample of n = 100 individuals with a mean income of \$34,000 was. In each case, the observed sample means are quite far in the extreme region of the sampling distribution, one quite far in the left region of the sampling distribution, and the other quite far in the right region of the sampling distribution.

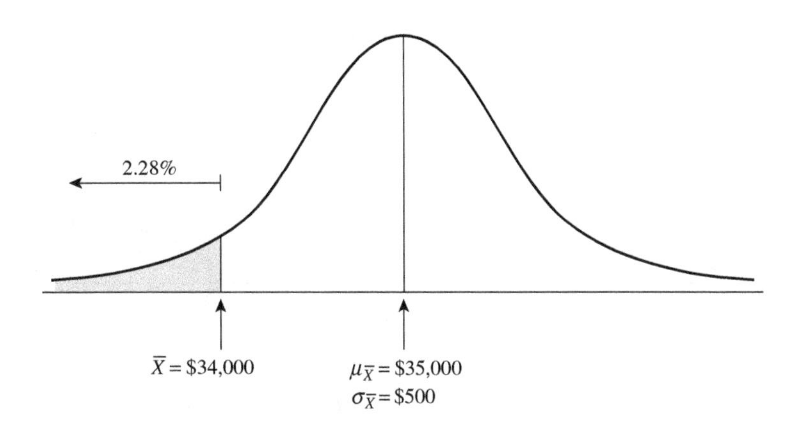

Figure 8.5 The Area Below $\bar{X}$ = \$34,000 on the Sampling Distribution

We could continue in this fashion indefinitely evaluating sample mean after sample mean, and trying to decide if a particular sample mean truly came from a given population's sampling distribution or not. We can never be sure, and we will always have a chance or probability of being wrong. We can quantify that probability of being wrong by converting the distance between the sample mean in question and the population mean to a *Z* score, and then seeing how much of the sampling distribution is equal to or beyond the sample mean in question.

At some point, we might decide to adopt a standard of proof based on how far to one extreme or the other in the sampling distribution a given sample mean would have to be located if the sample really did come from a given population. For example, we might adopt a standard of proof that if a sample mean in question would be in the extreme 5% of a population's sampling distribution, then our decision will be that the sample mean in question probably did not come from the population. Because the sampling distribution is a type of normal distribution, we could easily apply our 5% standard of proof by converting the distance between a sample mean in question and the mean of the population from which the sample might have been drawn into a *Z* score, and then checking to see if the resulting *Z* score falls into the extreme 5% of the normal distribution. If the resulting *Z* score is at or beyond the extreme 5% of the normal distribution then we will conclude that the sample mean is not likely to have come from the given population.

Conversely, if the distance between the sample mean in question and the population mean converted to a *Z* score that was not at or beyond the extreme 5% of the normal distribution, then we will conclude that the sample mean in question is likely to have come from the given population. If we adopt the extreme 5% of the normal distribution as our level of proof, then any sample mean that is far enough away from the population mean to equal a *Z* score of 1.645 or beyond toward the tail of the curve will be considered as being far enough away from the population mean to suggest that the sample mean did not come from the given population. Recall that the *Z* score 1.645 marks the boundary of the extreme 5% of the normal distribution on the positive side of the mean, and the *Z* score –1.645 marks the boundary of the extreme 5% of the normal distribution on the negative side of the mean.

As you will see in a limited way later in this chapter, and in a much more detailed fashion in Chapter 11, converting the distance from a sample mean to a population mean into a *Z* score is actually a type of statistical test. That particular test is called a "*Z* test," and the value resulting from the computation is called the observed **test statistic**. The 5% level of proof that we might adopt is called the **level of significance,** or sometimes the **alpha level**, because it is represented symbolically by the lower case Greek letter alpha "α." The *Z* score that represents the boundary on the normal curve associated with the level of significance is called the **critical value**, and the region of the normal curve beyond the critical value extending into the tail of the normal curve is called the **critical region**.

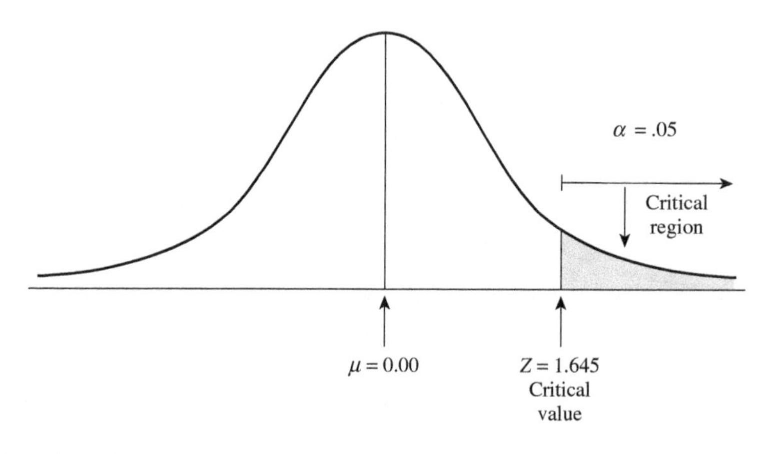

Figure 8.6 The Critical Value and Critical Region Associated with an Alpha = .05 Level of Significance

In fact, a standard of proof of the extreme 5% of the normal distribution is quite commonly used in behavioral science research, but it is not the only standard that is used. Other common standards are the extreme 10% of the distribution (not quite as high a standard of proof as the 5% standard), the extreme 2 1/2% of the distribution, the extreme 1% of the distribution, and the extreme .1% of the distribution (a very high level of proof that is less commonly used than the others). In each case, the size of the standard of proof represents our risk of being wrong in our conclusion if we decide that a particular sample mean did not come from a given population. Our method to apply each standard of proof is similar in each case. We simply need to convert the distance between a sample mean in question and the population mean into a Z score, and then see if the resulting Z score is at or beyond the chosen standard of proof on the normal distribution. For example, a 10% standard of proof would require a Z score of 1.28 or greater; a 5% standard of proof would require a Z score of 1.645 or greater; a 2 1/2% standard of proof would require a Z score of 1.96 or greater; a 1% standard of proof would require a Z score of 2.33 or greater; and, a .1% standard of proof would require a Z score of 3.08 or greater (assuming in each case that we were referring to the positive side of the distribution).

Hypotheses and Types of Relationships

At this point, we have actually gone quite far in the process of testing a hypothesis, but we have done it intuitively with the examples above. Before we formally state the steps involved in hypothesis testing, it will be useful to review the concept of a hypothesis, and to see how a hypothesis can be examined in the context of behavioral science research. In Chapter 1, a **hypothesis** is defined as a statement of relationship between two variables. In most cases, we are interested in discovering evidence of a causal relationship between two variables. In other words, we want to know "what causes what?" The variable presumed to be the cause is referred to as the independent variable, and the variable presumed to be the effect is referred to as the dependent variable. As you should recall from Chapter 1, we usually represent the independent variable as "*X*," and the dependent variable as "*Y*." The relationship between two variables is often easier to see when the variables are graphed on a Cartesian coordinate system consisting of an "*X*" (horizontal) axis, and a "*Y*" (vertical) axis. The resulting scatterplot of the two variables reveals the type of relationship between the two variables. You will see some detailed examples of using a Cartesian coordinate system to create a scatterplot of two variables in Chapter 9 on correlation, and how to interpret the relationship between the two variables.

Two variables may be related in several different ways. A positive relationship is one in which an increase in the level of the independent variable (the cause) results in an increase in the level of the dependent variable (the effect). In a positive relationship, the two variables move in the same direction. If there is an increase in the independent variable, there will be a corresponding increase in the dependent variable; conversely, if there is a decrease in the independent variable, there will be a corresponding decrease in the dependent variable. A classic example in behavioral science research is the relationship between education and income. Generally speaking, your level of education will determine (or cause) your income, or alternatively we can state, your income level is caused by your level of education. Other pairs of variables that are usually positively related include the following:

Independent Variable	Dependent Variable
Height	Weight
Severity of crime	Length of sentence
Level of anxiety	Blood pressure

Oftentimes, two variables may have a negative relationship. In the case of a negative relationship, as the level of the independent variable increases, the level of the dependent variable decreases. And conversely, as the level of the independent variable decreases, the level of the dependent variable increases.

Generally speaking, among older adults, as age increases, one's level of physical agility decreases. Other pairs of variables that would be expected to exhibit a negative relationship include the following:

Independent Variable	Dependent Variable
Price of an item	Quantity demanded
Severity of crime	Likelihood of probation
Blood alcohol content	Driving ability

In this case of education and income, the "level of education" is the independent variable, and "level of income" is the dependent variable. Our hypothesis might be stated as follows: As the level of education increases, the level of income increases. In this case, we have specified a positive relationship. One way that we might find evidence in support of our hypothesis is to select a sample from the population and compare the level of education and income on a case-by-case basis. In fact, we will use this type of strategy in the next chapter. An alternative approach to demonstrate evidence for our hypothesis might be to select a sample with an extremely high or extremely low level of education and compare its mean income to the known mean income of the population in general. For example, we could compare the mean income of a sample of individuals with a college education with the known mean income of the population in general. Individuals with a higher than average education would be expected to have a higher than average income. Alternatively, we might compare the mean income of a sample of individuals with less than a high school education with the mean income of the population in general. Individuals with a lower than average education level would be expected to have a lower than average income level. Both strategies are consistent with the expectation of a positive relationship; when education is high, income will be high; when education is low, income will be low.

There are times in behavioral science research when our hypotheses will not specify a positive or a negative relationship. That does not mean that two variables are not related, but that we simply do not know how they are related. For example, we might be interested in seeing how the mean income of a sample of recent college graduates who majored in one of the behavioral sciences differs from the mean income of all recent college graduates. Undoubtedly, the mean income of recent college graduates who majored in the behavioral sciences will be higher than the mean income of some recent college graduates, but it will likely be lower than the mean income of some of the other recent college graduates. In this case, we are interested in seeing how the mean income of recent college graduates who majored in the behavioral sciences is different from the mean income of all recent college graduates, but we are not in a position to specify if the difference will be positive or negative.

A One-Tail Test of the Hypothesis or a Two-Tail Test of the Hypothesis

If you think back to the examples we used above when trying to decide if a sample mean was likely to have come from a given population, we based our decision on how different the sample mean was from the population mean, or stated another way, we based our decision on how far into the tail of the sampling distribution the sample mean was located. As you saw, sometimes the sample mean in question could be considered quite different from the population mean because it was much larger than the population mean (i.e., it was in the extreme right tail of the sampling distribution), and sometimes the sample mean in question could be considered quite different from the population mean because it was much smaller than the population mean (i.e., it was in the extreme left tail of the sampling distribution). If we combine the idea of a sample mean being considered different from a population mean because it is either in the extreme right-hand tail or the extreme left-hand tail of the sampling distribution with the idea of a hypothesis, which implies a positive relationship, a negative relationship, or no specific relationship, we have the basis for the distinction that is made in statistical analysis between a **one-tail test** of a hypothesis and a **two-tail test** of a hypothesis.

A one-tail test of the hypothesis requires us to make an assumption about the direction of the relationship stated in the hypothesis. If we use the example of comparing a sample mean to a known population mean, a one-tail test of the hypothesis requires us to expect that the population represented by the sample mean will be either larger than the known population mean, in which case we expect the sample mean to be located in the extreme right tail of the sampling distribution, or we expect the population represented by the sample mean to be smaller than the known population mean, in which case we expect the sample mean to be located in the extreme left tail of the sampling distribution. A one-tail test of the hypothesis may refer to either tail of the sampling distribution, the key point is that the hypothesis leads us to expect the sample mean to show up in one tail or the other tail.

Our hypothesized relationship between education and income is an example of a one-tail hypothesis. If we choose to test this idea by comparing the mean income of a sample of college graduates with the mean income of the population in general, then we have a one-tail test because we expect the mean income of the sample of college graduates to be found in the extreme right-hand tail of the sampling distribution. That is, we are expecting the mean income of the sample of college graduates to be greater than the mean income of the population in general. If we choose to test the relationship between education and income by comparing the mean income of a sample of individuals with less than a high school education to the mean income of the population in general, then we still have a one-tail test because we are expecting the mean income of the sample of individuals with less than a high school education to be found in the extreme left-hand tail of the sampling distribution indicating a mean income far below that of the population in general. In each case, we have a one-tail test because we not only expect the sample mean to be different from the population mean, but we also have an expectation of direction leading us to look in either the right-hand tail of the sampling distribution or the left-hand tail of the sampling distribution.

Our hypothesis concerning the relationship of the mean income of the population represented by a sample of recent college graduates who majored in one of the behavioral sciences compared to the known mean income of all recent college graduates would be an example of a two-tail test of the hypothesis. We expect the mean income of the sample of recent behavioral science graduates to be different from all recent college graduates, but we do not have any statement of how it will be different. The mean income of recent behavioral science graduates can be different by being higher than the mean income of all recent college graduates (in which case we would be looking for the sample mean to be in the right-hand tail of the sampling distribution), or it can be different by being lower than the mean income of all recent college graduates (in which case we would be looking for it to be in the left-hand tail of the sampling distribution). An expectation of difference without direction will require us to look at both extremes of the sampling distribution, hence, a two-tail test.

Not all research in the behavioral sciences is guided by a formal statement of a hypothesis, but when a hypothesis is stated, it represents an expectation by the researcher as to what the outcome of the research will be. The expectation of the results is usually based on a logical analysis of the variables involved, combined with a review of the results of previous research on the topic. Once we have a formal statement of a hypothesis, which may or may not imply the direction of a relationship, we are still faced with the question of, "How do we provide evidence in support of our hypothesis?"

Devising a Research Strategy

Evidence in support of a hypothesis is provided by the collection of data, and the subsequent statistical analysis during the research process. Researchers are faced with the important questions of what types of data will be collected, and how will the data be analyzed to determine the accuracy of the hypothesis in question. For example, the hypothesis stating our expected positive relationship between education and income is a logical one, but to determine if there is evidence to support it, we must devise a research strategy that will allow us to test the hypothesis.

In the case of education and income, there is more than one research strategy that may be used to demonstrate support for the hypothesis. Three common strategies are (1) to examine groups who differ in the level of the independent variable to see if they also differ in levels of the dependent variable in the expected manner, (2) to look for evidence of an association between the independent and dependent variables, and (3) to examine a sample representing an extreme group with respect to the independent variable (a group with a high level of education or a group with a low level of education) to see if the group's mean income differs from the known population mean income in the expected direction.

One way we could implement the first research strategy would be to select a sample from the population of interest, and then to create subgroups based on their level of the independent variable (education). In this case, we might create two groups: those with a high level of education and those with a low level of education. We would then compare the two groups with respect to their level of the dependent variable (income). For example, we could compare the mean income of the high education group to the mean income of the low education group. Because we expect a positive relationship between the two variables, we would expect the mean income of the high education group to be higher than the mean income of the low education group.

The second strategy requires us to select a sample from the population of interest, and then to compare the level of education and the level of income on a case-by-case basis to see if the two variables move together. With education and income, we are expecting a positive relationship, so we would expect individuals with a high level of education to have a high level of income, individuals with a moderate level of education to have a moderate level of income, and individuals with a low level of education to have a low level of income. The statistic we would use to measure the degree of association between the independent variable and the dependent variable is called a correlation coefficient. There are several types of correlation coefficients, and the appropriate choice is usually based on the way in which the independent and dependent variables have been measured. Several types of correlation coefficients will be examined in the next chapter, and we will begin formally testing hypotheses related to them.

The third research strategy where we focus on a group representing an extreme position on the independent variable (either high education or low education), and then compare that group's position with respect to the dependent variable (income) to the population in general is most similar to the examples that we have seen above. Notice that in this case, our hypothesis would not involve an explicit statement of relationship between the two variables of education and income. In some cases, such as this one, we may test hypotheses regarding a single variable, although a relationship to another variable is implied. For example, if we were to test the relationship between education and income by selecting a sample of college educated individuals, and then comparing their mean income to the known mean income of the population in general, our hypothesis that would actually be tested is that the mean income of the population represented by the sample will be higher than the mean income of the population. On the surface it might appear that we are simply testing a hypothesis concerning the single variable of income (and technically we are), but implicit in the process of selecting a sample of college educated individuals is a hypothesized relationship between education and income. Testing hypotheses by examining differences between means is another common approach in the behavioral sciences, and there are several statistical tests used for that purpose, which we will examine in Chapter 11.

Up to this point, we have been using the term "hypothesis" to mean a stated relationship between two variables, or in some cases, a statement of expectation concerning a single variable that implies an underlying relationship with another variable. There is one final aspect to hypothesis testing that we need to examine before we pull all of these ideas together and proceed with a complete example of the process of hypothesis testing. The final aspect we need to address is the fact that hypothesis testing is conducted indirectly involving the concept of the null hypothesis.

Hypothesis Testing Is Conducted Indirectly: The Research Hypothesis and the Null Hypothesis

The conventional strategy for testing a statistical hypothesis involves two related hypotheses. The first is termed the research hypothesis, which is symbolized by H_1. The **research hypothesis** is a formal statement reflecting the researcher's expectations concerning the question under investigation. Up to this point in the chapter, whenever the term "hypothesis" has been used it has been generally synonymous with the term "research hypothesis." Conventionally, the research hypothesis is not actually subjected to statistical testing. The reason for this involves a subtle exercise in logic. Suppose we want to show that college educated individuals have a higher mean income than the population in general, and we plan to demonstrate this by selecting a sample of college graduates and comparing their mean income to the known mean income of the population in general. Our research hypothesis might be stated as follows:

H_1: The mean income of college graduates will be higher than the mean income of the population in general.

We could also state the research hypothesis symbolically using μ_{cg} to represent the mean income of the population represented by the sample of college graduates, and μ to represent the mean income of the population in general. Stated symbolically, the research hypothesis would be as follows:

$$H_1 : \mu_{cg} > \mu$$

One might think that the next logical step would be to select a sample of college graduates, compute their mean income, and compare it to the known mean income of the population in general in a fashion similar to what we have done earlier in the chapter by converting the difference between the two means to a *Z* score, which would indicate how likely it is that the sample mean would have come from the population's sampling distribution. As logical as that may seem, we do not proceed in that manner for two related reasons: the problem of sampling variation and the difficulty in proving something.

You have seen in Chapter 2 on sampling how it is possible to draw a large number of samples from even a relatively small population. A population of $N = 40$ will yield a total of 658,008 different samples of size $n = 5$. You have also seen that some of those samples are better than others; that is, some have a sample mean that is closer to the mean of the population from which they were drawn than others. The fact that there is sampling variation from sample to sample introduces some uncertainty into the research process. Even if we find that our sample of college graduates has a higher mean income than the population in general, how do we know that we had a "good" sample of college graduates (i.e., a sample with a mean that is very close to the true population mean of all college graduates). Maybe we just happened to get a sample of college graduates with exceptionally high incomes, and the true mean income of all college graduates is actually no different than the true mean income of the population in general. If someone believed that to be the case, it is unlikely that we would convince them otherwise with the results of a single sample.

The difficulty of making a convincing case based on the results of a single sample relates to the second problem I mentioned, which is how difficult it is to actually prove anything. If someone is not convinced based on the results of one sample, would they be convinced if we could demonstrate similar results from two different samples? Would it take similar results from three samples, or perhaps ten or twenty samples? In a sense it does not matter, because in almost all cases, we only draw a single sample when conducting research. Even if our audience would be satisfied with similar results from only 10 samples, we are still going to have results from the single sample that we actually draw from the population. The solution to this problem is that just as it is very difficult to prove something, it is actually quite easy to disprove something. If you think about it, you will realize that we can disprove a statement with a single contrary case.

Being able to disprove a contrary case is what the **null hypothesis** is all about. The null hypothesis, symbolized by H_0, is a specially crafted statement, which is based on the composition of the research hypothesis. The null hypothesis is a statement of no relationship designed to state a position contrary to what is stated in the research hypothesis. Instead of subjecting our research hypothesis to statistical test, we actually test the null hypothesis. Given our previous example of a research hypothesis:

H_1: The mean income of college graduates (cg) will be higher than the mean income of the population in general.

We would construct the null hypothesis to state:

H_0: The mean income of college graduates will be equal to the mean income of the population in general.

The null hypothesis can also be stated symbolically as in the form:

$$H_0 : \mu_{cg} = \mu.$$

We would then proceed as before, except that we are actually going to test the null hypothesis. We would convert the distance between the sample mean and the known population mean to a *Z* score, and if the resulting *Z* score is far enough into the extreme region of the normal distribution, then we will reject the null hypothesis. We still have a risk of being wrong in our conclusion based on the standard of proof that we have adopted, but the risk of being wrong will always be present no matter which hypothesis we evaluate. The major point is that if the null hypothesis is rejected as not being likely, then its alternative must be likely, and the alternative of the null hypothesis is the research hypothesis.

A Summary of the Steps in Testing a Statistical Hypothesis

As indicated at the beginning of this chapter, much of the material in this chapter can be quite challenging because it requires you to integrate a great deal of the previous material. It has taken us a while to get to this point, but we are now ready to summarize the steps involved in testing a statistical hypothesis. Following that, we will demonstrate the process with two examples.

Steps in testing a statistical hypothesis

1. State the research hypothesis H_1, and the null hypothesis H_0.
2. Examine the research hypothesis H_1, and determine if a one-tail test or a two-tail test is in order:
   a. If the research hypothesis suggests a difference and a direction (a positive or a negative relationship), then a one-tail test is in order;
   b. If the research hypothesis suggests only a difference, then a two-tail test is in order.
3. Determine the appropriate statistical test to use.
4. Select a level of significance at which to conduct the test.
5. Identify the critical value of the test statistic.
6. Compute the value of the test statistic using the data.
7. Compare the observed (computed) value of the test statistic to the critical value of the test statistic, and reach a decision regarding H_0:
   a. If the observed value of the test statistic falls on or in the critical region, then you REJECT H_0, and ACCEPT the alternative research hypothesis H_1.
   b. If the observed value of the test statistic fails to fall on or in the critical region, then you FAIL TO REJECT H_0.

Some Key Points to Keep in Mind

Keep in mind that you determine if a one-tail test or a two-tail test is in order by examining the research hypothesis. The null hypothesis is a statement of no relationship (or no difference), and can always appear as a two-tail test. The research hypothesis will always suggest a difference of some type (e.g., a difference between a population mean represented by a sample mean and a known population mean). If only a difference is stated, then a two-tail test is in order. If a difference plus a direction is stated, then a one-tail test is in order. A statement of direction would lead us to expect that the population mean represented by the sample mean will be higher than or lower than the known population mean, not just different from it.

Further, it is the null hypothesis that is subject to the statistical test, and the correct decision is to reject the null hypothesis, or to fail to reject the null hypothesis. This may seem like a minor point, but the terminology is important, and is very similar to the outcome of a criminal trial. In a criminal trial, assuming that the jury is able to reach a decision, the outcome is either "guilty," or "not guilty," it is not "guilty," or "innocent." There is a great deal of difference between a finding of "not guilty," and a finding of "innocent." The jury examines the evidence and decides that there is sufficient evidence of guilt beyond a reasonable doubt (similar to our level of significance), or there is not. The jury never finds the defendant "innocent." Similarly, the statistician examines the evidence regarding the null hypothesis and finds sufficient evidence to reject the null hypothesis, or fails to reject the null hypothesis. If the null hypothesis is rejected, then support for the alternative (the research hypothesis) is implied. However, it is not appropriate to use the terminology of "reject," or "fail to reject" with respect to the research hypothesis, or to use some other terminology such as to "accept" the null hypothesis. The null hypothesis is the one subjected to the statistical test, and the two appropriate conclusions are to either reject the null hypothesis or fail to reject the null hypothesis.

Two Examples of the *Z* test

Let's look at two examples of hypothesis testing using the *Z* test. Keep in mind that these examples are for the purpose of illustrating the steps involved in hypothesis testing, and not to give you a complete understanding of how and when to use this particular version of the *Z* test. It, along with several variations of the *Z* test, will be covered in detail in Chapter 11.

1. We will begin with the research hypothesis stated below:

   H_1: College graduates will have a higher mean income than the population in general.

   Or stated symbolically:

   $H_1 : \mu_{cg} > \mu$

   The null hypothesis can be stated as follows:

   H_0: College graduates will have a mean income equal to the population in general.

   Or stated symbolically:

   $H_0 : \mu_{cg} = \mu$

2. The research hypothesis implies a direction (we expect the mean income of college graduates estimated by a sample to have a higher mean income than the known mean income of the population in general), so a one-tail test of the hypothesis is in order.
3. We will be comparing a sample mean, $\bar{X}$, to a population mean, μ, so a *Z* test is an appropriate choice.
4. We will test the null hypothesis at the $\alpha = .05$ level of significance (this level is chosen arbitrarily, but the .05 level of significance is commonly used). Choosing the $\alpha = .05$ level of significance means that if the observed *Z* score based on the distance between the sample mean and the population mean is in the extreme .05 or 5% of the normal distribution's right tail (the critical region), then we will reject the null hypothesis. Otherwise, we will fail to reject the null hypothesis.

5. The critical value (the Z score that serves as the boundary of the extreme 5% of the normal distribution, or the critical region) is $Z_{.05} = 1.645$. (Note that the critical value of the test statistic is often identified by using the level of significance as a subscript.)
6. We will compute the value of the test statistic using the following formula for the Z test, and data.

$$Z = \frac{\overline{X} - \mu}{\sigma_{\overline{X}}}$$

Observed data:
Population mean income $\mu = \$28{,}800$; with a standard deviation $\sigma = \$7{,}200$
Sample of $n = 64$ college graduates observed to have a mean income $\overline{X} = \$30{,}825$
Note: standard error is equal to $\$7{,}200/\sqrt{64} = \$900$

Computing the value of the test statistic yields the following:

$$Z = \frac{30{,}825 - 28{,}800}{900}$$

$$= \frac{2025}{900} = 2.25$$

7. Comparing the observed value of the test statistic, $Z = 2.25$ to the critical value of the test statistic, $Z_{.05} = 1.645$ indicates that the observed value of the test statistic is well into the critical region of the normal curve as illustrated below.

Our decision is to reject the null hypothesis, and to accept the alternative. Our statistical analysis of the data indicates that a **statistically significant** difference exists between the mean income of college graduates and the mean income of the population in general.

One very important point to keep in mind when conducting a one-tail test of the hypothesis is that the sign of the test statistic is just as important as the absolute value of the test statistic. In order for us to reject the null hypothesis, the observed value of the test statistic not only needed to be equal to or greater than $Z_{.05} = 1.645$, but the value of the test statistic also needed to be positive in this case. As you can see in figure 8.7, the critical region is in the right side of the normal curve (the positive side). If the observed value of the test statistic had been negative in this particular case, we would fail to reject the null hypothesis no matter how large the observed value of the test statistic was. For example, an observed test statistic of $Z = -3.50$ would not be in the critical region, so we would fail to reject the null hypothesis.

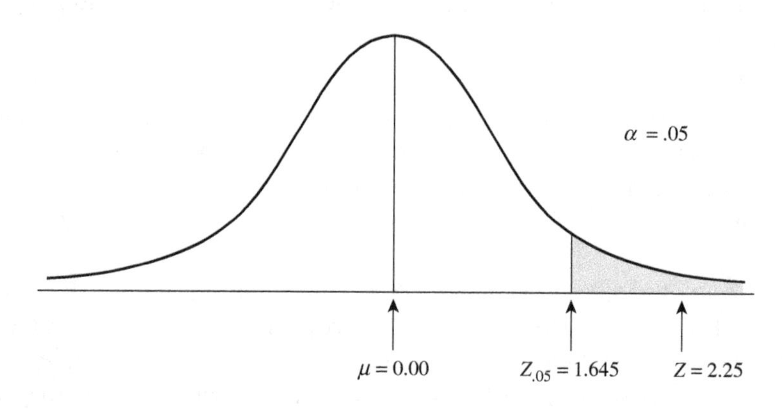

Figure 8.7 The Observed and Critical Values of *Z*

A Second Example (with an Important Twist)

In this second example, we will again be comparing an observed sample mean to a known population mean, but there is an additional concept to be introduced. Let's formally consider the hypothesis discussed earlier concerning the mean income of recent college graduates who majored in one of the behavioral sciences compared to the mean income of all recent college graduates.

1. We can state the research hypothesis as follows:

H_1: Recent college graduates who majored in one of the behavioral sciences (bs) will have a mean income different from that of the population of all recent college graduates.

Or stated symbolically:

$$H_1 : \mu_{bs} \neq \mu$$

The null hypothesis can be stated as follows:

H_0: Recent college graduates who majored in one of the behavioral sciences will have a mean income equal to that of the population of all recent college graduates.

Or stated symbolically:

$$H_0 : \mu_{bs} = \mu$$

2. The research hypothesis does not imply a direction (we expect the mean income of the population estimated by a sample of behavioral science majors to have a mean income *different* from the known population mean income of all recent college graduates, but we do not know how it will be different), so a two-tail test of the hypothesis is in order.

3. We will be comparing a sample mean, $\overline{X}$, to a population mean, μ, so a Z test is an appropriate choice.

4. We will test the null hypothesis at the $\alpha = .05$ level of significance. Because this will be a two-tail test of the hypothesis, we will have two critical regions (one in each tail of the normal curve), and two critical values of the test statistic (one to mark the boundary of each critical region). If the observed value of the test statistic falls on or in either of the critical regions, then we will reject the null hypothesis. Otherwise, we will fail to reject the null hypothesis.

5. A two-tail test of the hypothesis requires two critical values, one positive and one negative to mark the beginning of each of the two critical regions appearing in each tail of the distribution. We saw earlier that the critical value $Z = 1.645$ marked the boundary of the critical region for a one-tail test conducted at the $\alpha = .05$ level. However, for a two-tail test at the $\alpha = .05$ level, we cannot use $Z = -1.645$ and $Z = 1.645$ as the critical values because those values would leave us with .05 of the normal curve in each tail of the distribution, or an overall risk of being wrong of .10. *When conducting a two-tail test at a given alpha level, we must split the risk of being wrong and put half in each tail.* So for a two-tail test of the hypothesis conducted at the $\alpha = .05$ level, we want each critical region to contain one-half of .05, or .025. The proper critical values, therefore, are the Z scores that provide us with .025 in each tail of the normal curve, or $Z = -1.96$ and $Z = 1.96$. (Note: The appropriate critical values for a Z test for both one-tail and two-tail tests at commonly used alpha levels are provided at the end of Table 1 containing areas under the normal curve.)

6. We will compute the value of the test statistic using the following formula for the Z test, and data.

$$Z = \frac{\overline{X} - \mu}{\sigma_{\overline{X}}}$$

Observed data:
Population mean income of all recent college graduates $\mu = \$30{,}500$; with a standard deviation $\sigma = \$7{,}800$
Sample of $n = 100$ recent college graduates who majored in one of the behavioral sciences are observed to have a mean income $\bar{X} = \$31{,}850$
Note: standard error is equal to $\$7{,}800/\sqrt{100} = \$780$

Computing the value of the test statistic yields the following:

$$Z = \frac{31{,}850 - 30{,}500}{780}$$
$$= \frac{1350}{780} = 1.73$$

7. Comparing the observed value of the test statistic, $Z = 1.73$ to the critical value of the test statistic, $Z_{.05} = 1.96$ indicates that the observed value of the test statistic is not in either critical region of the normal curve as illustrated below in figure 8.8. *Our decision is to fail to reject the null hypothesis*. Our statistical analysis of the data indicates that there is not a statistically significant difference between the mean income of college graduates who majored in one of the behavioral sciences, and the mean income of the population of all recent college graduates. Although it is true that the data indicate the sample of recent college graduates who majored in one of the behavioral sciences has a mean income higher than the mean income of the population of all recent college graduates, the difference between the two groups could easily be the result of sampling variation.

Up to this point, we have established a foundation for hypothesis testing, and we have examined two detailed examples including both a one-tail test and a two-tail test using a statistical test called the Z test. Keep in mind that our use of the Z test in these examples was for the purpose of providing a complete example of the process of hypothesis testing beginning with the statement of the research hypothesis and progressing through to the decision to reject or fail to reject the null hypothesis. We will be examining the actual purpose of the Z test, and other versions of it, in more detail in Chapter 11.

Types of Error in Hypothesis Testing: Type I and Type II Error

We have seen that the level of significance, or alpha level, at which a statistical test is conducted represents the amount of error we are subject to when reaching a decision regarding the null hypothesis. To

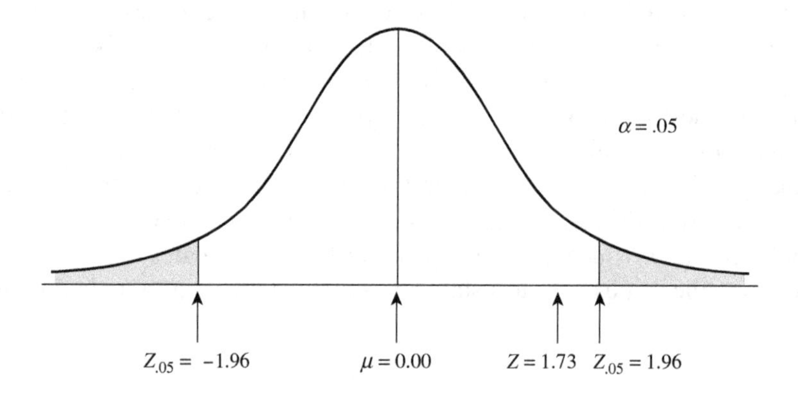

Figure 8.8 The Critical Regions for a Two-Tail Z Test at the $\alpha = .05$ Level of Significance

be more precise, the error associated with hypothesis testing consists of two types: **Type I error**, and **Type II error**. Type I error is the error made when we reject the null hypothesis when the correct decision should have been to fail to reject the null hypothesis. Stated differently, Type I error occurs when there is no significant difference in reality, but we conclude that there is based on our analysis of the available data. Type I error is the error associated with the level of significance or alpha level that we choose when conducting a statistical test. For example, when we chose an alpha level of $\alpha = .05$ above, we were declaring the extreme 5% of the normal distribution as the critical region, and any test statistic that appears in the critical region represents a statistically significant result even though there is a chance that we could be wrong 5% of the time. While we are always subject to Type I error, the good news is that we can control the likelihood of Type I error by our choice of the level of significance we use when conducting a statistical test. The lower the level of statistical significance (the lower the alpha level, or the more rigorous the standard of proof), the lower the risk of committing a Type I error.

Type I error can only occur when the correct decision should be to fail to reject the null hypothesis. Of course, if the correct decision is to fail to reject the null hypothesis and we do reject the null hypothesis, then we have made a correct decision. The probability of making a correct decision when the correct decision should be to fail to reject the null hypothesis is equal to $(1-\alpha)$.

Type II error is the error made when we fail to reject the null hypothesis when the correct decision should have been to reject the null hypothesis. Stated differently, Type II error occurs when a significant difference exists in reality, but we fail to reach that decision based on our analysis of the data available to us. Type II error is referred to as beta (β) error. Type II error can only occur when the correct decision should be to reject the null hypothesis. We have a probability of making the correct decision under this scenario as well. The probability of rejecting the null hypothesis when the correct decision is to reject the null hypothesis is equal to $(1-\alpha)$. Correctly rejecting the null hypothesis is referred to as the **power of a statistical test**. The consequences of the alternative conclusions that may be reached regarding the null hypothesis are summarized in Figure 8.9 below.

Most discussions of statistical analysis focus on Type I error rather than Type II error. One major reason for this is that Type I error is so much easier to control. Type I error is equal to the alpha level that we choose, so we have direct control over Type I error. Naturally, we want to minimize our risk of making a Type I error, but we also want a statistical test that is powerful, that is a test that maximizes our chance of rejecting the null hypothesis when it should be rejected. The only problem is that the size of the alpha level that we choose has a direct impact on our risk of making a Type II error. The more rigorous our level of significance, the smaller the alpha level, the greater the risk we take of failing to reject the null hypothesis when it should be rejected. A small alpha level increases the size of beta, or Type II error, and the larger the size of beta, the lower the power of the statistical test (review Figure 8.9 again to see the relationships involved between alpha, beta, and the likelihood of reaching the correct decision for H_0).

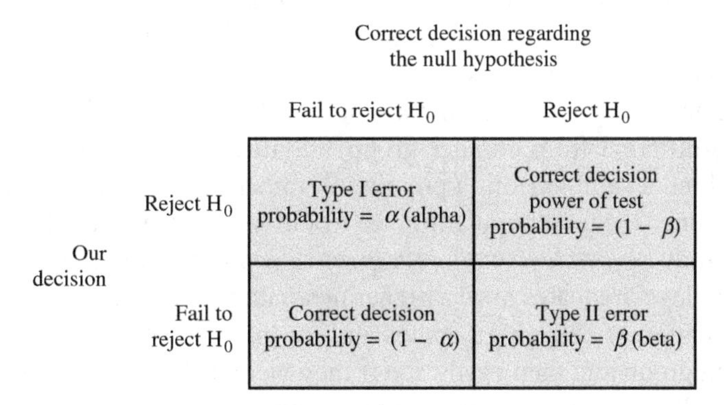

Figure 8.9 Possible Conclusions Regarding a Test of H_0

The size of the alpha level is not the only factor that has an impact on the size of beta, or Type II error. Other factors and their impact on Type II error include the following:

1. The distance between the true population mean and our hypothesized estimate of it (the smaller the actual distance, the greater the likelihood of Type II error)
2. The sample size "*n*" (the smaller the sample size, the greater the likelihood of Type II error)
3. The size of the population variance σ^2 (the larger the population variance, the greater the likelihood of Type II error)
4. The use of a one-tail or a two-tail test of significance (the use of a two-tail test of significance results in a greater likelihood of Type II error)

It is possible to compute the power of a statistical test, and consequently the size of Type II error, and to demonstrate how the likelihood of Type II error varies with changes in each of the conditions identified above. However, the computation of the power of a statistical test, and a detailed analysis of the various influences on Type II error is best left to a more advanced course in statistical analysis. At this stage, our focus is best left on Type I error, or the probability of rejecting the null hypothesis when the correct decision should be to fail to reject the null hypothesis, which we can control directly by our choice of an alpha level.

Why Do We Need Statistical Tests, and What Does a Finding of Statistical Significance Really Mean?

Most of the remainder of this text will cover a number of statistical procedures designed to test various statistical hypotheses. The statistical tests are designed for different types of research strategies, and different levels of measurement of the variables under investigation. We will focus on which statistical test is appropriate under a given set of circumstances, and the proper interpretation of the results. *The reason we conduct statistical tests, and the proper interpretation of a statistically significant result are two of the most commonly misunderstood concepts in statistics.* We will consider each of these two important issues in turn.

The Need For Statistical Tests

Recall from Chapter 1 that measured characteristics from a population are called parameters, and that measured characteristics from a sample are called statistics. Parameters are exact values that describe a population. Because a parameter is the result of an examination of all members of the population, there is one and only one value that a population parameter will equal. For example, if you think of the members of your statistics class as a population, and you want to compute the mean age of the class, there is only one value that the class mean will equal. On the other hand, we could take any number of samples from the class, and each sample could have a different mean age depending on which members from the population were included in the particular sample.

Statistical hypothesis testing is all about looking for differences. Oftentimes the question comes down to "Is one group different from another group?" Statistical tests are needed because we usually conduct research by taking a sample from a population rather than by examining the entire population itself. If we always examined the total population, we would not need statistical tests. *To see if two groups are different, we would only have to look at their respective population parameters*. For example, suppose we wanted to know if college graduates have a higher mean income than the population in general. Given data for the entire population of college graduates and for the population in general, we could compute a mean income for each group and then easily see if they were different. Life would be that simple, and no statistical testing would be necessary. The two mean incomes would either be the same or one group would be higher than the other.

Statistical testing is only necessary if we are trying to determine if two populations are different on the basis of sample information. When we select a sample from a given population, we have only one of many possible samples that could be obtained. We can never be sure which sample we have from a population's sampling distribution. The sample we actually draw may have a mean exactly equal to the population mean, or it may be somewhat higher or somewhat lower than the actual population mean. Because of the sampling variation involved, we cannot simply assume that the sample mean is exactly equal to the population mean from which it was drawn. For that reason, it would not be wise to compare a population mean estimated from a sample to a known population mean, and to decide if the two are different by simply looking at the two values. Instead, we will employ a statistical test and adopt a standard of proof to help us decide if the two values are sufficiently different to suggest a statistically significant result.

What Does a Finding of Statistical Significance Mean?

We have seen two complete examples of hypothesis testing in this chapter. In the first example, we rejected the null hypothesis, and found a statistically significant result. In the second example, we failed to reject the null hypothesis, and concluded that we did not have a statistically significant result. *But what does statistical significance really mean?* The best single word answer is that statistical significance means "difference." Let's assume that we have conducted a Z test to compare a mean estimated from sample mean to a population mean. A finding of statistical significance implies that there is enough of a difference between the sample mean and the population mean to indicate that a similar difference would have been found if we had been able to examine both population means directly instead of using a sample mean to estimate one of the populations.

In Chapter 11, we will examine another type of Z test that is used to compare two sample means. What does a statistically significant result indicate in that case? It means that the two sample means are different enough to suggest that the two population means they represent would also be different in a similar fashion. If the observed difference is not statistically significant, it means that the difference between the two sample means is small enough that it could represent sampling variation. That is, that the two samples are more likely to have come from the same population even though they are not exactly equal to one another.

How do we decide if an observed difference is a statistically significant one or not? We adopt a standard of proof (an alpha level, or level of significance), choose the appropriate statistical test, analyze the data, and determine if the test statistic falls in or on the critical region. If the test statistic falls in or on the critical region, then we reject the null hypothesis and conclude that we have a statistically significant result. If the test statistic does not fall in or on the critical region, then we fail to reject the null hypothesis and conclude that the observed difference is not statistically significant. Statistical significance just means that the observed difference is large enough to suggest that a similar difference would exist between the two population parameters had we been able to examine both of them directly.

What Does Statistical Significance NOT Mean?

Just as important as what statistical significance means, is the related question of "What does statistical significance not mean?" Statistical significance means that there is evidence of "difference," but it does not mean that the difference is an important one. Different disciplines approach the underlying concept here in a variety of ways such as saying that a finding is "important," or "theoretically important," or "material," or "substantial," but statistical significance does not imply any of those things. Statistical significance just means that the results of a statistical test are such that a real difference probably exists between the population parameters. When comparing two mean incomes, it is possible that an observed difference of only one dollar will be statistically significant! Is that difference an important one? That is not a statistical question; that is a theoretical or philosophical question. All we can say is that it is likely that the two population means are different by one dollar as well.

Summary of Key Points

This chapter has been an important one, and in some ways, a challenging one. To fully understand the material in this chapter, you have to integrate several important concepts from the chapters on sampling, the normal distribution, and probability theory. At this point, you should have a working knowledge of the logic of hypothesis testing including the concepts of a hypothesis, a research hypothesis, a null hypothesis, a one-tail and a two-tail test, the level of significance, the value of a test statistic, the critical value, the critical region, how to reach a decision regarding the null hypothesis, Type I error, and Type II error. If you have been able to do that successfully, you will have a good foundation for the next three chapters dealing with various types of commonly used inferential statistics procedures.

Hypothesis—A statement of relationship between variables.

Research Hypothesis (H_1)—A hypothesis stating the researcher's expectation regarding the relationship between variables.

Null Hypothesis (H_0)—A statement of no relationship between variables crafted to state the opposite of what the researcher expects to find. The null hypothesis is subjected to statistical testing.

One-Tail Test—A test of the hypothesis used when the research hypothesis states a direction (a positive relationship, or a negative relationship) for the relationship between variables.

Two-Tail Test—A test of the hypothesis used when the research hypothesis only states an expectation of difference, and does not state a direction for the relationship.

Level of Significance (Alpha Level)—The amount of error one is prepared to accept in hypothesis testing when reaching a decision.

Test Statistic—The value resulting from a statistical test used to test the null hypothesis.

Critical Region—The area in the distribution associated with a test statistic corresponding to the level of significance chosen for a statistical test.

Critical Value—The value of the test statistic marking the beginning of the critical region. The observed value of the test statistic must equal or exceed the critical value to reject the null hypothesis.

Statistical Significance—A result indicating that a difference observed in sample data is sufficiently large to suggest that a similar difference exits in the respective population parameters.

Type I Error—The error of rejecting the null hypothesis when the correct decision is to fail to reject the null hypothesis. Type I error is equal to the level of significance chosen for the hypothesis test. Type I error is often referred to as alpha error.

Type II Error—The error of failing to reject the null hypothesis when the correct decision is to reject the null hypothesis. Type II error is often referred to as beta error.

Power of a Test—The power of a statistical test is the probability of rejecting the null hypothesis when the correct decision is to reject the null hypothesis.

Inferential Statistics—Statistical procedures designed to allow us to make generalizations (inferences) regarding population parameters based on an analysis of sample statistics.

Questions and Problems for Review

1. Explain why we test the null hypothesis rather than the research hypothesis.
2. Read an article from a professional journal in your field, and identify the research hypothesis and the null hypothesis. Identify the type of statistical analysis used, and summarize the results. (Hint: there may be more than one research hypothesis, and the null hypothesis may not be formally stated in the article. Try to find an article that tests for a significant difference

between means, rather than one presenting a more complex testing of a "structural equation model.")

3. Explain the difference between a one-tail test and a two-tail test.
4. Identify three pairs of variables: one for which you would expect a positive relationship, one for which you would expect a negative relationship, and one for which you would expect a relationship but cannot state a direction.
5. State the research and null hypotheses for the relationships you identified in question 3 above.
6. Evaluate the following research hypotheses to determine if a one-tail test, or a two-tail test is in order.
   A. Individuals of an older age have a different level of physical agility than individuals of a younger age.
   B. Level of education is negatively related to level of prejudice.
   C. Males have a higher mean income than females.
   D. Females and males differ on the verbal skills.
7. Given the following information where μ_s represents a population mean estimated by a sample, and μ represents a known population mean:

   H_1: $\mu_s \neq \mu$
   H_0: $\mu_s = \mu$
   $\alpha = .05$
   $Z = -1.97$

   Identify the critical value of the test statistic, and reach a decision regarding H_0.
8. Given the following information where μ_s represents a population mean estimated by a sample, and μ represents a known population mean:

   H_1: $\mu_s < \mu$
   H_0: $\mu_s = \mu$
   $\alpha = .05$
   $Z = 2.08$

   Identify the critical value of the test statistic, and reach a decision regarding H_0.
9. A researcher expects that students using this statistics text will score higher than average on a standardized test of statistics knowledge than students in general. Given the data below:

   Mean score of all students taking the standardized test, $\mu = 75$, $\sigma = 16$
   Mean score of a sample of $n = 100$ students who used this text, $\bar{X} = 79$

   A. State the research hypothesis and null hypothesis in symbolic terms.
   B. Test the null hypothesis at the $\alpha = .01$ level.
   C. Reach a decision regarding H_0, and interpret the results.
   (Hint: A *Z* test is appropriate here. You can find the critical value of the test statistic at the end of Table 1 in the appendix.)
10. How could you have evaluated the researcher's hypothesis from question 9 above if you had access to population data for all students in general, and for all students using this text? Would the *Z* test be an appropriate choice? Why or why not?

Part IV Inferential Statistics

Part IV Inferential Statistics is the final section in the text, and consists of Chapters 9–13. Chapter 9 presents several types of correlation coefficients that are useful in telling us if there is an association between two variables. Chapter 10 deals with linear regression that allows us to use the level of an independent variable to predict the level of a dependent variable. Chapter 11 deals with statistical tests for means, and includes several statistical tests used to tell us if two means are statistically significantly different. Chapter 12 presents a discussion of one-way analysis of variance that allows us to examine a group of means for the presence of statistically significant differences. Finally, Chapter 13 presents a number of nonparametric statistical techniques that allow us to make inferences about a population based on the examination of a sample without meeting some of the rigorous assumptions required of the parametric statistical procedures covered in Chapters 9–12.

CHAPTER 9

Correlation

Key Concepts

Measures of Association
Pearson Correlation Coefficient
Point-Biserial Correlation Coefficient
Spearman Correlation Coefficient
Scatterplot
Statistical Significance
One and Two Tail Tests
Variance Explained
Partial Correlation

Introduction

The topics discussed in this chapter build on much of the material from the previous chapters including the different types of data from Chapter 1, the graphic display of variables from Chapter 3, and the idea of explaining variance from Chapter 5. One of the things that sets this chapter apart from the others is that we are beginning to deal with statistical techniques that examine two variables at the same time. More specifically, we will be analyzing data to determine if two variables are related in some way.

Correlation coefficients are statistics that indicate the existence and type of a relationship between two variables. You may recall from Chapter 4 that measures of central tendency were just statistics or numbers that were intended to tell us in a very efficient way what was central or typical of an entire set of scores, and that measures of variation from Chapter 5 were just statistics that were intended to tell us how much the scores tended to differ from a central point such as the mean. Correlation coefficients are similar in that they are just statistics or numbers that indicate how strongly and in what manner two variables are related to each other. In this chapter, we focus on three correlation coefficients: the **Pearson correlation coefficient (Pearson r)**, the **Point-biserial Correlation Coefficient (r_{pb})**, and the **Spearman Correlation Coefficient (Spearman r_s)**. All three of these correlation coefficients are similar in some respects, but they are appropriate for different situations depending on the type of data one is examining. We will also be discussing the idea of a partial correlation coefficient in this chapter, but first it will be useful to review a few key ideas from previous chapters.

A Brief Review of Levels of Measurement

Behavioral science research is often aimed at discovering causal relationships between two variables. You may recall that by convention we usually represent the independent variable (or presumed cause) by the letter "*X*," and the dependent variable (or the presumed effect) by the letter "*Y*." Variables included in our research may be measured in a number of different ways, but the resulting data will fall into one of four categories that we refer to as the scales of data or levels of measurement. A variable may be measured at either the nominal level, the ordinal level, the interval level, or the ratio level.

Nominal and ordinal level data are both categorical. Nominal level data are the simplest, and consist of only categories with no other underlying dimension of measurement. Nominal level data include such variables as sex, college major, state of birth, or race. Ordinal level data are also represented by categories, but there is an underlying dimension of measurement or sense of progression as you move from one category to the next. Variables that are often collected at the ordinal level include college classification, T-shirt size, or order of finish in a race. We are still measuring a variable with categories, but there is a natural or logical progression as you move from one category to the next.

Interval and ratio level data are measured on a continuous scale. Interval level data require a continuous scale, and the fact that the amount of change represented between any two adjacent points on the scale represent the same amount of change in the variable any place along the scale. Ratio level data are measured on a similar continuous scale, but the scale must also have an absolute zero point that represents a complete absence of what is being measured. Income is often measured at the ratio level. When someone reports that they had an income of $0.00, it indicates that they earned "absolutely" no money.

Choosing the Proper Correlation Coefficient

The point of this brief review over types of data is that the level at which two variables have been measured is the chief determinant of which correlation coefficient is appropriate in a given situation. The most frequently used correlation coefficient is the Pearson Product Moment Correlation Coefficient, or the Pearson r. The Pearson r is appropriate for investigating the relationship between two variables that have been measured at the interval level or above. Both variables do not necessarily have to be measured at the same level. One variable may be at the interval level, and the other may be at the ratio level, but neither should be measured at the nominal or ordinal level. The point-biserial correlation r_{pb} and the Spearman r_s are used less often than the Pearson r. The Spearman r_s is appropriate for two ordinal level variables, and in this case both variables must be measured at the ordinal level. The point-biserial r_{pb} is appropriate when the dependent variable is measured at the interval or ratio level, and the independent variable is measured as a "nominal" level dichotomy. A variable is dichotomous when it has only two possible values. The variable "sex" represents a true nominal level dichotomy. The variable is at the nominal level, and there are only two possible values—"male" or "female." The point-biserial correlation coefficient may be used in some cases when the independent variable is not a true nominal level dichotomy, as long as the independent variable is treated as one. For example, we may have a variable such as college major which is at the nominal level, but is not a dichotomy since there are many more than two possible outcomes. In this case, we would have to dichotomize the variable or collapse all possible values into two categories. One solution might be to put behavioral science majors into one category, and all other majors into the second category. I placed "nominal" in quotation marks above, because one is not necessarily restricted to using nominal level data for the independent variable with a point-biserial correlation coefficient. An ordinal level variable such as college classification could be used as an independent variable, but we would still have to dichotomize the variable in some fashion to create two categories such as "upper division students" (juniors and seniors), and "lower division students" (freshmen and sophomores), or the categories of "freshmen" and "nonfreshmen," or even "juniors" and "nonjuniors."

Bivariate Data Plots—Graphing Two Variables to Reveal the Relationship between Two Variables

In order for two variables to be related, there must be evidence that the variables "covary," or move together in some systematic fashion. In Chapter 3, we examined several ways to graphically display a single variable, and briefly introduced the concept of graphing two variables. In this chapter, we expand the idea of graphic display by looking at the presentation of two variables on a Cartesian coordinate system, and examine what types of causal relationships are implied by various bivariate (two variable) patterns.

The Cartesian coordinate system, consisting of two axes intersecting at a right angle to each other, is a useful method of displaying the interaction between two variables on a single graph. Each of the two axes represents a different variable, and conventionally the horizontal axis is used for the independent variable (the X variable), and the vertical axis is used for the dependent variable (the Y variable). What makes the system useful is that not only can we see the distribution of each variable alone, but we can also see how the two variables are related to each other. The resulting scatterplot (the pattern created by the data) is a valuable tool for examining the data for evidence of a cause and effect relationship.

We will demonstrate the method of plotting two variables on a Cartesian coordinate system by using two variables that are frequently encountered in behavioral science research: education (X) and income (Y). Hypothetical data for a sample of $n = 25$ individuals are presented below, and then plotted on the coordinate system in Figure 9.1. Each of the $n = 25$ individuals is plotted on the coordinate system by placing an asterisk (*) where each person's education intersects with their income level. For example, individual seven in the sample with 6 years of education, and 45 thousand dollars in income is represented by the asterisk in bold in Figure 9.1. The other 24 members of the sample are placed on the graph in a similar fashion.

The bivariate data presentation can provide us with a sense of how each variable is distributed by examining the spread of each variable along its own axis. For example, educational attainment among the sample ranges from a low of 6 years to a high of 22 years, and income among the sample ranges from a low of 10 thousand dollars to a high of 80 thousand dollars. While it is useful to be able to see how each of the two variables is distributed, what is of greater use is the ability of a scatterplot to show us how the two variables are related to each other. The relationship between the two variables can be seen by looking at the overall pattern of the 25 data points. If you examine the distribution of the data, you see that individuals with lower education levels also tend to have lower income levels. Similarly, individuals with higher education levels tend to have higher incomes. Two variables such as education and income that tend to move in the same direction are said to have a positive relationship. In a positive relationship, when the independent variable is low, the dependent variable tends to be low, and conversely, when the independent variable is high, the dependent variable tends to be high. As the independent variable increases, the dependent variable will increase, or conversely, as the independent variable decreases, the dependent variable will decrease. In a positive relationship, the independent and dependent variables tend to move in the same direction. When two variables with a positive relationship are graphed, the resulting scatterplot will appear much like the scatterplot of education and

Table 9.1 Data on Education and Income

Individual	Education (in years)	Income (× $1,000)
1	12	20
2	12	40
3	16	45
4	8	10
5	14	50
6	16	70
7	6	45
8	10	15
9	16	60
10	18	25
11	10	10
12	20	25
13	19	60
14	22	80
15	12	25
16	9	30
17	18	70
18	14	40
19	18	45
20	20	65
21	10	40
22	8	65
23	12	15
24	6	20
25	18	10

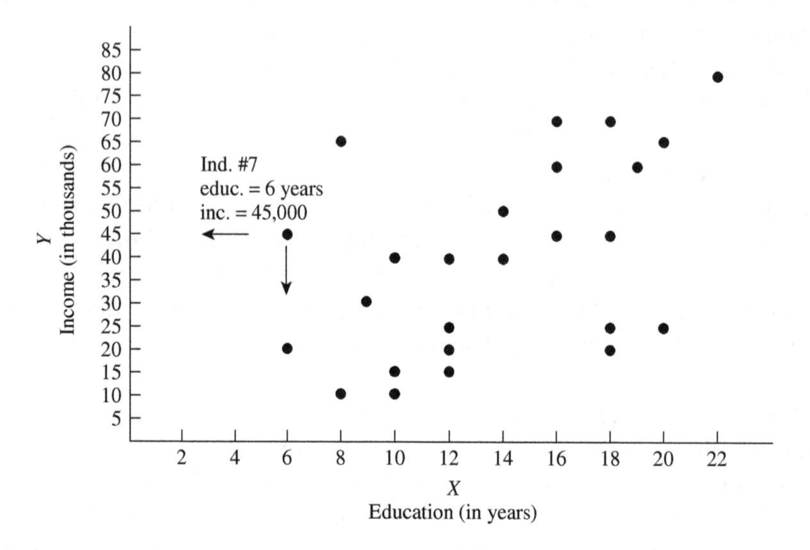

Figure 9.1 Scatterplot of Income and Education

income in Figure 9.1. In general, the farther an individual is to the right on the scale of the independent variable, the higher that same individual will be on the scale of the dependent variable. Of course there are a few exceptions; we can see some cases where individuals with low levels of education have high incomes, and a few cases where individuals with high levels of education have low incomes. If two variables have a perfect positive relationship, all of the data points in the scatterplot will form a straight line with a positive slope, that is, as you move to the right on the independent variable you will also move up on the dependent variable.

Two variables that tend to move in opposite directions are said to have a negative relationship. The two variables are still related, they just move in opposite directions. As you move to the right on the independent variable, you would move down on the dependent variable. Two variables with a negative relationship are an automobile's speed measured as miles per hour (X), and an automobile's gas mileage measured as miles per gallon (Y). In general, as an automobile's speed increases, the miles per gallon will decrease. Figure 9.2 presents a scatterplot of how the relationship between miles per hour and miles per gallon would appear.

The scatterplot depicting the relationship between miles per hour (X) and miles per gallon (Y) is typical of what you would see when two variables have a strong negative relationship. Note that the data points almost appear in a straight line with a negative slope. The farther you move to the right on the independent variable, the lower you are on the scale of the dependent variable. In addition to a positive relationship or a negative relationship, it is also possible that two variables will have no relationship to each other.

When two variables have no relationship, the scatterplot of the data points will appear in a random order which is sometimes referred to as a "shotgun" pattern much like the pattern of shotgun pellets striking a target. Figure 9.3 depicts a typical scatterplot of two variables with no relationship such as IQ and shoe size. The random pattern evident in Figure 9.3 is often referred to as a "shotgun" pattern, because it resembles the pattern that the pellets form on a target when a shotgun is fired. The pattern indicates no relationship between the two variables. Some people with small feet have low IQs and some people with small feet have high IQs. Similarly, some people with large feet have low IQs and some people with large feet have high IQs. In this case, knowing someone's shoe size will not help us predict their IQ.

When two variables have no relationship to each other, there will be no discernible pattern in the scatterplot. As is the case in Figure 9.3, at any given point on the scale of the independent variable you might find a value at any given point on the scale of the dependent variable. A scatterplot is not the only way in which the relationship between two variables may be presented. Oftentimes, data are presented in a table referred to as a contingency table. In Chapter 13, we will examine the way a contingency table is constructed, and you will see how the same patterns evident in a scatterplot are also evident in a contingency table.

We have seen that variables may covary in several different ways such as a positive relationship between education and income, a negative relationship between miles per hour and miles per gallon, and even no relationship such as between IQ and shoe size. Consider one more pair of variables, age and death rates, and imagine how they might covary. You might be tempted to think that age and death rates are positively related in that as one gets older one's risk of death gets higher. In some respects that is true,

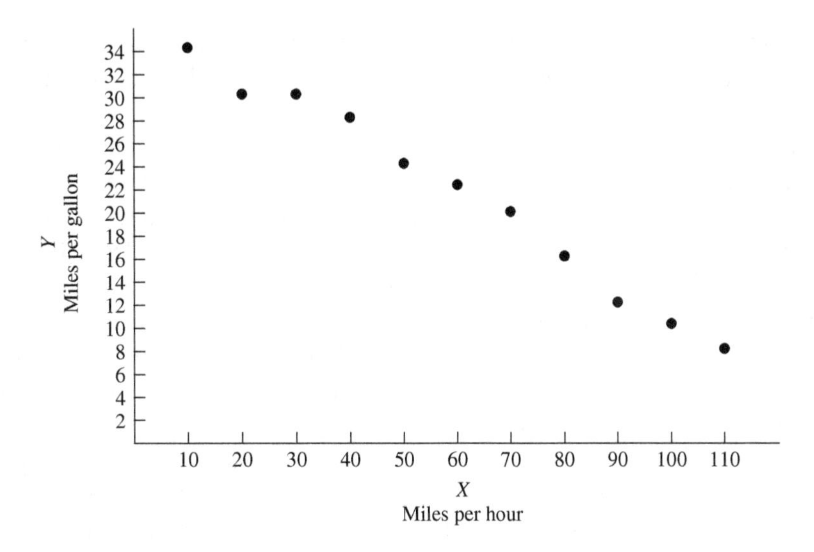

Figure 9.2 Scatterplot of Miles per Hour and Miles per Gallon

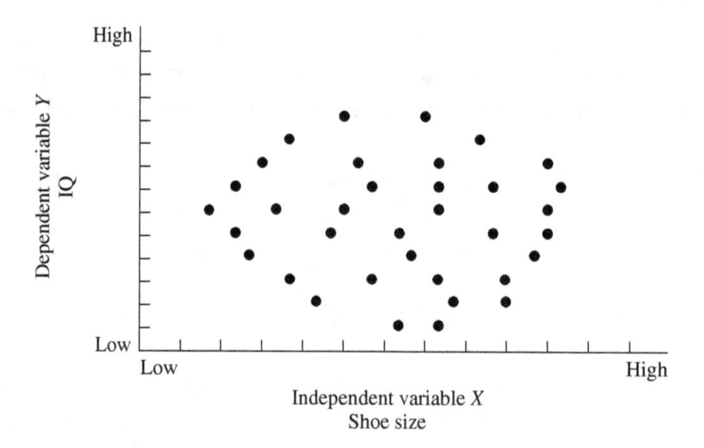

Figure 9.3 Scatterplot of I.Q. and Shoe Size

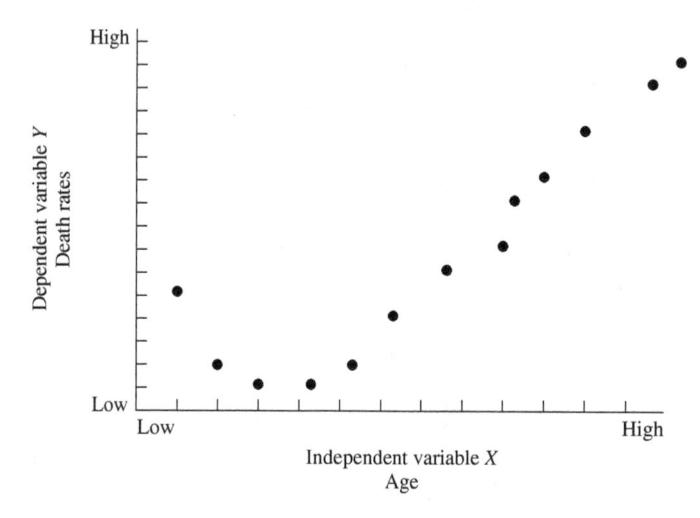

Figure 9.4 Scatterplot of Death Rates and Age

but the actual relationship between age and death rates is not quite that simple. The scatterplot presented below reflects the actual relationship between age and death rates in the United States, although not to exact scale.

Contrary to what you might have expected, the risk of death does not increase steadily with age. In fact, the risk of death is quite high in the first year of life, then falls off dramatically during youth and adolescence, and then begins to increase again. The relationship between age and death rates cannot be described as positive or negative; the relationship between these two variables is best described as curvilinear. Oftentimes, curvilinear relationships will take the form of a "U" shaped or "J" shaped curve. The relationship begins as a negative one, then the curve flattens, and then begins to take on the form of a positive relationship. In other cases, the curvilinear relationship may exhibit an inverse "U" shaped or "J" shaped curve, in which case the relationship begins by appearing positive, then the curve flattens out, and then takes on the form of a negative relationship.

Curvilinear relationships are just as important as linear relationships, and the fact that a relationship between two variables does not suggest a linear or straight line relationship does not invalidate it. However, the absence of a linear relationship does have one important consequence. The correlation coefficients covered in this chapter are only appropriately applied to variables exhibiting a linear relationship (either one that is positive or one that is negative).

An additional point to stress is that correlation coefficients only indicate the presence or absence of an association between two variables, but the suggestion of an association is not proof of a causal relationship between the two variables. In the language of formal logic we would say that evidence of an association between two variables is a necessary condition for a causal relationship, but not a sufficient condition. For one variable to be the cause of another it is necessary that the two variables have an association, or covary, but covariation alone is not sufficient to prove a causal relationship.

In some cases, what appears to be an association between two variables is not due to a causal relationship between the two variables that we examine at all, but is in fact the result of each of the two variables being related separately to a third variable. We can refer to the third variable as "*Z*" to go along with our convention of referring to the independent and dependent variables as "*X*" and "*Y*," respectively. When the third variable "*Z*" changes, it results in a corresponding change in each of the two other variables, "*X*" and "*Y*". In such cases, it will appear as if "*X*" and "*Y*" are related because they are moving together in response to changes in "*Z*," but the covariation between "*X*" and "*Y*" is not evidence of a true relationship between "*X*" and "*Y*." Two variables appearing to be related because of their common relationship to a third variable are said to have a false or spurious relationship. An example of a spurious relationship would be the apparent relationship between temperature and crime. Generally speaking, more crime is committed in the warmer months than in the colder months. In this case, changes in the temperature (X) might appear to be responsible for changes in the crime rate (Y); however, it is unlikely that the weather actually causes crime. What is more likely is the presence of a third variable that is related to each of the other two variables. One likely candidate for a third variable (Z) would be the amount of time that individuals spend outdoors in the presence of others. Individuals spend less time outdoors when the temperature is low during the winter months, and the result is less exposure to the risk of certain types of crime. As the temperature increases during the spring and summer months, individuals spend more time outdoors and increase their exposure to the risk of certain crimes.

It is important to control for the effects of a potential third variable when attempting to establish the presence of a true causal relationship between two variables. Controlling for the effects of a third variable may be done in a number of ways, but one common method is to use another type of correlation coefficient called the partial correlation coefficient. The partial correlation coefficient is based on the Pearson r, and is specifically designed to examine the relationship between two variables while controlling or excluding the effects of a third variable. The partial correlation coefficient is examined at the end of the chapter after we see how the bivariate, or two-variable, correlation coefficients in the chapter are used to analyze data for the presence of a basic relationship between two variables.

The Pearson Correlation Coefficient

The Pearson correlation coefficient is the most widely used measure of association in behavioral sciences research. The statistic takes its name from its developer, Karl Pearson, who officially named the statistic the Pearson Product Moment Correlation Coefficient, but it is more commonly referred to as the Pearson r. The Pearson r is appropriate for examining two interval or ratio level variables for evidence of a linear relationship (i.e., a relationship that can be represented by a straight line). Examples of pairs of variables that might be appropriate include: education and income, height and weight, amount of time studying and grade on an exam, city size and the crime rate, and any number of others.

When the Pearson r is computed for a pair of variables, the resulting correlation coefficient will range between –1.00 and +1.00. The absolute value of the Pearson r and its sign (positive or negative) are each important pieces of information.

$$\text{Possible Range of the Pearson } r = (-1.00 \le r \le 1.00)$$

The sign of the Pearson r indicates the direction or type of relationship between the two variables. A positive Pearson r indicates a positive relationship; that is, as the "X" variable increases, there will be a corresponding increase in the "Y" variable. A negative Pearson r indicates a negative relationship meaning that an increase in the "X" variable results in a corresponding decrease in the "Y" variable. A correlation coefficient equal to zero indicates no relationship between the two variables.

The absolute value of the correlation coefficient indicates the strength of the relationship between the two variables. When two variables are perfectly correlated, their scatterplot will appear as a straight line, and the absolute value of the correlation coefficient will equal 1.00. A correlation coefficient of +1.00 indicates a perfect positive relationship, and correlation coefficient of –1.00 indicates a perfect negative relationship.

Logic of the Pearson *r*

One challenge to creating a statistic to indicate the degree of association between two variables is the fact that any two variables may be measured on scales where the range of the measurement may vary widely. For example, the variable "Grade Point Average" measured on a ratio scale can vary from a low of 0.00 to a high of 4.00. The variable "Income" can be measured on a ratio scale from a low of 0.00 to a high in the millions or even billions of dollars. How can we create a statistic that can measure the degree of association present between two variables measured on such diverse scales? The answer lies in putting both variables on the same scale; a scale that can be used as a common basis of comparison. The Standard Normal Distribution (from Chapter 6) provides such a scale. In fact, the logic underlying the computation of the Pearson r is to examine where two variables tend to be in relation to each other when their raw scores are converted to Z scores. When we examine two variables "X" and "Y" for a group of individuals and find that the Z scores for the two variables are the same distance from the mean, then the variables are perfectly correlated; a positive correlation results if the Z scores are on the same side of the mean, and a negative correlation results if the Z scores are on opposite sides of the mean.

Calculating a Pearson r by converting raw scores into Z scores, and then determining the average correspondence of the two Z scores is needlessly cumbersome. A much simpler computational formula may be used as illustrated below.

A Computational Formula for the Pearson r

$$r_{yx} = \frac{(n * \Sigma XY) - (\Sigma X)(\Sigma Y)}{\sqrt{(n * \Sigma X^2) - (\Sigma X)^2}\sqrt{(n * \Sigma Y^2) - (\Sigma Y)^2}}$$

where

r_{yx} is the correlation between the two variables (by convention we usually list the dependent variable first when using the full notation, but in most cases we simple us the lowercase "r" to represent the correlation coefficient),

n is the sample size,

ΣXY is the sum of the cross-products (each X value times the corresponding Y value for each case in the sample),

(ΣX) is the sum of the *X*s,
(ΣY) is the sum of the *Y*s,
ΣX^2 is the sum of the *X*s squared (each *X* value squared, and then summed),
$(\Sigma X)^2$ is the sum of the *X*s, quantity squared,
ΣY^2 is the sum of the *Y*s squared (each *Y* value squared, and then summed), and
$(\Sigma Y)^2$ is the sum of the *Y*s, quantity squared.

The calculation of the Pearson *r* using the computational formula is straight forward and reasonably simple, but one note of caution is worth mentioning. The Pearson *r* formula is another example of a statistical formula that contains both the "sum of the *X*s squared" term ΣX^2, and the "sum of *X*s quantity squared" term $(\Sigma X)^2$. These terms are easy to confuse with each other, and doing so will result in an incorrect computation for the Perason *r*; however, it is an easy error to spot since reversing the two terms will result in a negative term in the denominator under the radical sign which is not a logical possibility.

The numerator is calculated by subtracting the product of the sum of the *X*s times the sum of the *Y*s from the product of the sample size "*n*" times the sum of the cross products (the sum of each *X* value times its corresponding *Y* value). The denominator contains two similar terms; one for the *X* values and one for the *Y* values. The denominator is calculated by subtracting the sum of the *X*s quantity squared from the product of the sample size *n* times the sum of the *X*s squared (each *X* value squared, and then those values summed), and then taking the square root. The resulting value is then multiplied by a similar term for the *Y* values, which is calculated by subtracting the sum of the *Y*s quantity squared from the product of the sample size *n* times the sum of the *Y*s squared (each *Y* value squared, and then those values summed), and then taking the square root. The product of the two values, one based on the *X*s and the other based on the *Y*s, completes the computation of the denominator.

The Pearson *r* value is then obtained by dividing the numerator by the denominator. Keep in mind that a negative Pearson *r* can only be the result of a negative value in the numerator, if you have a negative value for one or both of the terms in the denominator you have an error in your calculations.

Computing the Pearson *r*

Data for two variables, *X* and *Y*, for a sample of $n = 7$ cases are presented below along with a scatterplot of the data.

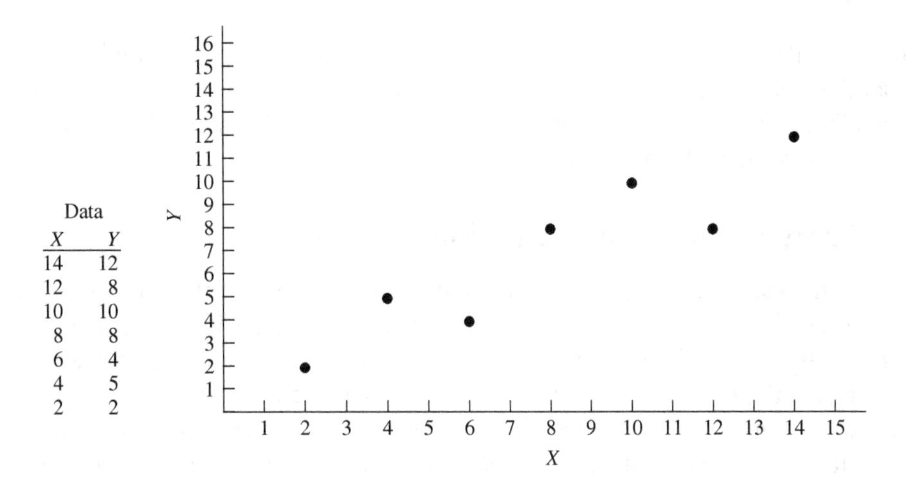

Data

X	*Y*
14	12
12	8
10	10
8	8
6	4
4	5
2	2

Figure 9.5 Scatterplot for Variables *X* and *Y*

Table 9.2 Data for Variables X and Y

X	Y	XY	X^2	Y^2
14	12	168	196	144
12	8	96	144	64
10	0	100	100	100
8	8	64	64	64
6	4	24	36	16
4	5	20	16	25
2	2	4	4	4
$\Sigma X = 56$	$\Sigma Y = 49$	$\Sigma XY = 476$	$\Sigma X^2 = 560$	$\Sigma Y^2 = 417$

While the data presented in the scatterplot do not form a straight line, there is evidence of a strong linear trend, and we can also see that the relationship between the two variables is positive.

The data for this example are presented below with appropriate columns for each of the terms we will need for the formula for the computation of the Pearson r.

$$r = \frac{(n \times \Sigma XY) - (\Sigma X)(\Sigma Y)}{\sqrt{(n \times \Sigma X^2) - (\Sigma X)^2}\sqrt{(n \times \Sigma Y^2) - (\Sigma Y)^2}}$$

$$r = \frac{(7 \times 476) - (56)(49)}{\sqrt{(7 \times 560\ ) - (56)^2}\sqrt{(7 \times 417) - (49)^2}}$$

$$r = \frac{3332 - 2744}{\sqrt{3920 - 3136}\sqrt{2919 - 2401}}$$

$$r = \frac{588}{\sqrt{788}\sqrt{518}}$$

$$r = \frac{588}{28 \times 22.76}$$

$$r = \frac{588}{637.27} = .923$$

The value of the Pearson r of .923 indicates a very strong positive association between the two variables. Now that the value of the Pearson r has been calculated, there are two additional procedures that we would normally conduct. First we want to test the value of the Pearson r for statistical significance, and second, we want to calculate the value of r^2.

Testing the Pearson *r* for Statistical Significance

The concept of statistical significance has been mentioned in several of the previous chapters, and has been discussed in detail in Chapter 8. As you should recall, the exact interpretation of a "statistically significant result" varies depending on the situation, but is often tied to the fact that we are interested in making inferences about population parameters based on observed sample statistics. In the case of the Pearson r, we are examining the size of the observed correlation between two variables for a sample of size "n" individuals, but we wish to state a conclusion concerning the nature of the correlation that would be found if we had examined data for the entire population. The correlation between the two variables for the entire population is symbolized by the Greek letter (ρ), read as rho, following the convention of using Greek letters to represent population parameters and English letters to represent sample statistics.

If we had data for the entire population, we know that an observed correlation of 0.00 indicates no relationship between the two variables, and an observed correlation of plus or minus 1.00 indicates a perfect relationship between the two variables; however, there are two problems commonly encountered in the actual practice of behavioral science research. First, we usually do not have data for the entire population, and second, most observed Pearson r values will be somewhere in-between the possible extreme values of 0.00 to plus or minus 1.00. These two problems effect the question of statistical significance in the following ways. You should recall from Chapter 2 that a single population can be the source for an incredibly large number of samples, and that a great deal of variation can exist among some of these samples even though they come from the same population. As a result, a correlation of a particular size based on the observation of a single sample may or may not be a good indicator of the "true" correlation that exists in the total population. In other words, reaching a conclusion regarding the nature of a population based on the observation of a sample always introduces some risk or probability of error.

An additional problem stems from the fact that a particular sample's observed correlation does not result in a value of either 0.00 indicating no correlation at all, or a value of plus or minus 1.00 indicating a perfect correlation. The issue here is how do we interpret a correlation that lies somewhere in-between the possible extreme values? For example, an observed correlation coefficient for a particular sample may not be equal to exactly 0.00, but may be only slightly different than 0.00. Would a correlation coefficient of .02 indicate the presence of a meaningful association between two variables, or would we be more correct to regard a correlation of .02 as not being substantially different than 0.00? What about an observed correlation coefficient of .09, or .15, or .23? The key question becomes, "How much larger than 0.00 does an observed correlation coefficient between two variables in a sample have to be before we are willing to assume that the actual correlation between the two variables in the population is not equal to 0.00?" It is precisely this question that is the subject of the interpretation of a test of statistical significance for a Pearson correlation coefficient.

The question that we are examining when we test a Pearson correlation coefficient for statistical significance is simply, "Is the observed correlation between two variables in the present sample different enough from 0.00 to suggest that a similar correlation exists between the two variables in the total population?" If we come to the conclusion that a particular correlation coefficient is statistically significant, it simply means that we have good reason to believe that a similar correlation would exist between the two variables in the total population; a statistically significant result means nothing more than that. It does not mean that the underlying relationship between the two variables is of some theoretical significance. The actual process of testing a Pearson r correlation coefficient for statistical significance is extremely easy, and is accomplished through the use of a table of critical values, which is presented in the appendix.

To test a correlation for statistical significance, we will follow the convention of testing the null hypothesis as discussed in the previous chapter. As you should recall, the logic of testing the null hypothesis, or statement of no relationship, is based on the idea that it is easier to disprove a statement than it is to prove a statement. In the case of the Pearson correlation coefficient, the null hypothesis that we test is the statement that rho, the correlation coefficient between the two variables in the population, is equal to 0.00, or in other words, the null hypothesis is stating that there is no relationship between the two variables in the population. Our corresponding research hypothesis is that there is a relationship between the two variables. If the value of the observed Pearson r (the correlation coefficient calculated from our sample data) is as large or larger than the appropriate critical value from the table, then we will reject the null hypothesis which in turn suggests support for the research hypothesis.

Table 3 in the appendix contains the critical values for the Pearson correlation coefficient for both a one-tail and a two-tail test of the null hypothesis at both the .05 and .01 levels of significance for selected degrees of freedom. A two-tail test of the null hypothesis is appropriate when the research hypothesis does not suggest a direction for the expected relationship between the two variables, and a one-tail test of the null hypothesis is appropriate when the research hypothesis does suggest a direction for the expected relationship between the two variables.

In our first example, we are examining the relationship between two unidentified variables that we have referred to simply as "*X*" and "*Y*." We certainly have no reason to expect any type of direction in the relationship, so a two-tailed test of the hypothesis would be appropriate in this case. By the way, do not make the mistake of thinking that a one-tail test of the hypothesis would be in order here because you can see evidence of a positive relationship in the scatterplot. In the actual practice of behavioral science research, the research hypotheses are stated prior to examination of the data. Expectations regarding the direction of a relationship are based on our reviews of previous research in the area under study, or on the basis of a logical argument.

Testing the null hypothesis of no relationship between the two variables is as simple as comparing the size of our observed result to a critical value from the table. Now that we have determined that a two-tailed test is in order we are ready to identify the critical value from Table 9.3. Conclusions based on statistical testing always involve a risk of being wrong due to variability that can exist from sample to sample from a given population. The **level of significance**, or alpha level, is the risk of being wrong for a given test. Table 9.3 contains critical values for the two more commonly used alpha levels, the .05 and the .01 levels of significance. A test of statistical significance at the .05 level indicates that our rejection of the null hypothesis has a .05 or 5% chance of being wrong, and a test of statistical significance at the .01 level indicates that our rejection of the null hypothesis has a .01 or 1% chance of being wrong. The final step in identifying the proper critical value from the table is calculating the appropriate degrees of freedom. As you may recall from Chapter 5, we lose one degree of freedom for each population parameter we are estimating. With the Pearson *r* we lose two degrees of freedom, one for the "*X*" variable and one for the "*Y*" variable. Degrees of freedom then are equal to our sample size ($n - 2$). The proper critical value is identified by moving to the appropriate row of the table based on the degrees of freedom, and then moving to the appropriate column depending on which level of significance we have chosen.

Brief Review of the Steps in Testing a Pearson *r* for Statistical Significance

1. Select the two variables to examine.
2. State the research hypothesis and the null version of the hypothesis.
3. Determine if a one-tail or a two-tail test of significance is in order based on the presence or absence of directionality in the research hypothesis.
4. Select a level of significance at which to conduct the test.
5. Identify the critical value from the table using ($n - 2$) degrees of freedom.
6. Compute the Pearson correlation coefficient for the sample.
7. Compare the observed value of the Pearson *r* (what you have computed) to the critical value (from the table).
8. Reach a conclusion regarding the null hypothesis: for a two-tailed test reject the null hypothesis if the absolute value of the observed Pearson *r* is as large or larger than the critical value from the table; for a one-tailed test reject the null hypothesis if the absolute value of the observed Pearson *r* is as large or larger than the critical value from the table and if the sign of the observed Pearson *r* is consistent with the expectation stated in the research hypothesis.

We would proceed through these steps as follows using the data from our first example. Since we have no reason to expect any direction in the relationship between "*X*" and "*Y*," our research hypothesis would be as follows:

H_1: The population correlation coefficient rho between *X* and *Y* is not equal to zero.

The null version of the hypothesis would then be:

H_0: The population correlation coefficient rho between *X* and *Y* is equal to zero.

Oftentimes, the research and null hypotheses will be stated in symbolic terms such as:

$$H_1 : \rho \neq 0$$
$$H_0 : \rho = 0$$

A two-tail test of the null hypothesis is in order in this case because there is no suggestion of a direction for the correlation expected between "*X*" and "*Y*" in the research hypothesis. We will arbitrarily choose the level of significance, $\alpha = .01$ to test the null hypothesis. The table provides critical values at both the .05 and .01 level, and both are perfectly acceptable levels of significance. Some statistical tables may provide critical values for levels of significance ranging from a low of .001 (a probability of 1 in 1,000 of being incorrect in rejecting the null hypothesis) to a high of .10 (a probability of 1 in 10 of being incorrect in rejecting the null hypothesis). It is highly unlikely that you will ever see an alpha level above the .10 level actually used in behavioral science research.

Our sample size in this example is $n = 7$, so the degrees of freedom are df = 5. Turning to the table of critical values for a two-tail test, we will find the appropriate critical value at the point where the row representing df = 5 intersects with the column for the .01 alpha level. The appropriate critical value in this case is plus or minus .874.

Note that the Pearson *r* table of critical values only lists the positive critical values. Remember that a two-tail test always has a critical region in each tail of the sampling distribution curve with a positive critical value marking the beginning of the critical region to the right, and a negative critical value marking the beginning of the critical region to the left. A one-tail test only has a single critical region that may be located in either the positive side or the negative side of the curve. When predicting a negative correlation you should assign the proper sign to the critical value.

Our observed Pearson correlation coefficient is $r = .923$ which is larger than the critical value, so the correct decision is to reject the null hypothesis and to accept the research hypothesis. The Pearson correlation coefficient is statistically significant at the $\alpha = .01$ level. Another way to interpret the statistically significant result is to say that there is only one chance in 100 that we would observe a sample where "*X*" and "*Y*" would appear to be correlated as they are in this case when in fact there is no correlation between the two variables in the population.

Computation and Interpretation of r^2

Now that we have determined that the observed Pearson correlation coefficient is statistically significant, we are ready to proceed to the last step in the process, which is the computation and interpretation of the r^2 value. Squaring the Pearson correlation coefficient provides us with some very important information. The computation is quite easy, we simply square the value of the observed Pearson correlation coefficient; or in this case $r^2 = (.923)^2 = .852$.

The resulting value, .852, indicates the proportion of the variance in "*Y*," the dependent variable, that can be explained or accounted for by "*X*," the independent variable. In this case, we see that .852 or 85.2% of the variance in "*Y*" can be explained by "*X*." Explaining variance is an important concept in statistical analysis, and it is one that was introduced in Chapter 5 on "Measures of Variation." You may recall the game where we were trying to guess the score on an examination paper drawn at random from an entire set of papers. (As I recall I had made a grand prize available that might have been a trip to lovely Cabo San Lucas. Unfortunately, we had no winners.) At any rate, what we discovered was that in the absence of any other information, the best strategy you can adopt when trying to predict the value of a dependent variable for a sample of individuals is to always guess the observed mean. We also saw that in some cases knowing the value of an individual's score on a related independent variable prior to predicting their value on the dependent variable can sometimes improve your prediction. It is this improvement in being able to predict a dependent variable by having knowledge of an independent variable that we refer to as explaining variance.

For any given dependent variable the total variance available to be explained is 1.00 or 100%. In our first example, we were able to explain 85.2%, which is quite good. In many ways it is the amount of variance explained in a relationship that is of more interest than the fact that a given Pearson correlation coefficient is statistically significant. There is a subtle, but important, point to be made here. If a Pearson correlation coefficient is not statistically significant, then we fail to reject the null hypothesis that states that the population correlation coefficient is equal to zero. If we cannot reject the null hypothesis, then it does not matter how large the r^2 value is. A nonsignificant Pearson r indicates that there is no evidence of a relationship between "X" and "Y," so we cannot reliably use knowledge of "X" to help us predict the dependent variable, "Y." However, we might find that an observed Pearson correlation is statistically significant, but that does not necessarily mean that the independent variable will be of much help in predicting the dependent variable. A statistically significant result is not always an important result!

How can a correlation coefficient be statistically significant, but not be very important?

Suppose that a Pearson correlation coefficient is computed for two variables, "X" and "Y" for a sample of $n = 47$ individuals, and that the resulting Pearson r is computed to be $r = .30$. Assuming a two-tailed test of the null hypothesis at an alpha level of $\alpha = .05$ and degrees of freedom df = 45 would result in a critical value of plus or minus .288. The absolute value of our observed Pearson correlation of $r = .30$ is larger than the critical value, so we would reject the null hypothesis and conclude that we have a statistically significant result. However, statistical significance does not tell us that the relationship is important, it only tells us that the relationship is a real one that is likely to exist in the population from which this sample was drawn (subject to the risk of being wrong 5 times out of 100 in this case). It is the r^2 value that really indicates the importance of the relationship, and in this case the r^2 value, $(.30)^2$, is only .09 indicating that only 9% of the variance in "Y" can be explained by the independent variable "X." A 9% improvement in predictive ability is better than nothing, but a 9% reduction in variance for "Y" would not be viewed as a major improvement.

You should keep in mind that statistical significance and the amount of variance explained are two completely different issues. They are related only in the sense that a correlation coefficient that is not statistically significant has no predictive ability. A finding of statistical significance just means that a particular result observed in the sample provides evidence of a similar relationship existing in the population (subject to the risk of error associated with the size of the alpha level, or level of significance, that we have chosen). Evidence of a statistically significant relationship is best thought of as a real relationship, but not necessarily an important one. Statistical significance is a necessary condition for a relationship to be important, but it is not a sufficient condition. To reinforce the point, you might want to scan the various critical values for the Pearson r correlation coefficient provided in the table. One of the things that you should notice is that the critical values become smaller as the sample size becomes larger. The reason is quite simple; as you select a larger and larger sample from a population, you have a smaller and smaller chance of making an error when using the sample statistics as estimates of the population parameters (assuming that the samples are selected in a random manner). In fact, when the sample equals the size of the population itself, then the issue of statistical significance becomes meaningless. There is no sampling error and the observed correlation between two variables is real. The correlation still may not be important, but if it is based on the observation of the entire population then there is no question that it exists in reality.

Some General Guidelines for the Interpretation of Correlation Coefficients

In your own research you will be in a position to compute the appropriate correlation coefficient, test the resulting value for statistical significance, and interpret the strength of the relationship by computing the r^2 value. In other situations, you may be reviewing reports of the research done by others, and the table below presenting some general guidelines for the interpretation of correlation coefficients may be of use.

Table 9.3 Guidelines for the Interpretation of Correlation Coefficients

$\pm r$	Strength of Relationship	r^2 Percent of Variance Explained
.75–1.00	Strong	56%–100%
.50–.74	Moderate to High	25%–55%
.25–.49	Low to Moderate	6.3%–24%
.00–.24	Weak	.0%–5.8%

A Pearson *r* Example Using Education and Income

The next example of a statistical analysis using the Pearson correlation coefficient involves the variables of education and income, which are frequently examined in behavioral science research. In this case, we want to calculate the Pearson correlation coefficient between the independent variable education in years and the dependent variable income in thousands of dollars. Income is reported in thousands of dollars rather than dollars to make the calculations less tedious, but the size of the Pearson *r* would be the same in either case. Some hypothetical data for a sample of $n = 20$ individuals are presented below along with the intermediate calculations we will need to compute the Pearson *r*. As an initial step, we will display the data in a scatter plot to visually look for evidence of a linear trend in the data.

Table 9.4 Data on Education and Income

Education (in years)	Income (×1,000)	XY	X^2	Y^2
8	15	120	64	225
12	20	240	144	400
16	40	640	256	1600
10	15	150	100	225
12	25	300	144	625
8	15	120	64	225
8	10	80	64	100
10	10	100	100	100
12	20	240	144	400
12	15	180	144	225
18	50	900	324	2500
16	45	720	256	2025
18	25	450	324	625
20	85	1700	400	7225
20	90	1800	400	8100
12	30	360	144	900
22	60	1320	484	3600
16	50	800	256	2500
16	45	720	256	2025
14	35	490	196	1225
$\Sigma X = 280$	$\Sigma Y = 700$	$\Sigma XY = 11430$	$\Sigma X^2 = 4264$	$\Sigma Y^2 = 34850$

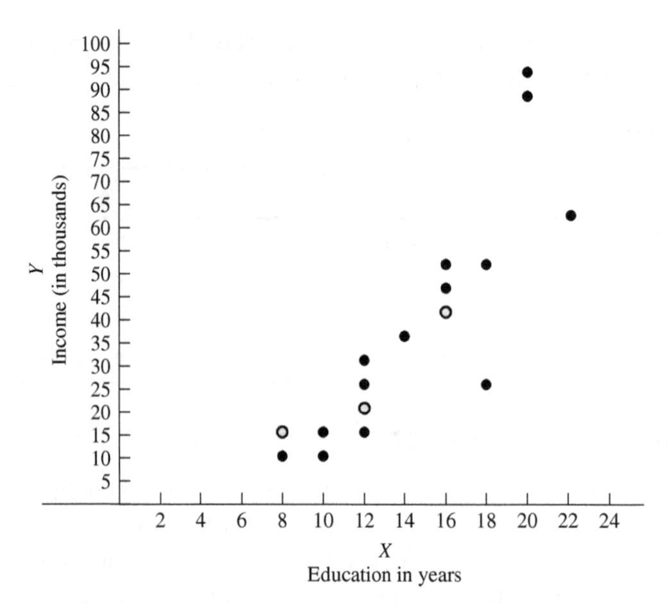

Figure 9.6 Scatterplot of Income and Education

The data have a clear linear trend, and there is evidence of a strong positive relationship. We will calculate the value of the Pearson r by substituting the appropriate values from the intermediate calculations into the formula below.

$$r = \frac{(n \times \Sigma XY) - (\Sigma X)(\Sigma Y)}{\sqrt{(n \times \Sigma X^2) - (\Sigma X)^2}\sqrt{(n \times \Sigma Y^2) - (\Sigma Y)^2}}$$

$$r = \frac{(20 \times 11430) - (280)(700)}{\sqrt{(20 \times 4264) - (280)^2}\sqrt{(20 \times 34850) - (700)^2}}$$

$$r = \frac{228600 - 196000}{\sqrt{85280 - 78400}\sqrt{697000 - 490000}}$$

$$r = \frac{32600}{\sqrt{6880}\sqrt{207000}}$$

$$r = \frac{32600}{82.95 \times 454.97}$$

The final step is to divide our numerator by the denominator to calculate the Pearson r value for the data.

$$r = \frac{32600}{37739.76} = .864$$

The Pearson r value of .864 appears to be reasonably strong, but we will need to test the correlation for statistical significance. In this case, we will perform a one-tail test of the hypothesis since we would reasonably expect a positive relationship to exist between education and income. We will technically be testing the null form of the hypothesis, which is a statement of no relationship, and I have arbitrarily chosen the $\alpha = .01$ level of significance. The research and null versions of the hypothesis are stated symbolically below.

$$H_1 : \rho > 0$$
$$H_0 : \rho = 0$$

The appropriate critical value obtained from the table for a one-tail test with df = 18, and $\alpha = .01$ is +.516. Since our observed Pearson *r* value of .864 is as large or larger than the critical value, we will reject the null hypothesis and accept the research hypothesis.

Our final step is to compute the r^2 value and to interpret the result.

$$r^2 = (.864)^2 = .75$$

The resulting r^2 value indicates that education explains 75% of the variance in income. Or stating the result the way we have previously, if we wanted to predict the income for a particular member of a sample and we knew nothing else, then our best strategy would be to guess the mean income of the sample. Doing so would result in some error or variance; however, if we are given information on the person's educational level prior to guessing, we can reduce the error in our prediction by 75%. Just exactly how we would use the information on educational level to better predict income is one of the major topics in the next chapter on linear regression.

The Point-Biserial Correlation Coefficient

The second correlation coefficient we will be examining is called the point-biserial correlation coefficient, r_{pb}. The point-biserial correlation coefficient is appropriate in cases where the dependent variable is measured at either the interval or ratio level, and the independent variable is a nominal level dichotomy, or a variable that can be converted to a dichotomy as was discussed earlier. A classic application of the point-biserial correlation coefficient would be examining data for a relationship between sex and income. In this case, the independent variable, sex, is a true nominal level dichotomy, and the dependent variable, income, can be measured at the ratio level. A situation where we do not actually have an independent variable that is a true nominal level dichotomy, but where we can convert the independent variable to a dichotomy would be to apply the point-biserial correlation coefficient to search for a relationship between undergraduate classification and grade point average. In almost all cases, undergraduate classification is measured as an ordinal level variable with four possible attributes: freshman, sophomore, junior, or senior; however, we could dichotomize the variable by creating two categories of under classmen and upper classmen, or freshman and nonfreshman, or any other classification scheme that results in two mutually exclusive categories.

The Logic of the Point-Biserial Correlation Coefficient

The logical basis of the point-biserial correlation coefficient is quite simple. Just as was the case with the Pearson *r*, with the point-biserial we are most interested in being able to explain variation in some dependent variable, which has been measured at the interval or ratio level. However, unlike the Pearson *r*, our independent or predictor variable is not a similarly measured interval or ratio level variable, but is instead a nominal level dichotomy—or in more direct language, the independent variable can be thought of as group membership in one of two possible groups. In the case of trying to predict income, we would be in the usual situation of using the sample's mean income as our best guess (the guess that would give us the least amount of variance) in the absence of any other information. The point-biserial correlation coefficient will provide us with a measure of how much variance in the dependent variable we can explain by knowing to which of two groups an individual belongs. In the case of sex and income, we would be measuring how much variance in income can be explained by knowing the person's sex prior to making our best guess on income.

There are several similarities between the point-biserial correlation coefficient r_{pb} and the Pearson correlation coefficient r. First, the range of the point-biserial correlation coefficient is from −1.00 to +1.00, just as that of the Person r that we have just examined.

$$\text{Range of the Point-Biserial } r_{\text{pb}} = (-1.00 \leq r_{\text{pb}} \leq 1.00)$$

Second, the interpretation of the value of the observed r_{pb} is similar to that of the Pearson r with a value of plus or minus 1.00 indicating a perfect relationship between the independent and dependent variables, and an observed value of 0.00 indicating no relationship between the independent and the dependent variables. However, unlike the Pearson r, the sign of the point-biserial correlation coefficient does not indicate anything about the direction of the relationship between the independent and dependent variables. In a nominal level variable, there is no underlying dimension of measurement among the various groups, so the order of the groups is rather arbitrary. As a result, the sign of the point-biserial correlation coefficient is a function of which group we arbitrarily call Group 1 and which group we arbitrarily call Group 2. If we reverse the order of the two groups, we will also reverse the sign of the point-biserial correlation coefficient.

There are two additional similarities between the point-biserial correlation coefficient and the Pearson r. The point-biserial correlation coefficient may be tested for statistical significance in the same manner as the Pearson r, and by using the same table of critical values as the Pearson r. In the case of the point-biserial correlation, we will use the total sample size (the number of individuals in Group 1 plus the number of individuals in Group 2) – 2 in determining the degrees of freedom. Finally, computing the square of the point-biserial correlation coefficient r^2_{pb}, will provide us with a measure of the amount of variance in the dependent variable that can be explained by the independent variable.

The Formula for the Point-Biserial Correlation Coefficient

The formula for the point-biserial correlation coefficient is as follows:

$$r_{\text{pb}} = \frac{\overline{Y}_1 - \overline{Y}_2}{S_y} \times \sqrt{p \times q}$$

where

$\overline{Y}_1$ is the mean of the dependent variable for Group 1,
$\overline{Y}_2$ is the mean of the dependent variable for Group 2,
S_y is the standard deviation of the dependent variable for the entire sample (Group 1 and Group 2),
p is the proportion of the sample in Group 1 (the number of people in Group 1 divided by the total sample size),
q is the proportion of the sample in Group 2 (the number of people in Group 2 divided by the total sample size).

Note: Because the entire sample must be a member of either Group 1 or Group 2, the values of "p" and "q" will equal 1.00 or 100% of the total sample.

To illustrate the computation and interpretation of the point-biserial correlation coefficient, we are going to analyze some data that we saw previously in Chapter 5 on measures of variation. The following data on the incomes of college and noncollege graduates were used to introduce you to the very important concept of "explaining variance."

We used these data to show that the total variance of the dependent variable "income" was $S^2 = 365.65$, and that one way to think about variance was that it represented the average amount of squared error you would have when using the sample's mean income value to predict the income of all of the individuals in the sample. We also saw that using the mean as the prediction was the best strategy, that is, there was no other income value you could choose as a prediction that would result in a lower amount of average squared error

Table 9.5 Data on College Graduate Status and Income

College Graduate	Income (× 1,000)	
YES	30	
NO	20	College Graduate's Mean = 43.4
YES	50	Noncollege Graduate's Mean = 15.6
NO	15	S^2 Variance of Income = 365.65
NO	12	
YES	75	
YES	40	
NO	15	
YES	22	
NO	16	

as long as no other information about the sample was available to you. However, we also saw that if you knew if a member of the sample was a college graduate or not prior to predicting income that you could do much better in your prediction. If you used the college graduate's mean as your prediction when you knew the sample member was a college graduate, and the noncollege graduate's mean when you knew the sample member was not a college graduate, then your overall prediction, the average amount of squared error, was reduced from 365.65 to 172.44. The reduction of 193.21 points of variance (365.65 – 172.44) represented a reduction of .528 or 52.8% of the variance in income when using college graduate or noncollege graduate status as a predictor variable. One final word about these data from Chapter 5, you might recall that this entire process of demonstrating the amount of variance in income that can be explained by college graduate or noncollege graduate status was quite tedious and time consuming. Now we are going to accomplish the same thing in a much easier fashion by making use of the point-biserial correlation coefficient.

Computing the Point-Biserial Correlation Coefficient

$$r_{pb} = \frac{\overline{Y}_1 - \overline{Y}_2}{S_y} \times \sqrt{p \times q}$$

College graduate mean $\overline{Y}_1 = 43.4$

Noncollege graduate mean $\overline{Y}_2 = 15.6$

Standard deviation of income $S_y = 19.122$

(S_y is obtained by taking the square root of the variance, S^2 which was given)

p the proportion in Group 1 = 5/10 = .50

q the proportion in Group 2 = 5/10 = .50

$$r_{pb} = \frac{43.4 - 15.6}{19.122} \times \sqrt{.5 \times .5}$$

$$r_{pb} = \frac{27.8}{19.122} \times \sqrt{.25}$$

$$r_{pb} = 1.454 \times .5$$

$$r_{pb} = .727$$

Testing the Point-Biserial Correlation Coefficient for Statistical Significance

The observed point-biserial correlation coefficient r_{pb} of .727 can be tested for statistical significance just as the Pearson r was by finding the appropriate critical value from the Pearson r tables. The null hypothesis for the point-biserial correlation coefficient assumes no relationship between the independent and dependent variables (in the population from which our sample was selected), and a one-tail or two-tail test of the null hypothesis may be in order depending on the presence of a statement of the direction of the relationship in the research hypothesis.

In the case of examining the relationship between income and college graduate status we would reasonably expect that college graduates would have higher incomes than noncollege graduates, and that this difference would help us better predict income level. The issue of direction of the point-biserial correlation can be a little confusing, and is not so easily interpreted as the direction of the Pearson r, where the direction of the relationship and the sign of the observed correlation coefficient are related. With the point-biserial correlation coefficient, the sign of the observed correlation coefficient is simply a matter of which group we arbitrarily decided to call Group 1 and which group we arbitrarily decided to call Group 2. In our example, we have used the college graduates as Group 1, so our research and null hypotheses concerning the population correlation coefficient rho (ρ pb) would be stated as follows:

$$H_1 : \rho_{pb} > 0$$

$$H_0 : \rho_{pb} = 0$$

At this point, we would proceed just as we did earlier with the statistical test of the Pearson r. We will obtain a critical value from the appropriate section of the Pearson r tables for either a one-tail or two-tail test of the hypothesis using $N-2$ degrees of freedom (where N represents the total sample size of Group 1 and Group 2) for our desired level of significance. In this case, we will use a level of significance of $\alpha = .01$, with eight degrees of freedom for a one-tail test, and find that the appropriate critical value is $r_{pb\alpha} = .716$.

Our observed point-biserial correlation coefficient of .727 is as large or larger than the critical value, so we will reject the null hypothesis and accept support for the research hypothesis. Again, a finding of statistical significance when testing the correlation coefficient just tells us if we have reason to believe that a relationship similar to what we observe in the sample is also likely to be found in the population from which it was drawn. It is still the amount of variance in the dependent variable that can be explained by the dependent variable that is of more importance. It is the amount of variance explained that tells us how important the relationship is.

To determine the amount of variance explained in the dependent variable by the independent variable, we simply square the value of the obtained point-biserial correlation coefficient. The procedure is the same as that for the Pearson r, and the interpretation is the same as well.

$$r_{pb}^2 = (.727)^2 = .528$$

Squaring the value of the observed point-biserial correlation coefficient yields a value of .528, indicating that 52.8% of the variance in income can be accounted for by knowing if a given individual is a college graduate or not. Incidentally, this is the same result we obtained in Chapter 5 when we used the more tedious process of using the appropriate group mean as our prediction for income rather than the overall group mean, but the point-biserial correlation provides us with a much simpler solution.

Using the Point-Biserial Correlation Coefficient to Measure the Relationship between Sex and Physical Dexterity

A second example of the point-biserial correlation coefficient will examine data on sex and physical dexterity. The scores on a physical dexterity test will be examined for a total sample of $n = 50$ males and females to see if sex can help predict physical dexterity. We have no reason to hypothesize an advantage on physical dexterity for one sex or the other, so our research hypothesis will reflect a two-tail test, which we will test at the $\alpha = .05$ level.

$$H_1 : \text{pb} \neq 0$$
$$H_0 : \text{pb} = 0$$

$$r_{\text{pb}} = \frac{\bar{Y}_1 - \bar{Y}_2}{S_y} \times \sqrt{p \times q}$$

Male mean score on dexterity test $\bar{Y}_1 = 77.4$
Female mean score on dexterity test $\bar{Y}_2 = 80.2$
Standard Deviation of dexterity test $S_y = 5.95$
p the proportion in Group 1 (male) = 20/50 = .40
q the proportion in Group 2 (female) = 30/50 = .60

$$r_{\text{pb}} = \frac{77.4 - 80.2}{5.95} \times \sqrt{.4 \times .6}$$
$$r_{\text{pb}} = \frac{-2.8}{5.95} \times \sqrt{.24}$$
$$r_{\text{pb}} = .471 \times .49$$
$$r_{\text{pb}} = -.231$$

As before, the next step is to test the observed correlation coefficient for statistical significance. We have a two-tail test situation with an $\alpha = .05$, and degrees of freedom ($df = 50 - 2$) of 48. The appropriate critical value taken from the Pearson r tables is $r_{\text{pb}\alpha} = \pm.288$. Notice that in the absence of a critical value for exactly 48 degrees of freedom in the table, we must choose between using the critical value for 45 degrees of freedom or 50 degrees of freedom. Even though our actual degrees of freedom are closer to 50, we use the critical value associated with 45 degrees of freedom because that is the more conservative choice. The critical value associated with $df = 45$ is more conservative because it is larger than the critical value associated with $df = 50$, thus making it harder to reject the null hypothesis.

A comparison of the absolute value of our observed point-biserial correlation coefficient ($r_{\text{pb}} = -.231$) to the critical value from the table ($r_{\text{pb}\alpha} = \pm.288$) indicates that the correlation coefficient is not statistically significant at the $\alpha = .05$ level. A correlation between sex and physical dexterity of the size we observed in our sample of $n = 50$ could easily be the result of sampling error. Further, even if the observed correlation coefficient had been statistically significant, it would not have provided evidence of a very important relationship because the r^2_{pb} value of .053 indicates that only approximately 5% of the variation in physical dexterity could be accounted for by the sex of the individual.

The Spearman Rank Order Correlation Coefficient

The final bivariate correlation coefficient we will examine in this chapter is the Spearman rank order correlation coefficient r_s (also known as Spearman's rho) which is appropriate in situations involving two ordinal level variables. The steps involved in applying the Spearman rank order correlation coefficient are similar to what we have done previously with the Pearson r and the point-biserial r_{pb} in that we will compute the test statistic, and then test the value for statistical significance using a critical value obtained from the appropriate table. Both a one-tail and a two-tail test of the hypothesis may be conducted depending on the presence or absence of directionality in the research hypothesis. Finally, as with the other two correlation coefficients we have examined, the range of the Spearman r_s is between 0.00 and plus or minus 1.00, with a value of 0.00 indicating no relationship between the independent and dependent variables and a value of plus/minus 1.00 indicating a perfect relationship.

$$\text{Range of the Spearman } r_s = (-1.00 \le r_s \le 1.00)$$

One major difference between the Spearman r_s and the previous two correlation coefficients we have examined is that because the dependent variable is at the ordinal level it is not appropriate to compute a variance, and as a result, there is not a "variance explained" interpretation with the Spearman r_s.

The Logic of the Spearman Rank Order Correlation Coefficient

The Spearman r_s is appropriate in situations where both the independent and dependent variables are measured at the ordinal level, and are expressed as rank order values. For example, to use the variable "classification" with the categories of freshman, sophomore, junior, or senior in a Spearman r_s computation would require the assignment of a rank order to each of the four categories such as 1, 2, 3, or 4. The value of the correlation coefficient is based on the degree of association of the ranks on the independent variable with the ranks on the dependent variable. A Spearman r_s value of 1.00 would indicate that the highest rank on the dependent variable is associated with the highest rank on the independent variable, the second highest rank on the dependent variable is associated with the second highest rank on the independent variable, and so on through to the lowest rank on the dependent variable being associated with the lowest rank on the independent variable, or a perfect positive association. A Spearman r_s value of 0.00 would indicate a random association of ranks between the independent and dependent variable, or no association. A Spearman r_s value of –1.00 would indicate that the highest rank on the dependent variable is associated with the lowest rank on the independent variable, and so on through to the lowest rank on the dependent variable being associated with the highest rank on the independent variable, or a perfect negative association.

The Formula for the Spearman Rank Order Correlation Coefficient

The formula for the Spearman rank order correlation coefficient is as follows:

$$r_s = 1 - \frac{6 \times \Sigma d^2}{n(n^2 - 1)}$$

where

1 is a constant,
6 is a constant,
Σd^2 is the sum of the squared difference between ranks on the independent and dependent variables,
n is the sample size, and
$(n^2 - 1)$ is the square of the sample size minus 1.

Note in the computations that the value of 6 times the sum of the differences squared is divided by the product of "n" times $(n^2 - 1)$, and it is that result that is subtracted from the constant "1" in the formula. The value "1" is not part of the numerator, but is instead a separate term.

Computing the Spearman Rank Order Correlation Coefficient

The computation of the Spearman r_s is relatively simple and straight forward, and will be demonstrated with a typical situation where one might want to apply the statistic. Data are presented below for a sample of $n = 8$ individuals with the "X" or independent variable representing rank on job satisfaction and the "Y" or dependent variable representing rank on worker productivity. The worker with the highest level of job satisfaction is ranked "1," the worker with the second highest level of job satisfaction is ranked "2," and so on through to the least satisfied worker ranked "8." Similarly, the most productive worker is ranked "1," the second most productive worker is ranked "2," and so on through to the least most productive worker ranked "8." In this case, we might reasonably hypothesize that the more satisfied a worker is with his or her job the more productive that the worker will be. Because direction is implied, we are expecting a positive correlation between the two ranks, a one-tail test of the hypothesis is in order. Both the research and null hypotheses are presented below in symbolic form, and we will test the null hypothesis for statistical significance at the $\alpha = .05$ level of significance.

$$H_1 : \rho_s > 0$$
$$H_0 : \rho_s = 0$$

The computation begins by obtaining the difference in rank between the independent and dependent variables. With this particular statistic the sign of the difference between the ranks is not important because we will be squaring the difference, and the resulting values will all be positive. However, it is a good idea to get into the habit of subtracting in a consistent manner: either always subtract the "Y" variable rank from the "X" variable rank, or always subtract the "X" variable rank from the "Y" variable rank and be careful to assign the proper sign to the difference. In the examples here, I will be subtracting the "X" variable rank from the "Y" variable rank.

Table 9.6 Data on Rank on Job Satisfaction and Rank on Worker Productivity

X *Rank on Job Satisfaction*	*Y* *Rank on Worker Productivity*	*d* *difference between ranks*	*d*² *difference squared*
1	2	1	1
2	3	1	1
3	1	−2	4
4	4	0	0
5	8	3	9
6	7	1	1
7	5	−2	4
8	6	−2	4
			$\Sigma d^2 = 24$

$$r_s = 1 - \frac{6 \times \Sigma d^2}{n(n^2 - 1)}$$

$$r_s = 1 - \frac{6 \times 24}{8\ (64 - 1)}$$

$$r_s = 1 - \frac{144}{504}$$

$$r_s = 1 - 0.29$$

$$r_s = .71$$

Testing the Spearman Rank Order Correlation Coefficient for Statistical Significance

Testing the observed Spearman r_s value of 0.71 for statistical significance follows the same general procedure as before with the Pearson r, but the Spearman correlation coefficient has its own table and critical values. As before, we will turn to the appropriate section of the table depending on whether a one-tail or two-tail test of the hypothesis is in order, and we have a choice of either an $\alpha = .05$, or $\alpha = .01$ level of significance. The only major difference between the Spearman r_s table of critical values and the Pearson r table that we used previously is that the Spearman table does not require any computation of degrees of freedom; we find the appropriate critical value by choosing the row of the table representing the sample size "n," and the column representing the appropriate level of significance.

In our example, we are performing a one-tail test at the $\alpha = .05$ level of significance with a sample size "n" = 8. Turning to the table of critical values for the Spearman r_s, we find the appropriate critical value is $r_{s\alpha} = .643$. Our observed Spearman correlation coefficient of $r_s = .71$ is larger than the critical value so we will reject the null hypothesis, and claim support for the research hypothesis. There is a statistically significant relationship between rank on job satisfaction and rank on worker productivity at the $\alpha = .05$ level of significance. Again, a finding of statistical significance in this case just means that based on our observation of $n = 8$ individuals, we have sufficient reason to believe that a similar relationship between job satisfaction and worker productivity would be found in the population from which our sample was drawn. Beyond that there may or may not be any special significance to the relationship. Also note that with the Spearman r_s we do not have basis for a reduction in variance interpretation as we did with our previous correlation coefficients.

An Example of the Spearman r_s Involving Tied Ranks

One common problem encountered in the application of a Spearman r_s correlation coefficient is a situation where two or more cases in the sample have the same value on one or both of the variables resulting in a tie at a particular rank value. While it is possible to compute a Spearman r_s with data containing tied ranks, it is generally a better idea to resolve the tied score prior to computing the correlation. Resolving tied ranks is very easy, and just involves assigning the mean rank for the scores tied at a particular rank. For example, if two cases in the sample are tied at rank "1," they are in reality occupying ranks "1" and "2," and the solution is to assign each case the mean value of the ranks they are occupying (1.5 in this case). If three cases were tied at rank "5," they would be occupying the ranks "5," "6," and "7." We would resolve the tie by assigning each of the three cases the rank of "6."

You might wonder at this point what difference it makes if the Spearman r_s were calculated with all three cases having the tied rank of "5" as opposed to resolving the tied ranks and assigning the three cases the "resolved" rank of "6." Three cases tied at "5" or three cases tied at "6," what difference does it make?

The difference is that because the Spearman r_s is based on the squared difference between ranks, having several scores tied at a lower rank value (in this case "5" rather than "6") will result in an artificially inflated value for the correlation coefficient. Conclusions based on sample statistics rather than population parameters always carry a risk of error, and we do not want to go out of our way to add to the risk of being wrong. For this reason, we adopt a conservative strategy in statistical analysis wherever possible, and in the case of tied ranks it is more conservative (i.e., it will make it harder for us to reject the null hypothesis) to resolve tied ranks.

An Example of the Spearman Correlation Coefficient r_s with Rank on Physical Attractiveness and Rank on Perceived Intelligence

This next example involves two variables that are occasionally encountered in behavioral science research: physical attractiveness and perceived intelligence. The major research question is, "Does physical attractiveness have an effect on an individual's perceived intelligence?" A typical research strategy to investigate this question might focus on children, and a researcher would take photographs of a sample of children and have an independent group of judges rank the children on their perceived intelligence. If our sample consisted of $n = 9$ children, then the judges would assign a rank of "1" to the child perceived as most intelligent, "2" to the child perceived as second most intelligent, and so on through to "9" to the child perceived as being the least intelligent. The photographs would then be shown to a second panel of judges who would rank the children on physical attractiveness with a rank of "1" for the child perceived as most attractive and so on through to a rank of "9" for the child perceived as least attractive. The final ranks for the children on each variable would be determined by averaging the ranks from each judge. Such a procedure will often result in tied ranks which should be resolved prior to the computation of the Spearman r_s. To illustrate the effect that unresolved tied ranks can have on the size of a Spearman r_s correlation coefficient, the final example of the Spearman r_s correlation coefficient will be computed first without resolving the tied ranks, and then the analysis will be repeated with the tied ranks resolved.

Our research hypothesis is that physical attractiveness will have an effect on perceived intelligence. No direction is implied, so a two-tail test of the hypothesis is in order, which we will test at the $\alpha = .01$ level. Both the research hypothesis and the null hypothesis are stated symbolically below.

$$H_1 : \rho_s \neq 0$$
$$H_0 : \rho_s = 0$$

Table 9.7 Data on Rank on Physical Attractiveness and Rank on Perceived Intelligence

X Rank on Physical Attractiveness	*Y* Rank on Perceived Intelligence	*d* Difference between Ranks	d^2 difference squared
1	3	2	4
2	3	1	1
3	1	−2	4
4	2	−2	4
5	5	0	0
6	7	1	1
7	8	1	1
8	6	−2	4
9	9	0	0
			$\sum d^2 = 19$

$$r_s = 1 - \frac{6 \times \Sigma d^2}{n\ (n^2 - 1)}$$

$$r_s = 1 - \frac{6 \times 19}{9(81 - 1)}$$

$$r_s = 1 - \frac{114}{720}$$

$$r_s = 1 - .158$$

$$r_s = .842$$

Our observed Spearman correlation coefficient of r_s = .842 is based on computations without resolving the two tied ranks occupying positions "3" and "4" on perceived physical attractiveness. We will resolve the tied ranks by assigning a rank of "3.5" to each case, and then compute the value of the Spearman r_s again. Both observed values of the Spearman r_s will then be tested for statistical significance.

Table 9.8 Data on Rank on Physical Attractiveness and Rank on Perceived Intelligence with Tied Ranks Resolved

X Rank on Physical Attractiveness	*Y* Rank on Perceived Intelligence	*d* Difference between Ranks	*d*² Difference Squared
1	3.5	2.5	6.25
2	3.5	1.5	2.25
3	1	−2	4
4	2	−2	4
5	5	0	0
6	7	1	1
7	8	1	1
8	6	−2	4
9	9	0	0
			$\Sigma d^2 = 22.50$

$$r_s = 1 - \frac{6 \times \Sigma d^2}{n(n^2 - 1)}$$

$$r_s = 1 - \frac{6 \times 22.5}{9(81 - 1)}$$

$$r_s = 1 - \frac{135}{720}$$

$$r_s = 1 - .188$$

$$r_s = .812$$

With a sample size of $n = 9$ and a two-tail test of the hypothesis at the $\alpha = .01$ level of significance, we obtain a critical value of $r_{s\alpha} = \pm.833$. Our observed Spearman r_s obtained without resolving the tied ranks was $r_s = .842$ that is greater than the critical value and results in a rejection of the null hypothesis; however, when we resolved the tied ranks prior to computing the Spearman correlation coefficient, we obtained a result of $r_s = .812$. The difference may seem slight, but in the second case with the ties resolved our observed correlation coefficient was not in either critical region and we would fail to reject the null hypothesis.

Using a Correlation Coefficient to Control for the Effects of a Third Variable

Up to this point, we have examined the issue of a causal relationship between two variables, and have seen that demonstrating an association between two variables through the use of one of the three correlation coefficients presented in the chapter is a necessary first step toward establishing evidence of a cause and effect relationship between two variables. However, evidence of an association between two variables, that is, that they move together or covary is not sufficient to prove causality. Often two variables may appear to be related to each other when in fact they are each related to a third variable. It is their relationship to a third variable that makes it appear that the first two variables may be related to each other when in fact they are not. For example, there is a positive association between a woman's age at the birth of her first child and family income. Generally speaking, women who have their first birth at an early age have a lower family income, and women who give birth to their first child at an older age have a higher family income. While there is evidence of an association between the two variables, there is reason to believe that a true cause and effect relationship does not exist, or in the language of research, we have reason to believe that the bivariate relationship between age at first birth and family income is spurious or false. We can attempt to demonstrate that a relationship is spurious in a number of ways, and one common way is to compute a partial correlation coefficient.

The partial correlation coefficient is based on the Pearson r, and provides us with a measure of the association between two variables while controlling for any possible effects from a third variable. In the case of the relationship between age at first birth and family income, a likely third variable that might be related to both would be education. Theoretically we could argue that what appears to be a positive relationship between age at first birth and family income is really due to the fact that women who give birth at an early age are often forced to terminate their education. Women with lower levels of education are likely to have a lower family income. Conversely, women who give birth for the first time at an older age are more likely to have an opportunity to complete their education. Women who have completed their education are likely to have a higher family income. The partial correlation coefficient provides us with a mechanism to provide quantitative support for our theoretical argument; we will be able to examine the relationship between age at first birth and family income while controlling for the effects of education.

The Partial Correlation Coefficient

The partial correlation coefficient indicates the strength of the relationship between two variables while controlling for the effects of a third variable. When examining a bivariate or two variable relationship, we have been symbolizing the independent variable as "*X*," and the dependent variable as "*Y*." We will let the third variable whose effects we wish to control be symbolized by "*Z*." A partial correlation coefficient is typically represented as $r_{yx.z}$, which is read as "the correlation between "*Y*" and "*X*" controlling for "*Z*." In our case, age at first birth is "*X*," family income is "*Y*," and education is "*Z*." To compute the partial

Table 9.9 Pearson r Correlation Matrix

	(X) Age at First Birth	(Y) Family Income	(Z) Education
Age at first birth	1.00	.425	.588
Family Income	.425	1.00	.628
Education	.588	.628	1.00

correlation coefficient, or partial r, we simply need the bivariate correlation coefficients, the Pearson r values, for every combination of the three variables involved: r_{yx}, r_{xz}, and r_{yz}.

Oftentimes, a set of bivariate correlation coefficients is presented in a tabular format, and is referred to as a correlation matrix. The correlation matrix presents the bivariate correlation coefficient for each variable with every other variable, including itself, which is always a "perfect correlation" equal to 1.00. A correlation matrix for the three variables of age at first birth, family income, and education might appear as follows:

Notice that any variable's correlation with itself is 1.00, so the main diagonal in a correlation matrix as you move from the top left corner to the bottom right corner will consist of values of 1.00, where each variable intersects with itself. Also notice that the bottom half of the matrix, the section below the main diagonal, simply repeats the information found in the top half of the matrix above the main diagonal. The duplication is due to the fact that the correlation between "*X*" and "*Y*" is the same as the correlation between "*Y*" and "*X*," or in this case, the correlation between age at first birth and family income is the same thing as the correlation between income and the age at first birth.

Formula for the Partial Correlation Coefficient

To compute the partial correlation coefficient, we will make use of the Pearson correlation coefficients in the top half (or bottom half) of the correlation matrix. The formula for the partial correlation coefficient of "*X*" and "*Y*" controlling for "*Z*" ($r_{yx.z}$) is as follows:

$$r_{yx.z} = \frac{r_{yx} - r_{xz} r_{yz}}{\sqrt{1 - r_{xz}^2}\sqrt{1 - r_{yz}^2}}$$

where

r_{yx} is the Pearson r correlation between the "*X*" (independent) variable and the "*Y*" (dependent) variable,

r_{xz} is the Pearson r correlation between the "*X*" (independent) variable and the "*Z*" (control) variable,

r_{yz} is the Pearson r correlation between the "*Y*" (dependent) variable and the "*Z*" (control) variable,

r_{xz}^2 is the squared Pearson r correlation between the "*X*" and "*Z*" variable, and

r_{yz}^2 is the squared Pearson r correlation between the "*Y*" and "*Z*" variables.

If you examine the denominator of the formula, you will see that it contains terms of the square root of $(1 - r^2)$ for each of the two main variables with the control variable. As we have seen, r^2 represents the amount of variance explained in a bivariate relationship, so the quantity $(1 - r^2)$ represents the amount of variance that is left unexplained. In effect, the partial correlation coefficient is providing us with an estimate of the amount of variation in the independent variable "*Y*" that can be explained by the independent variable "X" after the variation in both "X" and "Y" attributed to the control variable "Z" has been removed. The computation is demonstrated below.

$$r_{yx.z} = \frac{.452 - (.588 \times .628)}{\sqrt{1-(.588)^2}\sqrt{1-(.628)^2}}$$

$$r_{yx.z} = \frac{.452 - .369}{\sqrt{1-.346}\sqrt{1-.394}}$$

$$r_{yx.z} = \frac{.056}{\sqrt{.654}\sqrt{.606}}$$

$$r_{yx.z} = \frac{.056}{.809 \times .778}$$

$$r_{yx.z} = \frac{.056}{.629}$$

$$r_{yx.z} = .089$$

The final value of the partial correlation coefficient of age at first birth and family income controlling for education, $r_{yx.z} = .089$ is much smaller than the original bivariate correlation observed between age at first birth and family income ($r = .45$ as seen in the correlation matrix). When the effects of education have been controlled, the correlation is very small and the resulting r^2 value of $r^2_{yx} = .008$ indicates that almost no variation in family income can be explained by age at first birth when the effects of education are controlled. We can compare this to the original r^2 value of the bivariate relationship that indicated that .181, or just over 18% of the variation in family income could be explained by age at first birth.

The partial correlation coefficient may be tested for statistical significance just as a Pearson r is using the same table of critical values and degrees of freedom of $df = n - 2$. The partial correlation coefficient is an effective tool in statistical analysis that helps us take an important step beyond the evidence of an association between two variables that is provided by the Pearson r, and moves us toward evidence of a true causal relationship by controlling for the effects of other variables.

Computer Applications

1. Select the variables Education (Educ) Respondent Income in Constant Dollars (Coninc) and Age (Age) from the GSS data set and conduct a Pearson correlation analysis. Interpret the results.
2. Select the variables Education (Educ) Respondent Income in Constant Dollars (Coninc) from the GSS data set and conduct a partial correlation analysis controlling for the effects of Age. Interpret the results.
3. Input the ordinal level data from one of the examples in the text and conduct a Spearman correlation analysis. Interpret the results.

How to do it

Load the GSS data set. Click on "Analyze," "Correlate," and "Bivariate" to open the bivariate correlations dialog box. Highlight the desired variables and select them by clicking on the appropriate direction arrow. Make sure that the Pearson procedure is selected, and then choose either a one-tail or a two-tail test of significance. Click on "OK" to complete the procedure.

Click on "Analyze," "Correlate," and "Partial" to open the partial correlations dialog box. Highlight the variables "Educ" and "Coninc" and move them to the selected "Variables:" box. Highlight the variable "Age" and move it to the "Controlling for:" box. Choose either a one-tail or two-tail test, and then click on "OK" to complete the procedure. Compare the results to the bivariate results above.

Use the "File" "New" "Data" command to clear the GSS data set and obtain a clear data editor screen. Input ordinal level data from the text and conduct a Spearman correlation analysis. Click on "Analyze," "Correlate," and "Bivariate" and select the variables for analysis. Remember that you will have to click on "Spearman" to obtain a Spearman analysis.

Summary of Key Points

Correlation coefficients are statistics that are intended to indicate the existence and type of a relationship between two variables. Correlation coefficients are important tools in establishing evidence of a causal relationship between two variables, but evidence of a correlation alone does not establish causality. Correlation coefficients may best be thought of as measures of association rather than measures of causation. The correlation coefficients can indicate that two variables are associated, or covary, and that is an important step toward eventually establishing evidence of causality. Further evidence of causality may be provided by a partial correlation coefficient which measures the degree of association between two variables while controlling for the possible influence of a third variable.

Pearson r—The Pearson r, or the Pearson Product Moment Correlation Coefficient, is the most widely used measure of association in behavioral science research. The Pearson r range is ($-1.00 \le r \le 1.00$), with values of plus or minus 1.00 indicating a perfect relationship between two variables, and a value of 0.00 indicating no relationship. The Pearson r is appropriate for two variables that are measured at the interval or ratio level. The square of the Pearson r indicates the amount of variation in the dependent variable "Y" that can be explained by the independent variable "X."

Point-Biserial r_{pb}—The point-biserial correlation coefficient is a measure of association appropriate for a dependent variable "Y" measured at the interval or ratio level and an independent variable "X" that is expressed as a dichotomy (only two values are possible). Like the Pearson r, the point-biserial r_{pb} range is ($-1.00 \le r_{pb} \le 1.00$) and the interpretation is similar except that the sign of the r_{pb} is a function of which group is considered Group 1 and which group is considered Group 2. Because the dependent variable is measured at the interval or ratio level, the square of the r_{pb} can provide an estimate of the amount of variation in the dependent variable "Y" that can be accounted for by the independent variable "X" which is membership in one of two possible groups.

Spearman r_s—The Spearman correlation coefficient is a measure of association for two variables measured at the ordinal level and expressed as rank values. Like the other correlation coefficients the range of the Spearman r_s is ($-1.00 \le r_s \le 1.00$) with an interpretation like that of the Pearson r. Because the dependent variable is not measured at the interval or ratio level, there is no r_s^2, or variance explained interpretation possible with the Spearman r_s.

Partial Correlation Coefficient—The partial correlation coefficient is based on the Pearson r, and provides evidence of an association between two variables while controlling for the possible effects of a third variable. The partial correlation coefficient may be treated just as the Pearson r with the same range of possible values, and the same interpretation for the r^2 value.

Cartesian Coordinate System—A set of two axes set at right angles to each other with the horizontal axis representing an independent variable and the vertical axis representing the dependent variable. Cartesian coordinate systems are used to display the relationship between two variables on a single graph.

Positive Relationship—A relationship between two variables in which the variables move in the same direction.

Negative Relationship—A relationship between two variables in which the variables move in opposite directions.

Level of significance—The level at which a statistical test is performed. The level of significance, also called the alpha level, is the risk of being wrong when rejecting the null hypothesis.

Statistical Significance—A test of statistical significance performed on the null hypothesis, a statement of no relationship between two variables in the population from which a particular sample was drawn, indicates if we have reason to believe that a particular association observed between two variables in a sample is likely to exist in the larger population from which the sample was drawn. An observed correlation coefficient that has an absolute value as large or larger than the indicated critical value from the appropriate table for a given correlation coefficient allows us to reject the null hypothesis, and indicates that the observed correlation is likely to exist in the larger population (subject to our risk of being wrong as indicated by the size of our α level).
One-Tail and Two-Tail Test—One-tail and two-tail tests of the null hypothesis depend on the presence or absence of a statement of direction in the research hypothesis. If the research hypothesis states a direction for a particular relationship, then the null hypothesis is tested with a one-tail test; otherwise, the null hypothesis is tested with a two-tail test.
Variance Explained—The amount of variance that can be explained in the dependent variable "*Y*" by the independent variable "*X*." The amount of variance that can be explained is usually taken as a measure of how important a particular relationship between two variables is.

Questions and Problems for Review

1. Compute the appropriate measure of association for the following data measured at the ordinal level. Test the resulting correlation coefficient at the $\alpha = .05$ level, assume a one-tail test of the hypothesis, and interpret the results.

X	*Y*
2	8
4	6
1	9
8	3
7	5
9	1
10	2
3	10
6	7
5	4

2. Why is it not appropriate to compute an estimate of variance explained with the data in question 1?
3. Compute the appropriate correlation coefficient given the data below. Test the observed correlation for statistical significance at the $\alpha = .05$ level, assume a one-tail test of the hypothesis and interpret the results. Estimate the amount of variance in "*Y*" that can be explained by group membership.

Sex	ACT Score
male	25
female	20
male	32
male	15
female	19
female	32

Sex	ACT Score
female	17
male	16
male	20
female	28

4. Demonstrate the reduction in variance in problem 3 that can be explained by group membership (hint: compare the reduction in variance obtained by using the appropriate group mean as your prediction compared to the original variance in the data).
5. Create a scatterplot for the data below; assume they have been measured at the ratio level. Describe the nature of the relationship you observe, and estimate what you would expect the correlation coefficient to equal (hint: do not forget to use the appropriate sign for the direction of the relationship you expect).

X	Y
10	8
15	5
30	15
25	18
40	20
18	12
8	5
12	10

6. Compute the appropriate correlation coefficient for the data in problem 5 assuming that they have been measured at the ratio level. Test the observed correlation coefficient at the $\alpha = .01$ level, assume a two-tail test of the hypothesis, and interpret the results. Provide an estimate of variance explained if appropriate.
7. Resolve the ties for the rank order data below, and compute the Spearman r_s correlation coefficient for the data. Test the observed correlation at the $\alpha = .05$ level of significance assuming a one-tail test of the hypothesis.

X	Y
1	10
2	7
3	8
4	8
5	5
6	3
7	6
7	4
7	2
10	1

8. Given the correlation coefficient matrix below:

	X	Y	Z
X	1.00	–.75	.60
Y	–.75	1.00	–.50
Z	.60	–.50	1.00

Compute the following partial correlation coefficients:

A. $r_{yx.z}$

B. $r_{xz.y}$

9. How do the results from the partial correlation coefficients in problem 8 differ from the bivariate correlations provided in the correlation matrix? Compare the amount of variance that is explained in the dependent variable in the bivariate relationship with the amount of variance that is explained when a control variable has been introduced. Does the use of a control variable suggest that the original bivariate relationships were spurious?

10. Compute the appropriate correlation coefficient given the following data. Test the correlation coefficient for statistical significance at the $\alpha = .01$ level, assume a two-tail test of the hypothesis and interpret the results. How much variation in "*Y*" can be explained by group membership?

$\bar{Y}_1 = 87.5$ $\bar{Y}_2 = 95.8$ $S_y^2 = 64$

$n_1 = 80$ $n_2 = 120$

CHAPTER 10

Linear Regression

Key Concepts

Linear Regression
Regression Equation
Least Squares
Regression Coefficient
Slope
Y Axis Intercept
Coefficient of Determination
Standard Error of the Estimate
Multiple Regression
Multiple *R* Squared

Introduction

Linear regression is a statistical procedure that allows us to predict the level of a dependent variable by developing an equation based on the relationship between the dependent variable and a single independent variable. The ability to use one variable as a predictor of another also tells us a great deal about the underlying relationship between the two variables. In the last chapter, we saw that the relationship or correlation between two variables was stronger the more closely the scatter plot of the two variables suggested a straight line. The regression equation that we develop is actually the equation for a straight line that best fits the data.

The values of the dependent variable that are predicted by the regression equation are the points along the line defined by the equation. Unless there is a perfect correlation between the two variables (either a Pearson $r = 1.00$, or $r = -1.00$), the actual values of the dependent variable will not lie in a straight line. As a result, the predicted values of the dependent variable will not exactly equal the actual values, because the predicted values always lie in a straight line. The difference between the predicted values and the actual values may be thought of as error. The regression equation that provides the best fit to the data will be the equation that produces the minimum amount of error when comparing the predicted values of the dependent variable with the actual values. We will examine this idea of measuring prediction error in more detail in a later section, but first let's review the equation for a straight line.

Equation for a Straight Line

A straight line is defined by the following equation:

$$Y = bX + a$$

where

Y = the values of the dependent variable, which are plotted on the vertical axis;
X = the values of the independent variable, which are plotted on the horizontal axis;
b = the slope of the line, which is the unit change in the Y variable for each unit change in the X variable; and,
a = represents the Y axis intercept, which is the point where the line intersects the Y axis.

By substituting some arbitrary values for "b" and "a" into the general form of the equation, we can create the equation for a specific straight line and then plot the line given some values for X. For example, if we let "b" equal 1.5 and "a" equal 2, the equation becomes:

$$Y = 1.5X + 2$$

Given the following values of X, we can solve for the corresponding values of Y.

X	$Y = 1.5X + 2$
2	$1.5(2) + 2 = 5$
4	$1.5(4) + 2 = 8$
8	$1.5(8) + 2 = 14$
0	$1.5(0) + 2 = 2$
−2	$1.5(-2) + 2 = -1$
6	$1.5(6) + 2 = 11$
−4	$1.5(-4) + 2 = -4$

Plotting the values of X and Y will demonstrate that the points do in fact lie in a straight line.

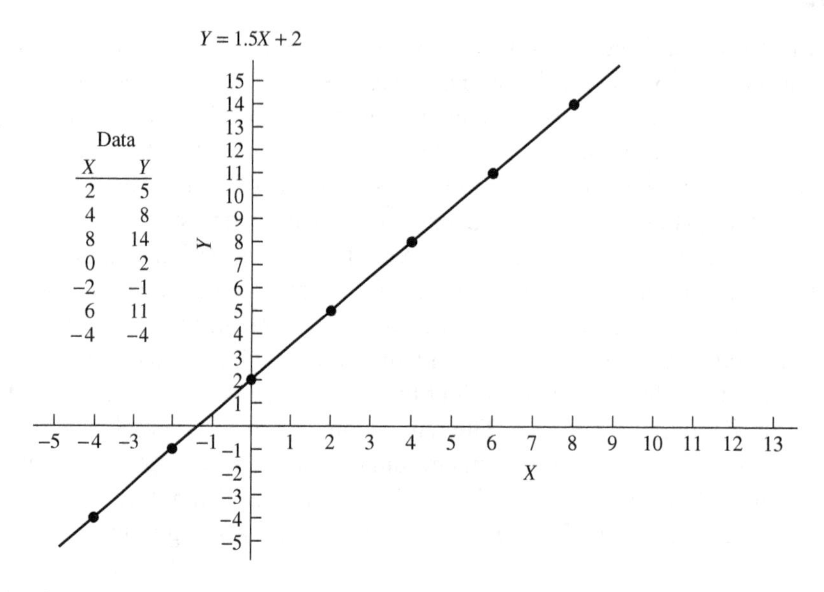

Figure 10.1 Straight Line Generated by the Equation $Y = 1.5X + 2$

The meaning of the value of "b," the slope of the line, and "a," the Y axis intercept, may be more clear now. A slope equal to 1.5 means that for each 1 unit change in X, there is a corresponding 1.5 unit change in Y. A move of 1 unit to the right on the X axis results in a move upward on the Y axis of 1.5 units. The value of "a" equal to 2 indicates that the line will cross the Y axis at 2 (note that the Y axis intercept is where X is equal to 0). See Figure 10.2 below.

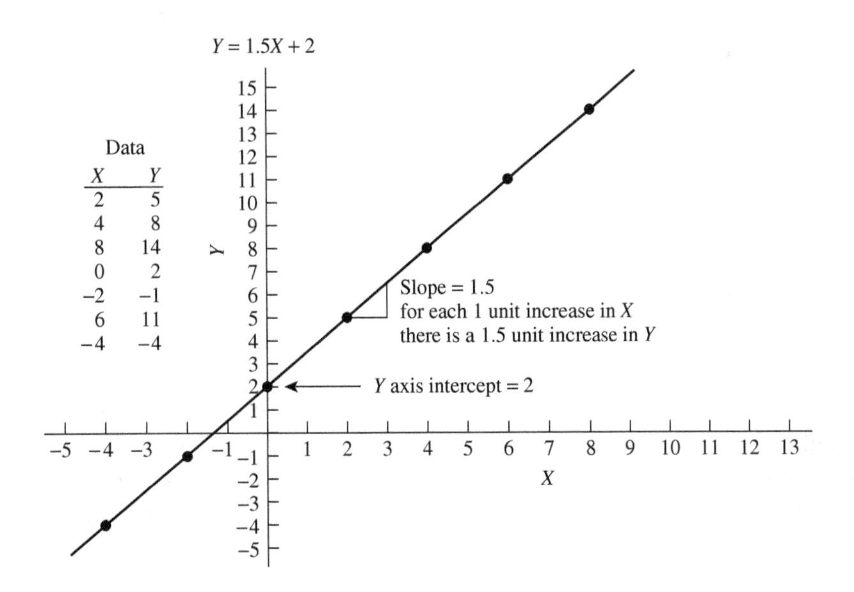

Figure 10.2 The Slope and Y Axis Intercept for $Y = 1.5X + 2$

Fitting a Straight Line to Describe a Linear Relationship

A linear relationship is one in which two variables covary or move with respect to each other in a consistent manner in a pattern suggestive of a straight line. As you recall from previous chapters, the relationship may be positive where the values of Y increase as the values of X increase, or the relationship may be negative where the values of Y decrease as the values of X increase. In either case, there is a linear trend in the data. Keep in mind that not all variables are related in a linear fashion. Some variables may not be related in any manner, and others may have a curvilinear relationship. However, if two variables are related in a linear fashion, it is possible to describe the relationship by means of an equation for a straight line. Consider the data plotted below in Figure 10.3.

While the data are not in a straight line, there is clearly a strong linear relationship between the two variables. We have seen these data before as an example of calculating the Pearson correlation coefficient in Chapter 8, and you may recall that we observed a correlation between X and Y of $r = .923$ suggesting a very strong positive relationship.

Our goal with the linear regression analysis is to generate an equation for a line passing through the $n = 7$ data points that will provide the best fit to the data. If you examine the scatter plot of the data, you might see a path for the regression line that to your eye looks like it might be a very good fit. I might see the ideal path in a slightly different location, and some third observer might suggest a slightly different location still. To find the best fit to the data, we must first define what we mean by the best fit.

Our definition for the line providing the best fit to the data is the line that **minimizes the sum of the squared deviations** of the actual Y values from the predicted values along the line. We will measure the deviation between the predicted Y values and the actual Y values by placing a line perpendicular to the X axis from the actual Y value to the predicted Y value along the potential regression line. The length of the

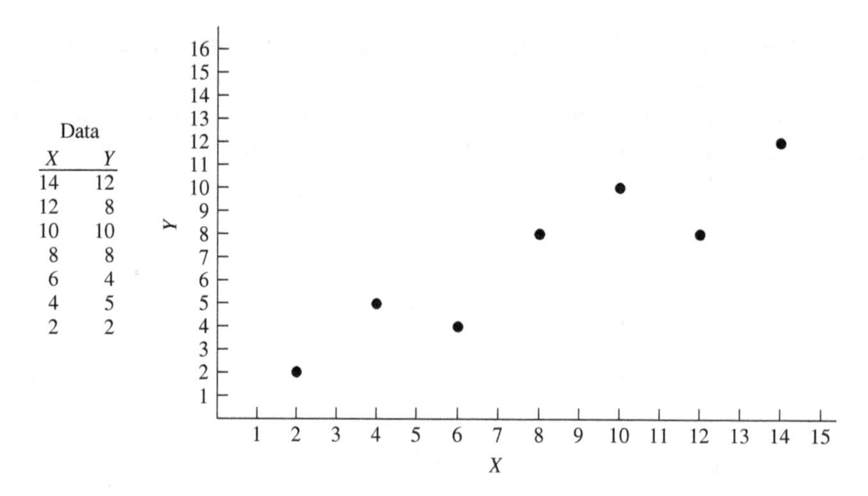

Figure 10.3 A Scatterplot of Seven Data Points

line from the actual *Y* value to the predicted *Y* value represents our error in the prediction of that particular value of *Y*. The scatter plot of our data is presented below in Figure 10.4 with a potential regression line placed through the data. I have arbitrarily placed the line so that it crosses the *Y* axis at $Y = 5$, and passes through the middle data point of ($X = 8$, $Y = 8$). Perpendicular lines from the actual *Y* value to the predicted *Y* value along the potential regression line illustrate the amount of error for this particular line.

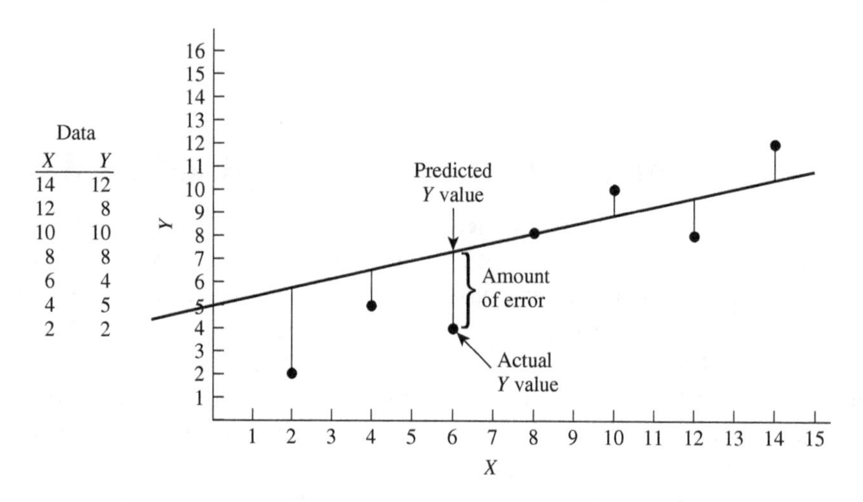

Figure 10.4 A Hypothetical Regression Line through the Data with Perpendicular Lines Demonstrating Error

Because some of the predicted values of *Y* will be too high and others will be too low, we will square the distance or deviations to keep the positive and negative distances from cancelling each other out. (This is same strategy we used in Chapter 5 in computing variance.) The sum of the **squared deviations** may then be computed as a measure of goodness of fit of the regression line to the actual data points. The best fit to the data will be the line that minimizes the sum of the squared deviations. Minimizing the sum of squares is also referred to as using the "least squares" criterion for determining the best fit to the data.

There is one point in the distribution of data through which the line providing the best fit will always pass. The best fit to the data will always pass through the point $(\overline{X}, \overline{Y})$, the point representing the means of the dependent and independent variables. You might think of the point of $(\overline{X}, \overline{Y})$ as being the center of the entire distribution. The scatter plot of the data is presented again below in Figure 10.5 with the intersection

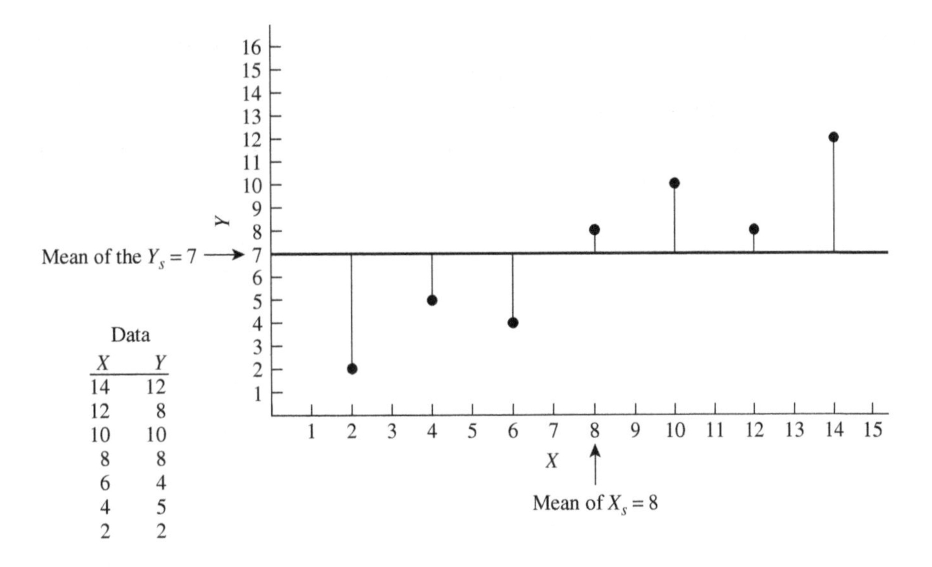

Figure 10.5 Using the Mean of *Y* to Predict Each *Y* Value

of $(\bar{X}, \bar{Y})$ represented by small Arrows. The horizontal line at $Y = 7$ indicates what our prediction line would look like if we were to base our prediction of *Y* on the mean without using any information from the *X* variable. The perpendicular lines drawn from each data point to the line at the mean indicates what our error would be in that case. If we were to square the size of each error (to eliminate the positive and negatives from canceling out each other), sum the errors, and then find the mean error by dividing by *n*, we would have the variance of the *Y* variable.

The best fit regression line will still pass through the intersection of $(\bar{X}, \bar{Y})$, but the effect of our linear regression analysis will be to provide a slope to the line that will minimize the error as much as possible.

The Regression Equation

The equation for a regression line is similar to the general formula for a straight line that we just reviewed. The **linear regression equation** may be written as:

$$Y' = bX + a$$

where

Y' = the predicted values of the dependent variable *Y* plotted on the vertical axis;
X = the values of the independent variable which are plotted on the horizontal axis;
b = the regression coefficient which is the slope of the regression line indicating the unit change in Y' for each unit change in *X*; and,
a = the *Y* axis intercept which is the point where the regression line intersects the *Y* axis.

The values of *X* are given; these are the values of the independent variable that we will have measured among our sample. To develop the regression equation, we must calculate the values of "*b*" (the regression coefficient) and "*a*" (the *Y* axis intercept).

Formula for "b," the Regression Coefficient

The regression coefficient may be calculated with the following formula:

$$b = \frac{n \times \Sigma XY - (\Sigma X)(\Sigma Y)}{n \times \Sigma X^2 - (\Sigma X)^2}$$

The formula for "b" is very similar to the one we used to compute the Pearson correlation coefficient in Chapter 9. The numerator is actually the same, and the denominator consists of a portion of one of the terms we used in the Pearson correlation formula.

Our data are presented below with appropriate columns for each of the terms we will need for the formula for the regression coefficient "b."

X	Y	XY	X^2
14	12	168	196
12	8	96	144
10	10	100	100
8	8	64	64
6	4	24	36
4	5	20	16
2	2	4	4
$\Sigma X = 56$	$\Sigma Y = 49$	$\Sigma XY = 476$	$\Sigma X^2 = 560$

$$b = \frac{7 \times 476 - (56)(49)}{7 \times 560 - (56)^2}$$

$$b = \frac{3332 - 2744}{3920 - 3136}$$

$$b = \frac{588}{784}$$

$$b = 0.75$$

The regression coefficient $b = .75$ (representing the slope of the regression line) indicates that there is a .75 unit change in the value of Y for each unit change in the value of X. The value of "b" is also used to solve for the value of "a," the Y axis intercept with the following formula:

Formula for "a," the Y Axis Intercept

$$a = \bar{Y} - b(\bar{X})$$

The values of the mean of the X and Y variables were previously provided, and were $\bar{X} = 8$ and $\bar{Y} = 7$. Solving for "a"

$$a = 7 - (0.75)(8)$$
$$a = 7 - 6$$
$$a = 1$$

The result indicates that our best fit regression line will cross the Y axis at $Y = 1$. Having solved for both "b" and "a," we can write the regression equation as follows:

$$Y' = 0.75X + 1$$

Given any value of X, we will be able to predict the value of Y through simple substitution into the equation. We will solve for some predicted values of Y shortly, but I want to make two important points before doing so. First, it is often desirable to draw the regression line through the scatterplot of the data. Since a straight line can be drawn once we know any two points along the line, we are already in a position to draw the regression line. We know one point the line will pass through is the Y axis intercept of $Y = 1$. We also know that the best fit line will pass through the intersection of $(\bar{X}, \bar{Y})$. So our regression line may be drawn through the scatter plot as indicated below in Figure 10.6.

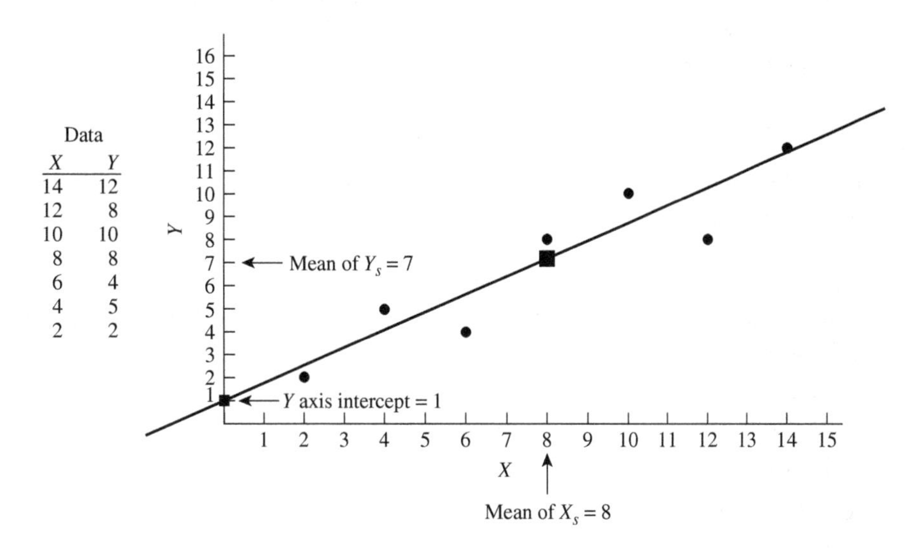

Figure 10.6 The Regression Line for the Seven Data Points

The second point to consider involves the interpretations of the values of "*a*" and "*b*." You will see when a second example of regression analysis is presented below that the value of "*b*" has a more significant meaning than just being the slope of the regression line. In many ways, the value of "*b*" helps us to understand the nature of the relationship between the dependent and independent variable in quantitative terms. Given the links between correlation analysis and linear regression, it may be tempting to associate the regression coefficient "*b*" with the Pearson correlation coefficient *r*. In fact, the two statistics are related, and we will see in a later section that one may calculate the value of the regression coefficient "*b*" from the Pearson correlation coefficient *r* and the standard deviations of the *X* and *Y* variables. It is also true that the sign of the regression coefficient will indicate the direction of the relationship between *X* and *Y* just as the sign of the correlation coefficient did. That is, a positive "*b*" indicates a positive relationship between *X* and *Y*, and a negative "*b*" indicates a negative relationship between *X* and *Y*. However, the regression coefficient and the correlation coefficient are not the same, and one important difference to keep in mind is that while the range of the Pearson correlation coefficient is restricted to $(-1.00 \leq r \leq 1.00)$, *there is no such limitation on the value of the regression coefficient* "*b*."

One other important consideration about the value of "*b*" is that just as with the case of the Pearson *r*, the value of the regression coefficient "*b*" can be tested for statistical significance. For example, a regression coefficient with a value of $b = 0.00$ indicates no relationship between *Y* and *X*. (Using the mean of the "*Y*" variable as our prediction would be just as effective as a regression line with a slope of $b = 0.00$.) In fact, as we saw earlier, using the mean of the "*Y*" variable as our prediction results in a line with a slope of 0.00. "How much larger than $b = 0.00$ does the observed value of b need to be before it is considered statistically significant?" is the question we are considering when we test the observed value of the regression coefficient for statistical significance. The statistical test is based on ratio of the size of the observed regression coefficient to its standard error. We will not be computing a test of statistical significance for the regression coefficients in this chapter, but you will see them reported in the output from statistical packages such as SPSS.

What about the value of "a"? Does the fact that "a" takes on a value of 1, or any value for that matter, have any special significance? In the general case, the value of "a" does not have any special meaning. There are some exceptions, but for the general case you should simply think of "a" as the point where the regression line crosses the Y axis. In fact, there is a special application of linear regression called **standardized linear regression** that results in a value of "a" = 0, that is, the regression line will always pass through the origin (the point $X = 0$, $Y = 0$). To perform a standardized linear regression analysis, just convert the values of the dependent and independent variables into standard scores (Z scores) prior to performing the regression analysis.

Predicting Values of *Y*

The predicted values of Y are the points along the regression line. Given any value of X, we can calculate the predicted value of Y by use of the regression equation:

$$Y' = 0.75X + 1$$

Let's take each of the values of X, calculate the predicted value of Y, and then compare the predicted Y value to the actual Y value.

X	$0.75X + 1 = Y'$	Actual Y
14	$0.75(14) + 1 = 11.5$	12
12	$0.75(12) + 1 = 10.0$	8
10	$0.75(10) + 1 = 8.5$	10
8	$0.75(8) + 1 = 7.0$	8
6	$0.75(6) + 1 = 5.5$	4
4	$0.75(4) + 1 = 4.0$	5
2	$0.75(2) + 1 = 2.5$	2

What is the point of regression analysis?

At this point we have plotted some data, calculated the values of "b" and "a" to determine the equation for a regression line that provides the best fit to our data, and computed the predicted values of Y for each of our observed values of X. Some fundamental questions may have occurred to you somewhere along the way. What is the point of all of this? Why do we even care about the predicted values of Y when we already have the actual values of Y, especially since our predicted values of Y have a degree of error associated with them? If all we cared about were the characteristics of our particular sample, then there really would not be very much point to linear regression analysis. We would in fact have the actual X and Y values, and would not have any use for predicted values of Y. There would still be some utility in using regression analysis in that it allows us to describe the relationship between X and Y in quantitative terms. However, very often we are interested in going beyond our actual data. For example, we know what the actual values of Y are when X is equal to 2, 4, 6, 8, 10, 12, or 14 (the values that we have observed in our small sample of $n = 7$). What if we want to know what Y is equal to when X is 5, or X is 11? Since we have no observations of Y at those values of X we do not know, but with the use of our regression equation we can predict the value of Y quite easily by substituting the given value of X into the regression equation.

$$\text{If } X = 5, \text{ predicted } Y' = 0.75(5) + 1$$

$$Y' = -3.75 + 1$$

$$Y' = 4.75$$

$$\text{If } X = 11, \text{ predicted } Y' = 0.75(11) + 1$$
$$Y' = 8.25 + 1$$
$$Y' = 9.25$$

Note that in both of the above examples, we are predicting values of Y for values of X that lie within our observed range of X for the sample (X ranges from $X = 2$ to $X = 14$). Theoretically we can predict a value for Y for any value of X. For example,

$$\text{If } X = 100, \text{ predicted } Y' = 0.75(100) + 1$$
$$Y' = 75 + 1$$
$$Y' = 76$$

However, for any given value of X, the validity of the resulting predicted value of Y decreases the more the value of X deviates from the actual range of the X values in the sample used to generate the regression analysis. In the current example, we can be relatively confident in predicting a value for Y for any value of X ranging from $X = 2$ to $X = 14$, but the more a value of X is below 2 or the more a value of X is above 14, the less valid will be our predicted value of Y. In some cases, using extreme values of X will result in a predicted value of Y that lies outside the range of possibility. For example, we might find that there is a positive relationship between family income and college grade point average based on the analysis of a typical sample of college students. However, if we use the resulting regression analysis to predict the grade point average of a new student from an extremely wealthy family, we might observe a predicted grade point average of 5.25, which is outside the range of possible values.

Another potential use of regression analysis is in situations when we may not have any information at all on our dependent variable Y, but may be able to turn to regression analysis for suitable predictions. Consider the case of an admissions director at a university who suspects that high school senior year grade point average (X) will be a good predictor of college freshman year grade point average (Y). The goal is to predict the academic performance of first year college students prior to admission, so that only those students with a reasonable chance of success will be admitted. The problem is that the decision to admit a student must be made before the student has an opportunity to demonstrate his or her academic ability at the college level. Linear regression analysis may provide a solution by building a regression equation using data from existing students. The existing students will have both a high school senior year grade point average and a college freshman year grade point average, and these data will be the source of the linear regression equation. Once the values of "b" and "a" have been computed and the equation is developed, we may then estimate the college freshman year grade point average of the student applicants by using their high school senior year grade point average in the linear regression equation. We do not have an actual college freshman year grade point average for the student applicants, but we will have a predicted freshman year grade point average. How accurate will the predictions be? That will depend on how good the regression model is, which is the topic we will examine next.

Assessing the Quality of the Regression Model

The Coefficient of Determination

We can assess the quality of a linear regression model by examining two statistics: **the coefficient of determination** and **the standard error of the estimate**. The concepts behind both of these statistics should already be familiar to you. We first encountered the coefficient of determination in the previous chapter on correlation. At that time we saw that the Pearson correlation coefficient r indicated the strength of a relationship between two variables, and that the coefficient of determination r^2 indicated the amount of variation in the dependent variable Y that could be accounted for by the independent variable X.

One of the most appealing aspects of presenting the material on linear regression is that it affords an opportunity to tie together several important concepts that you have encountered previously. Linear regression and the Pearson correlation coefficient are closely related, which is why the chapter on regression follows the one on correlation. We are also dealing with two other key ideas from Chapter 5: **variance** and the concept of **explaining or reducing variance**. By way of a brief review, a variable takes on different values. We can calculate the mean of the variable which indicates its central point, and then we can measure the amount of variation in the variable by calculating the variance, s^2. The variance represents the average squared deviation of each score from the mean for the distribution. We also saw that if we want to predict the value of a variable, our best strategy is to guess the mean in the absence of any other information. Of course we will be wrong some of the time, and one way to measure how much error we would have is to compute the variance, the amount of squared deviation or squared error of each actual score from our prediction of the mean. Finally, we saw in Chapter 5 that we may be able to improve our ability to predict a variable by using information about another variable. For example, we used our knowledge of the level of education to better predict income. We measured the amount of improvement in our ability to predict the dependent variable by measuring how much we were able to reduce the variance in the dependent variable; that is, how much variation is there when we measure the squared deviation from the actual score to our new predicted score based on information from the level of the independent variable compared to the previous amount of variance based on the squared deviation from the actual score to our prediction of always using the mean.

In the previous chapter, we saw that the Pearson correlation coefficient for the data set we have used for our linear regression analysis was $r = .92$. The coefficient of determination, r^2, is then .85. The value of the coefficient of determination indicates that when we use our knowledge of the independent variable X to predict the level of the dependent variable Y, we will be able to reduce or explain 85% of the variation in Y. Let's demonstrate how this reduction in variance is accomplished.

Our regression equation $Y' = .75X + 1$ enables us to predict the value of Y using information on X. We have previously calculated the predicted values of Y for each of our observed values of X. These data are displayed below along with calculations for the difference between the actual Y values and the predicted Y' values, the squared difference between the actual Y values and the predicted Y' values, and the squared difference between the actual Y values and the mean of the actual Y values.

Actual Y	Y'	$(Y - Y')$	$(Y - Y')^2$	$(Y - \bar{Y})^2$
12	11.5	.50	.25	25
8	10.0	−2.00	4.00	1
10	8.5	1.50	2.25	9
8	7.0	1.00	1.00	1
4	5.5	−1.50	2.25	9
5	4.0	1.00	1.00	4
2	2.5	−.50	.25	25
			$\Sigma = 11$	$\Sigma = 74$

The sum of the squared deviations of the actual Y values from the mean is 74, and if we divide that value by the sample size $n = 7$, we arrive at the variance:

$$s_y^2 = \frac{\Sigma(Y - \bar{Y})^2}{n}$$

$$s_y^2 = \frac{74}{7} = 10.57$$

The sum of the squared deviations of the actual Y values to the predicted Y' values is 11, and if we divide that value by the sample size $n = 7$, we have the average squared deviation of the actual Y values to the predicted Y' values:

$$s_{y'}^2 = \frac{\Sigma(Y - Y')^2}{n}$$

$$s_{y'}^2 = \frac{11}{7} = 1.57$$

You can think of the $s_{y'}^2$ as a type of modified measure of variance; a variance based on the estimated values of Y'.

As you can see, using our knowledge of the independent variable X to predict the value of Y has allowed us to reduce the variance from 10.57 to 1.57, or we can express the reduction as a proportion. The variance has been reduced by 9.0 points (10.57 – 1.57); expressing the reduction in variance as a proportion of the original variance yields:

$$\frac{9.0}{10.57} = .85$$

Or alternatively, we could simply find the proportional reduction in variance by

$$1 - \frac{s_{y'}^2}{s_y^2} = 1 - \frac{1.57}{10.57} = 1 - .15 = .85$$

which is the same value we observed earlier for the coefficient of determination $r^2 = .85$. So how can we use our knowledge of an independent variable to better predict a dependent variable? By analyzing the data with linear regression, and then using the resulting regression equation to predict values of the dependent variable (Y) based on our observation of the independent variable (X).

The Standard Error of the Estimate

In Chapter 5, when we dealt with variance, we also found it useful to compute the standard deviation; the same is true when working with the variance of the estimate $s_{y'}^2$. By taking the square root of the variance of the estimate, we can find the **standard error of the estimate** defined as the square root of the average squared deviation of each actual value of Y from the estimated value of Y'. The interpretation of the standard error of the estimate is very similar to that of a standard deviation.

A Conceptual Formula for the Standard Error of the Estimate

Conceptually, the formula for the standard error of the estimate is just the square root of what we were calling the modified variance $s^2_{y'}$. The formula could then be written as:

$$s_{y'} = \sqrt{\frac{(Y - Y')^2}{n}}$$

However, just as we saw in Chapter 5 that we had to surrender a degree of freedom when calculating the standard deviation and used the value $n - 1$ in the denominator of the formula for a standard deviation in some applications, so too do we have to surrender degrees of freedom when calculating the standard error of the estimate. Technically, we must surrender one degree of freedom for each estimated parameter in our regression equation. In the case of linear regression, we have two estimated parameters: the Y axis intercept "a" and the regression coefficient "b." Therefore, our formula for the standard error of the estimate (for simple linear regression) is correctly written as:

$$s_{y'} = \sqrt{\frac{(Y - Y')^2}{n - 2}}$$

One of the assumptions or requirements of the data to legitimately conduct a linear regression analysis is that for any given value of X, the actual Y values will be normally distributed around the estimated value of Y' on the regression line. This may be difficult to visualize since we have used examples with a small sample size to demonstrate the technique of linear regression. Imagine a regression line passing through a very large number of data points. Instead of there being one or perhaps two values of Y at each value of X as we have seen in the examples so far, there might be 50 or 100 values of Y. Naturally, the regression line will not pass through all of the data points. The assumption is that at any given value of X, the actual Y values will be normally distributed around the estimated value of Y'. That is at any given value of X, half of the actual Y values would be above the regression line, and half of the actual Y values would be below the regression line. Further, the actual Y values would fall away from the estimated Y' value on the regression line in a normal fashion.

Because the standard error of the estimate operates as a standard deviation for the difference between the actual Y values and the estimated Y' values, and the errors around the regression line (the difference between the actual Y values and the estimated Y' values) are normally distributed, we would expect approximately 68% of the errors (the difference between the actual Y values and the estimated Y' values) to fall within plus and minus one standard error of the estimate, and approximately 95% of the errors to fall within plus and minus two standard errors of the estimate (since in a normal curve approximately 68% of the distribution is within plus and minus one standard deviation, and approximately 95% of the distribution is within plus and minus two standard deviations).

Our two statistics then, the *coefficient of determination*, and the *standard error of the estimate* tell us a great deal about the quality of the regression model. The higher the coefficient of determination the greater the amount of variation in the dependent variable that is explained by the independent variable. The smaller the size of the standard error of the estimate, the closer the regression line is to the actual Y values.

Explained and Unexplained Variance

The coefficient of determination and the standard error of the estimate allow us to partition the variance of a dependent variable into two important components. The total variance can be expressed as the sum of the variance that is **explained** by the regression model, and the variance that remains **unexplained** by the regression model. It is easier to see the relationship among the total variation, the explained variation, and the unexplained variation if we just look at the **sum of squares** (the sum of the squared deviations) terms rather than dividing by n to calculate the variance terms. In that case, we have the total variation in the dependent variable referred to as the **sum of squares total**, the variation that is explained by the regression model referred to as the sum **of squares explained**, and the variation that remains unexplained referred to as the **sum of squares error**. Each of these terms is represented below symbolically:

$$\text{Sum of squares total} = \text{Sum of squares explained} + \text{Sum of squares error}$$

$$\Sigma(Y - \bar{Y})^2 \quad = \quad \Sigma(Y' - \bar{Y})^2 \quad + \quad \Sigma(Y - Y')^2$$

We can easily rearrange the terms to emphasize the sum of squares explained:

$$\text{Sum of squares explained} = \text{Sum of squares total} - \text{Sum of squares error}$$

$$\Sigma(Y' - \bar{Y})^2 \quad = \quad \Sigma(Y - \bar{Y})^2 \quad - \quad \Sigma(Y - Y')^2$$

These same terms may also be used to express the amount of the variance in the dependent variable that is explained by the regression model as a proportion of the total variance.

$$\text{Proportion of Sum of Squares Explained} = \frac{\text{Total Sum of Squares} - \text{Sum of Squares Error}}{\text{Sum of Squares Total}}$$

In this presentation, you can see that as the sum of squares error approaches 0, the proportion of variance explained will approach 1 indicating that 100% of the variance in the dependent variable is accounted for by the regression model. Conversely, as the sum of squares error approaches the size of the sum of squares total, the proportion of variance explained will approach 0 indicating that the regression model provides no improvement in predicting the dependent variable.

A Second Example: Using Education to Predict Income

Now that we have examined the basics of linear regression, let's look at a second example using the independent variable *education in years* to predict the dependent variable *income in thousands of dollars*. Below we have some hypothetical data that we will also display in a scatter plot (see Figure 10.7). Note that these are the same data we used in the previous chapter for an example of the Pearson r, and found that the correlation between education and income was $r = .864$.

Education (in years)	Income (× 1,000)	XY	X^2
8	15	120	64
12	20	240	144
16	40	640	256
10	15	150	100
12	20	300	144
8	15	120	64
8	10	80	64
10	10	100	100
12	20	240	144

Education (in years)	Income (× 1,000)	XY	X^2
12	15	180	144
18	50	900	324
16	45	720	256
18	25	450	324
20	85	1700	400
20	90	1800	400
12	30	360	144
22	60	1320	484
16	50	800	256
16	45	720	256
14	35	490	196
$\Sigma X = 280$	$\Sigma Y = 700$	$\Sigma XY = 11430$	$\Sigma X^2 = 4264$

Mean Education $\bar{X} = 14$ (years)
Mean Income $\bar{Y} = 35$ (thousand dollars)

The data appear to have a linear trend, although the points are far from a straight line. We will proceed with the regression analysis by first solving for the regression coefficient "*b*," and then for the *Y* axis intercept "*a*."

$$b = \frac{(n \times \Sigma XY) - (\Sigma X)(\Sigma Y)}{(n \times \Sigma X^2) - (\Sigma X)^2}$$

$$b = \frac{(20 \times 11430) - (280)(700)}{(20 \times 4264) - (280)^2}$$

$$b = \frac{228600 - 196000}{85280 - 78400}$$

$$b = \frac{32600}{6880}$$

$$b = 4.74$$

Having solved for our regression coefficient "*b*," we will find the value of the *Y* axis intercept "*a*."

$$a = \bar{Y} - b(\bar{X})$$
$$a = 35 - (4.74)(14)$$
$$a = 35 - 66.36$$
$$a = -31.36$$

With the values of "*b*" and "*a*," we may now write the regression equation.

$$Y' = 04.74X - 31.36$$

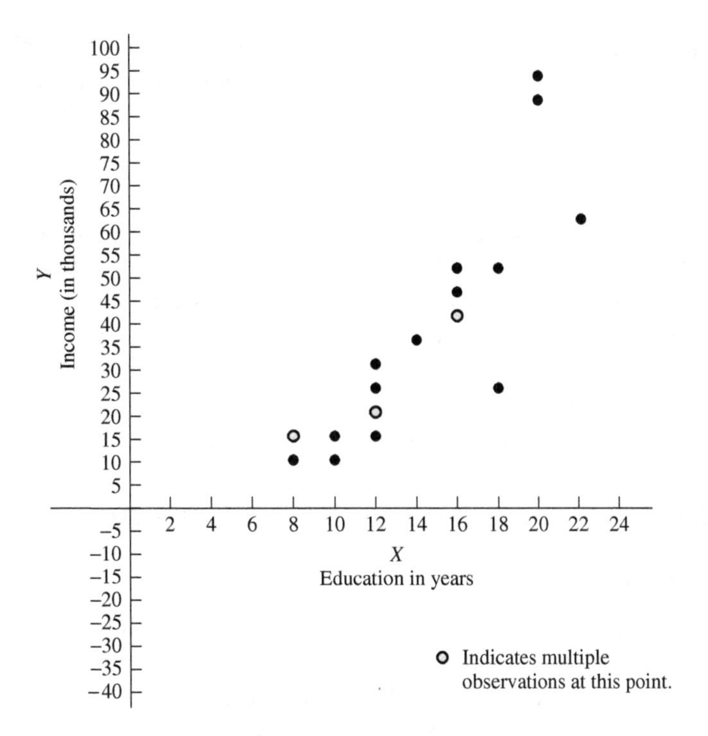

Figure 10.7 A Scatterplot of Income by Education

Given the value of the Y axis intercept and the means of the X and Y variables, we are able to position the regression line through the data; recall that the best fit line will always pass through the intersection of $(\bar{X}, \bar{Y})$ (see Figure 10.8).

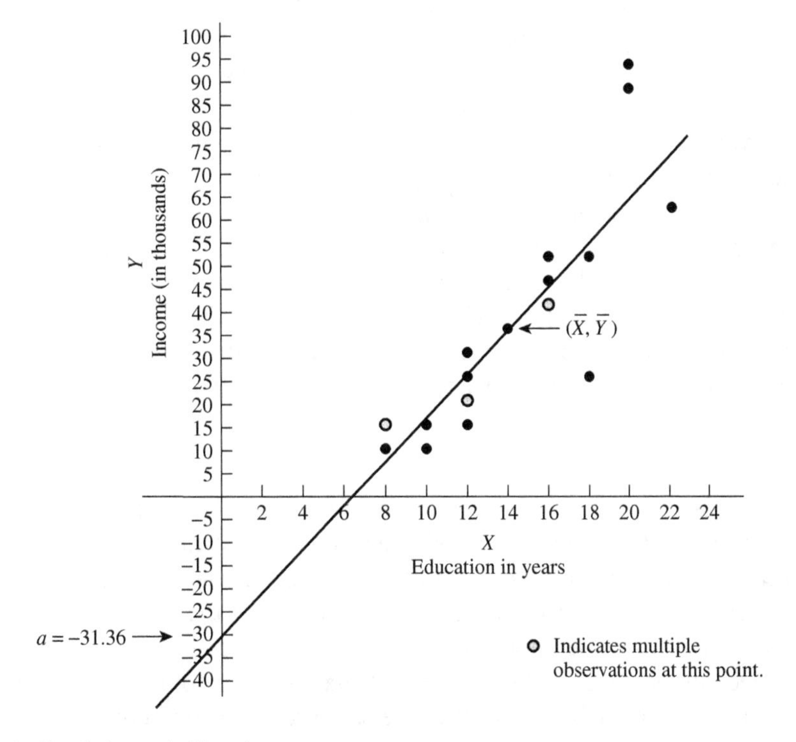

Figure 10.8 The Regression Line for Income by Education

Our regression line appears to provide a reasonably good fit to the data. The value of $r^2 = .75$ indicates that education explains 75% of the variation in income. Now, take another look at the regression equation that we have constructed:

$$Y' = 4.74X - 31.36$$

We saw previously that the regression coefficient "*b*" is the slope for the regression line, and that it could be defined as the unit change in *Y* for each unit change in *X*. With this definition in mind, think about how we have measured the *X* and *Y* variables. The independent variable *X* is education, measured in years. The dependent variable *Y* is income, measured in thousands of dollars. The value of "*b*," our regression coefficient, is 4.74, but what is that telling us about the relationship between education and income? Because education is measured in years, and income is measured in thousands of dollars, a regression coefficient of $b = 4.74$ indicates that **for each additional year of education we should expect an additional $4,740 in income.**

Standardized Regression Analysis and Outliers

Standardized Regression

The regression analysis of income and education provided an opportunity to place a more meaningful interpretation on the value of the regression coefficient "*b*." We may also briefly consider two other issues. I mentioned previously that the value of the *Y* axis intercept "*a*" generally does not have any special significance, and that it is possible to generate a regression equation where the *Y* axis intercept is zero by performing a standardized regression analysis. Standardized regression is performed by converting the raw scores to standard scores (or *Z* scores) prior to the regression analysis.

Our regression analysis of income and education resulted in a *Y* axis intercept of $a = -31.36$. Since the *Y* axis intercept occurs where the value of *X* is zero, our regression analysis indicates that we predict that an individual with zero years of education will have an income of **negative** $31,360. Logically we would not expect an individual with no formal education to earn a very large income, but to predict that the individual would have a negative income does not really make sense. Of course to some extent that would not matter because a value of $X = 0$ lies outside the observed range of *X* in the sample used to generate the regression equation. Still, for aesthetic reasons we might choose to perform a standardized regression analysis which would force the regression line to cross the *Y* axis at the origin. The result would be to predict that an individual with zero years of education will have zero income. While this prediction is also likely to be false, it is not as illogical as predicting a negative income.

You might want to perform a standardized regression analysis on these data by first converting each raw score to a standard score (recall that this is accomplished by subtracting the mean from each score, and then dividing by the standard deviation). The procedure is somewhat tedious to do by hand, so I would not recommend that you do so if you have access to a computer statistical analysis package of some sort. Should you perform a standardized regression analysis you will find that the regression equation becomes:

$$Y' = 0.865X.$$

In standardized regression analysis, the regression coefficient takes on a slightly different interpretation than the "*b*" resulting from a regular regression analysis. The standardized regression coefficient is referred to as a **beta** rather than a "*b*," and it is interpreted as the change in units of standard deviation in the *Y* variable for each unit change in the standard deviation of the *X* variable. Our observed beta = .865 in the present example would be interpreted as follows: for each increase of one standard deviation in education, we expect an increase of .865 standard deviations in income. It may not appear so at first, but the standardized regression analysis results in the same relationship between education and income as we saw in the original analysis.

For our data of $n = 20$ individuals, the standard deviation of education (X) is $s_x = 4.26$ years, and the standard deviation of income (Y) is $s_y = 23.34$ thousand dollars. The results of the standardized regression analysis tell us that a one standard deviation unit increase in education (4.26 years) will result in a 0.865 standard deviation unit increase in income. A .865 of a standard deviation of income is ($0.865 \times 23.34 = 20.19$) or 20.19 thousand dollars. Keep in mind the 20.19 thousand dollar increase in income is for a standard deviation unit increase in education, but a standard deviation unit of education is equal to 4.26 years. To express the relationship in terms of a single year increase in education, we simply need to divide 20.19 thousand dollars by 4.26 years ($20.19/4.26 = 4.74$), indicating a 4.74 thousand dollar per year increase in income for each one year increase in education which is exactly the result obtained earlier in the linear regression analysis.

It is important to note here that some statistical analysis packages offer an option of performing a linear regression analysis which forces the regression line to pass through the origin. This technique is not the same thing as performing a standardized regression analysis, and the results are not comparable to regular linear regression analysis, or to standardized regression analysis.

Outliers

Before we leave this example of analyzing income and education, there is one more important point to consider. When you examined the scatter plot of the data, you may have noticed that most of the data points form a relatively strong linear trend, but there are two points that seem different from the rest. Both occur when $X = 20$; in one case the associated Y value is 85, and in the other case the associated Y value is 90. These income levels are considerably higher than the mean value, and are over 20 thousand dollars higher than the next highest income level observed in the data.

Data points that lie far outside the normal range of a distribution are referred to as **outliers**. In many cases, the presence of an outlier will distort the results of a statistical analysis. Just as one or two extreme values may distort the value of a mean, one or two extreme values may distort the results of a regression analysis. Unusually high (or low) values will distort the regression analysis. The resulting regression line will still be the best fit possible, but the fit to the data would have been much better if the extreme values did not exist. In some cases, it is legitimate to declare the extreme values as outliers, and to remove them from the data. The resulting analysis will produce a regression line that is a much better fit to the remaining data.

There is no standard rule of thumb to declare a particular observation as an outlier. The most important issue is one's motivation for declaring an observation as an outlier. It is proper to remove an observation if its continued presence in the data set will distort the analysis and result in an inferior product. It is quite another situation to remove an observation because it does not support one's preconceived notion of what the data should look like. In the latter situation, it becomes improper to remove an observation just because its continued presence in the data set makes it harder for you to demonstrate statistically what you had intended.

An Alternative Method of Calculating "*b*"

I mentioned earlier that there was an alternative method of calculating the regression coefficient "*b*" based on the Pearson correlation coefficient and the standard deviations of X and Y. The alternative formula for "*b*" is as follows:

$$b = r \times \left(\frac{s_y}{s_x} \right)$$

where

b = the regression coefficient
r = the Pearson correlation coefficient between X and Y

s_y = the standard deviation of the Y variable
s_x = the standard deviation of the X variable

We have conducted a linear regression analysis on two data sets thus far in this chapter. In our first analysis we computed a regression coefficient of $b = .75$ by using the formula previously presented for "b." In the second analysis on the income and education data we computed a regression coefficient of $b = 4.74$ using the formula for "b." Summary information is presented below for the correlation coefficients and the standard deviations of the X and Y variables for both data sets.

Data Set	r	s_y	s_x
One	.923	3.25	4.00
Two	.864	22.75	4.15

With these summary data, and the alternative formula, we can solve for the regression coefficient "b."

Data set one:

$$b = .923 \times \left(\frac{3.25}{4.00} \right)$$

$$b = .923 \times .8125 = .7499 \text{ or } .75 \text{ when rounded}$$

Data set two:

$$b = .864 \times \left(\frac{22.75}{4.15} \right)$$

$$b = .864 \times 5.482 = 4.736 \text{ or } 4.74 \text{ when rounded}$$

In each case, we obtain the same results as before. With additional information on the means of the X and Y variables, we would be able to complete the analysis since the value of the Y axis intercept "a" is obtained from the two means and the value of "b."

Assumptions for Linear Regression

I made mention earlier that one of the assumptions or requirements necessary for legitimately conducting a linear regression analysis was that the errors around the regression line are normally distributed. There are a few other assumptions. The most obvious assumption is that there is a linear trend in the data. Again, not all relationships between two variables may be adequately expressed by a straight line. Linear regression analysis assumes that a linear relationship is present in the data.

Linear regression analysis also assumes the presence of at least interval level data for both the independent and dependent variables. Finally, a condition called **homoscedasticity** is assumed in the data. When homoscedasticity is present, the Y values will be equally distributed about the regression line across all values of X. The alternative is heteroscedasticity that results when the Y values are tightly distributed around the regression line at some values of X, and broadly distributed around the regression line at other values of X.

A final assumption for linear regression concerns the sample size. For most applications in the social sciences, the sample size is sufficiently large that there are no problems with performing a linear regression analysis. As a rule of thumb, one should have at least 10 observations or members of the sample for each independent variable included in the regression model. This assumption applies more to the case of multiple regression discussed below than linear regression since the latter only includes a single independent variable.

Two Important Alternatives to Linear Regression

There are several important alternatives to linear regression, and I want to mention two of them here. One of the assumptions of linear regression is that both the independent and dependent variables are measured at the interval level or better. Oftentimes, we may have a dependent variable that is at the nominal level. The most frequent case is when we are trying to predict a particular condition that is either present or absent. For example, we might be interested in predicting recidivism among recently released convicts, or we might be interested in predicting retention among college students. In each case there are only two alternatives. A released convict will commit another crime and go back to prison or not; a student will return for another year of college or not (for some of you those alternatives may sound very similar). Neither of these variables would be suitable for a typical linear regression analysis, but both might be suitable for a **logistic** or **logit** regression analysis. Logit regression does not assume a linear relationship, instead it assumes a logistic or curvilinear relationship. I will not be presenting an example of logistic regression here, but you should be aware that there are some alternatives to simple linear models in regression analysis that may be more appropriate for some applications.

The second alternative to linear regression is **multiple regression**. Actually, multiple regression is more of an extension of linear regression than an alternative to it. Whereas linear regression employs a single independent variable to predict a dependent variable of interest, multiple regression employs several independent variables. Since there is only one dependent variable there will still be only a single *Y* axis intercept, but a multiple regression analysis will have several regression coefficients—one for each of the independent variables. A generalized multiple regression equation for the simplest case of one dependent variable *Y*, and two independent variables X_1 and X_2 would take the following form:

$$Y' = b_1X_1 + b_2X_2 + a$$

The data could be plotted on a coordinate system consisting of one axis for each variable. In this case, there would be three axes: a vertical *Y* axis, and two horizontal *X* axes set at right angles to each other and to the *Y* axis. This simple case is the only one that we can visualize since having three axes at right angles to each other requires three dimensional spaces—where we all normally exist. It is possible to have a multiple regression equation with many more variables, but we cannot visualize it since the mathematical model involves more than three dimensions.

Our simple generalized multiple regression equation contains two regression coefficients or slopes. One regression coefficient would adjust the regression line up or down with respect to the *Y* axis, and the second regression coefficient would adjust the line to the left or right of the *Y* axis. To better visualize the effect, you can take any straight object such as a pencil or pen to represent a regression line, and hold it in front of you parallel to the floor. Now imagine a set of data points scattered about the line (you might think about a swarm of small insects as the data points, but they should not be moving). Multiple regression analysis will place the regression line such that it provides the best fit to the data just as linear regression analysis did, but in multiple regression the data points will surround the best fit line instead of just being above or below it in a single plane as in the case of linear regression. The resulting equation will have two regression coefficients or slopes. One slope will adjust the line upward toward the ceiling or downward toward the floor, while the other slope will adjust the line toward you or away from you.

A Conceptual Example of Multiple Regression

You are much more likely to encounter examples of multiple regression in the scholarly literature of your field than your linear regression. The models we employ in the behavioral sciences usually involve several variables, and to try to describe a complex relationship with a simple linear regression model in which we can introduce only a single independent variable will usually provide unsatisfactory results. However, it is not practical to present a computational example of multiple regression since the technique requires a great deal of tedious calculations, but we can present a typical situation that would require multiple regression, and report the results obtained from a computer aided analysis of the data.

Let's look at a typical use of multiple regression that one might see in an applied setting. Very often multiple regression analysis is used for site selection evaluations. A typical situation involves a corporation that has a number of established locations, and plans to open another outlet at one of several possible sites. The goal is to develop a suitable statistical model using multiple regression analysis that predicts annual sales for the existing locations, and then use the model to evaluate the possible sites for the new location. The process requires several important steps.

1. Identify potential independent variables that should enable us to predict the dependent variable—annual sales.
2. Build a database for the existing locations consisting of data on annual sales, and each of our possible independent variables.
3. Analyze the existing data with multiple regression to develop a suitable regression equation.
4. Collect data for the independent variables for each of the potential sites for the new location.
5. Use the regression equation to predict annual sales at each of the potential sites by plugging the data for the independent variables into the regression equation.
6. Reach a decision as to the best location.

In an actual application, we might run several different regression models trying different combinations of independent variables until we found the model that provides the best fit to the existing data. In this example, let's assume that we are evaluating four possible locations for a new fast food restaurant, and that we have found that the following three independent variables provided satisfactory results in our regression analysis of existing locations:

X_1—Population within a 5 mile radius
X_2—Number of competing restaurants within a 1 mile radius
X_3—Automobile traffic count per hour at the location

Let's assume that the corporation has a total of 200 existing restaurants. We would have to develop a database consisting of information on each of our three independent variables, and the dependent variable (annual sales) for each of the 200 locations. The database would appear as follows (note: annual sales are expressed in thousands of dollars):

	Y	X_1	X_2	X_3
Location	Sales	Population	Competition	Traffic
1	1250	3200	8	1000
2	2250	4500	5	1800
3	1120	2200	12	1500
4	850	1150	15	750
...	...	...	...	...
200	2000	3000	6	1900

Our database for all 200 locations would be analyzed with a multiple regression analysis. A typical multiple regression output from a computer statistical package might look as follows:

Multiple Regression Output	
Constant	1061.68
Standard Error of *Y* Estimate	586.50
R Squared	.62
X Regression Coefficients:	
X_1 .236 X_2 −105.56 X_3 .553	

With the information contained in the multiple regression output for the constant (the value of "*a*," the *Y* axis intercept) and the regression coefficients for each of our three independent variables, we are ready to write the regression equation:

$$Y' = .236X_1 - 105.56X_2 + .553X_3 + 1061.68$$

Note that the value of b_2 (the regression coefficient for the variable indicating the number of competitors within a 1 mile radius) is negative, while the values of b_1 (the regression coefficient for the variable indicating the population within a 5 mile radius) and b_3 (the regression coefficient for the variable indicating the traffic count per hour) are positive. We would expect such a result. Annual sales should be positively related to both population and traffic count since each of these variables represent the volume of potential customers. However, annual sales should be negatively related to competition in the area.

Notice that the multiple regression output contains the value of the standard error of the *Y* estimate (586.50) and a statistic called *R* Squared (.62). The standard error of the estimate is interpreted just as it was with linear regression. Since our *Y* variable, annual sales, is measured in thousands of dollars, our standard error of the *Y* estimate should be interpreted on that scale as well. The statistic indicates that on average the square root of the squared distance between the actual sales and the sales predicted by the regression model is 586.50 thousand dollars. We have not previously encountered the *R* Square statistic, but it is directly analogous to the coefficient of determination, r^2 in linear regression. The *R* Square statistic tells us the amount of variance in the dependent variable that is explained by all of our independent variables combined. In this case, we are able to explain 62% of the variation in sales by using population, competition, and traffic count as predictors.

Assuming that we are satisfied with our multiple regression model, we would now collect data for the three independent variables for our four possible sites for a new location. Our results might look like the following:

Potential Location	X_1 Population	X_2 Competition	X_3 Traffic
1	3100	12	1500
2	2500	7	1200
3	1200	8	900
4	2750	6	1800

We can then predict annual sales for each of the four possible sites by entering each site's data into the regression equation.

Predicted Annual Sales

$$Y'_1: .236(3100) - 105.56(12) + .553(1500) + 1061.68 = 1356.06$$

$$Y'_2: .236(2500) - 105.56(7) + .553(1200) + 1061.68 = 1576.36$$

$$Y'_3: .236(1200) - 105.56(8) + .553(\ 900) + 1061.68 = 998.10$$

$$Y'_4: .236(2750) - 105.56(6) + .553(1800) + 1061.68 = 2072.72$$

Clearly location 4 has the highest predicted annual sales of 2027.72 (thousands of dollars). Keep in mind that this does not mean that we should automatically choose location 4 as the best site. Multiple regression analysis, or any statistical technique for that matter, is just a tool to aid us in our decision-making process. What sorts of other factors might you suggest taking into account before a final decision is made? One important point is that the predicted sales are indicative of conditions as they exist currently. However, neighborhoods within a city change over time, and location 4 might be in an area that is in the process of urban decay. While it looks good on paper today, it might get worse and worse over time. Location 2 might not have the highest level of predicted sales today, but it might be a location in one of the high growth areas of the city. If so, its ability to generate sales will get stronger and stronger as time goes by. We also might want to consider the traffic flow in each area, not just the traffic volume. Just because an area has a high traffic volume does not mean that it will translate into increased sales. If access to a particular location is restricted potential customers will likely pass it by in favor of a location with easier access. The point is that there should always be more to the decision-making process than just the results of a statistical analysis.

Computer Applications

1. Load the GSS data set and conduct a simple linear regression analysis using Education (Educ) to predict Respondents Income in Constant Dollars (Coninc). Note the amount of variance explained by education. Write the regression equation.
2. Conduct a multiple regression analysis using "Coninc" as the dependent variable as above, but select three independent or predictor variables. Choose "Educ" as one of the independent variables, and two others that you think will help explain more variance in income. Indicate the improvement in the variance explained, and write the regression equation.

How to do it

Load the GSS data set. Click on "Analyze," "Regression," and "Linear" to open the linear regression dialog box. Highlight "Coninc" and select it as the dependent variable by clicking on the appropriate direction arrow. Highlight "Educ" and select it as the independent variable by clicking on the appropriate direction arrow. Click on "OK" to complete the procedure.

Click on "Analyze," "Regression," and "Linear" to reopen the linear regression dialog box. Highlight the two additional variables of your choice and add them to the selected list of independent variables. Click on "OK" to complete the procedure.

Summary of Key Points

Linear regression and its multivariate extension multiple regression are important and powerful tools in behavioral science research. As we have progressed through this chapter, we have encountered a number of key points that are summarized below.

The Equation for a Straight Line: $Y = bX + a$
The Linear Regression Equation: $Y' = bX + a$—A linear regression equation can be used to describe the relationship between two variables resulting in enhanced predictive ability.

The Regression Coefficient "*b*" is found by:

$$b = \frac{n \times \Sigma XY - (\Sigma X)(\Sigma Y)}{n \times \Sigma X^2 - (\Sigma X)^2}$$

or by the alternative formula:

$$b = r \times \frac{s_y}{s_x}$$

The *Y* Axis Intercept "*a*" is found by:

$$a = \bar{Y} - b(\bar{X})$$

Minimizing the sum of squares (the sum of the squared deviations from the actual *Y* values to the predicted *Y* values along a regression line)—The regression line providing the best fit to the data will be the line which minimizes the sum of squares. That line will always pass through the point $(\bar{X}, \bar{Y})$.
The Coefficient of Determination r^2—The coefficient of determination r^2 indicates the amount of variation in the dependent variable *Y* that is explained by the independent variable *X*. The value of r^2 is one way we can assess the quality of our regression model. In the case of multiple regression, multiple *R* Squared indicates the amount of variation explained by all of the independent variables included in the regression model.
The Standard Error of the Estimate—The standard error of the estimate serves as a standard deviation of the size of the error of the predicted values of *Y*. The size of the standard error also helps us assess the quality of our regression model.
Standardized Regression—Standardized regression is accomplished by converting the raw scores to normal scores prior to the regression analysis. One result is that the best fit regression line will always pass through the origin.
Assumptions of Linear Regression—Linear regression assumes data measured at the interval level or higher; the existence of a linear relationship in the data; normally distributed errors around the regression line; homoscedasticity (equal variances for the *X* and *Y* variables); and, approximately 10 cases in the sample for each independent variable included in the regression model.
Logistic Regression and Multiple Regression—Logistic regression and multiple regression are two important alternatives to simple linear regression. Logistic regression is appropriate when the dependent variable is not measured on an interval scale, and only takes on two values. Multiple regression is a valuable tool appropriate when we have several independent variables that we can use to predict a single dependent variable *Y*.
Outliers—Data points that lie far outside the normal range of other points in the distribution.
Homoscedasicity—Literally a condition of equal variance among subsets of the dependent variable *Y*. In linear regression homoscedasicity is present when the distances from the actual *Y* values to the regression line are similar across the entire range of the *X* axis. For example, the regression line would fit the *Y* data in a similar fashion across all values of *Y* as opposed to fitting the data very well at one end of the distribution (i.e. the higher values of *Y*), but not fitting the data at all well at the other end (i.e. the lower values of *Y*).

Explained Variance—Variation in the dependent variable Y that is due to (explained) by the independent variable X.

Unexplained Variance—Variation in the dependent variable Y that is not due to (not explained) by the independent variable X, and is therefore considered unexplained by the regression analysis.

Questions and Problems for Review

1. Plot the data provided below and describe the nature of the relationship. Is the relationship positive, negative, or close to zero? Do the data suggest a linear trend?

X	Y
8	2
7	3
9	1
5	10
10	0
2	12
5	9
7	7
6	8
12	1

2. Conduct a linear regression analysis on the data from problem 1.

A. Find the values of "b" and "a"; and, write the regression equation.
B. Draw the regression equation on your plot of the data.
C. Estimate the value of Y when:

$X = 20$
$X = -14$
$X = 3$

D. How much variation in Y is explained by the regression equation?

3. Read an article from a scholarly journal in your field that employs linear or multiple regression. Identify the dependent variable, and the independent variable(s). Note the values of "a" and "b" and write the regression equation. How much variation in Y is explained by the regression model?

4. Conduct a regression analysis given the information below:

Pearson correlation coefficient $r = .78$
Mean of the Xs = 25
Mean of the Ys = 30
Variance of the Xs = 25
Standard deviation of the Ys = 4

5. How much variation in Y is explained by the regression model for problem 4?

6. Write the general regression equation for a multiple regression analysis with 4 independent variables.

7. In the regression example earlier in the chapter where we predicted income by using education, we raised the possibility that two of the cases might be considered outliers ($X = 20$, $Y = 85$; and $X = 20$, $Y = 90$).

A. Declare those two cases as outliers (take them out of the analysis), and use the remaining $n = 18$ cases to conduct a new regression analysis.

B. Estimate the amount of variance education can explain in income in the new regression analysis (review the section of Chapter 9 on correlation if you need a reminder in computing a Pearson r correlation).

C. What impact did removing the two suspected outliers have on the overall analysis?

8. The data from problem 1 above have been converted to Z scores, and are presented below.

X	Y
.32	−.76
−.04	−.53
.67	−.98
−.74	1.08
1.02	−1.21
−1.79	1.53
−.74	.85
−.03	.38
−.39	.62
1.72	−.98

A. Plot the data, and compare the resulting scatter plot to the one produced for problem 1.

B. Perform a linear regression analysis on the data. Compute the value of "b" and "a," and write the regression equation.

C. Draw the regression line on the scatter plot, and compare the results to the regression line for problem 1.

9. Given the information below:

Pearson correlation coefficient $r = -.58$
Mean of the Xs = 12
Mean of the Ys = 8
Standard deviation of the Xs = 4.8
Standard deviation of the Ys = 6.25

A. Conduct a regression analysis, and write the regression equation.

B. What is your assessment of the quality of the regression model?

10. Plot the data for *X* and *Y* below.

X	*Y*
2	4
4	6
2	8
5	10
7	14
8	12
12	14
14	11
14	16
16	10
18	12
18	6
20	6
20	4

A. Do these data appear to be a good candidate for linear regression analysis? Why or why not?
B. Perform a linear regression analysis on the data. Compute the value of "*b*" "*a*," and write the regression analysis.
C. How much variation does the independent variable explain in the dependent variable?

CHAPTER 11

Hypothesis Tests for Means

Key Concepts

- *Z* Test for a Sample Mean and a Population Mean
- t-Test for a Sample Mean and a Population Mean
- t-Test for Two Independent Sample Means
- *Z* Test for Two Independent Sample Means
- *Z* Test for Two Proportions
- t-Test for Two Related Samples
- Research Hypothesis H_1
- Null Hypothesis H_0
- Alpha Level
- Critical Region
- Critical Value
- Test Statistic
- Statistical Significance

Introduction

The primary focus in this chapter is on tests of **statistical significance** and related procedures dealing with the mean. The procedures that we will cover in this chapter are some of the more common ones in use in behavioral science research, and if you have had any exposure to the results of research presented in professional journals, you will likely have seen applications of one or more of the techniques covered in this chapter. We will be testing hypotheses about means, and the first procedures we will examine will be situations where we want to compare a sample mean "$\overline{X}$," for a significant difference to a known population mean "μ." You should find the initial material very familiar because the first statistical test we will examine is the *Z* test that we used in Chapter 8 to demonstrate the procedure of hypothesis testing. From there, we will move to a discussion of the t-test used to compare a sample mean to a known population mean. The procedures for the *Z* test and the t-test are very similar with the difference being that the *Z* test is based on the normal distribution, and the t-test is based on the "t" distribution. You might recall that the concept of the "t" distribution was introduced briefly in Chapter 6 dealing with the normal distribution. This chapter will provide us with an opportunity to use the "t" distribution.

Following the discussion of the *Z* test and t-test for comparing a sample mean to a known population mean, we will examine several procedures used when the population mean μ is unknown. The first procedure is a variation of the *Z* test and t-test used to compare two sample means for a significant difference. Another related *Z* test will be examined that is used to compare two sample proportions for a significant difference. Finally, we will examine a version of the t-test that is used to compare two related samples for a significant difference. We will begin with a discussion of the *Z* test used to compare a sample mean for a significant difference from a known population mean following a brief review of the logic of hypothesis testing.

A Brief Review of the Logic of Hypothesis Testing

In Chapter 8, dealing with hypothesis testing we discussed the concept of a research hypothesis, symbolized by H_1, which represents a researcher's expectation for a particular research question. These hypotheses, or expectations, can come from our own logical analysis of a given question, or they may come from our knowledge of previous research efforts in our area of study. Also recall that it is actually the **null hypothesis**, symbolized by H_0, which is a statement of no relationship that we actually subject to statistical testing. We test the null hypothesis because it is much easier to disprove a statement than it is to prove a statement, and disproving the null hypothesis suggests support for its alternative, the **research hypothesis H_1**.

In Chapter 8, we also looked at some common strategies for providing evidence in support of a research hypothesis. One strategy that we discussed was comparing the mean for a sample to a known population mean. For example, to provide evidence in support of our expected positive relationship between education and income, we compared the mean income of a high education group (college graduates) to the mean income of the population in general. This type of research strategy can be tested through the use of a *Z* test that converts the distance between the sample mean $\overline{X}$ and the population mean μ to a *Z* score, which we can then place on the normal distribution. Our ability to use a *Z* test is based on the fact that the sampling distribution of the mean, the distribution of all possible samples of size "*n*" taken from a population, has the important characteristics of being normal in form, it is an example of a normal distribution, with a mean "$\mu_{\bar{X}}$" equal to the population mean "μ" from which the samples are drawn, and a standard deviation (called the standard error, "$\sigma_{\bar{X}}$") equal to the population's standard deviation "σ" divided by the square root of the sample size "*n*."

For example, if we know that the mean heart rate for the total population in general is $\mu = 72$ beats per minute with a standard deviation of $\sigma = 12$ beats per minute, and we were to take every possible sample of size $n = 36$, then the sampling distribution of all of the possible sample means would appear as a normal distribution with a mean $\mu_{\bar{X}} = 72$, and a standard error $\sigma_{\bar{X}} = 2$ (note: the standard error $\sigma_{\bar{X}} = 12/\sqrt{36}$) See Figure 11.1.

We then might be presented with the mean heart rate for a sample of $n = 36$ individuals, and asked to decide if it is likely that this sample of $n = 36$ individuals came from the general population or not. If you are wondering where the sample may have come from if it did not come from the general population, two possibilities are that the sample may have come from a population of highly trained athletes, in which case the sample's mean heart rate would be expected to be much lower than that of the general population, or the sample may have come from a population of sedentary individuals in which case the sample's mean heart rate would be expected to be higher than that of the population in general. While we can never be sure if the sample of $n = 36$ individuals came from the general population or not, we do know that if it did come from the general population, then the sample must be one of the samples in the general population's sampling distribution for all samples of size $n = 36$.

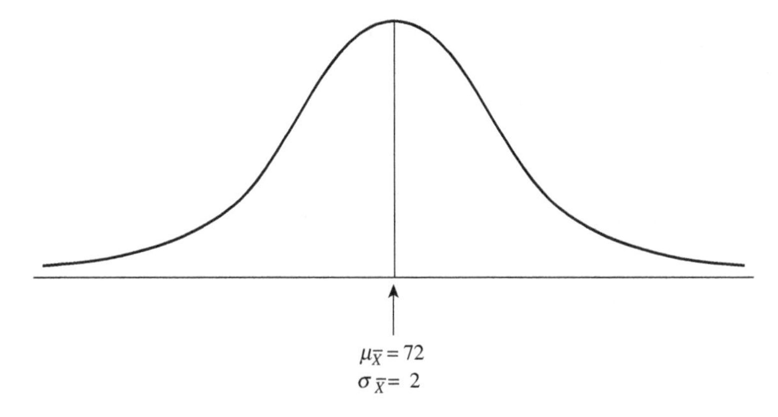

Figure 11.1 The Sampling Distribution of All Possible Samples of Size $n = 36$ Drawn from a Population with a Mean $\mu = 72$ and a Standard Deviation $\sigma = 12$

Given our knowledge of the structure of the normal distribution and of probability theory, we know that it is likely that any sample selected at random from a given population will have a sample mean that is reasonably close to the true population mean from which the sample was selected. We also know that the greater the distance between a given sample mean and a population mean, the less likely it is that the sample came from that particular population. It is more likely that the sample came from some other population, a population with a mean closer to the mean observed for the sample.

We saw that we can quantify our decision, or state it in terms of a specific probability, by converting the distance between the population mean and the sample mean we are examining into a *Z* score. We can then adopt a standard of proof regarding the likelihood that a sample mean came from a particular population's sampling distribution. For example, if the *Z* score indicating the distance between the sample mean in question and the population mean was in the extreme 5%, or perhaps the extreme 1% of a population's sampling distribution, then our decision will be that the sample mean in question probably did not come from the population.

Conversely, if the distance between the sample mean in question and the population mean converted to a *Z* score that was not at or beyond the extreme 5%, or the extreme 1% of the normal distribution, then we will conclude that the sample mean in question is likely to have come from the population. If we adopt the extreme 5% of the normal distribution as our level of proof, then any sample mean that is far enough away from the population mean to equal a *Z* score of 1.645 or beyond toward the tail of the curve will be considered as being far enough away from the population mean to suggest that the sample mean did not come from the population. Recall that the *Z* score of 1.645 marks the boundary of the extreme 5% of the normal distribution for a one-tail test of the null hypothesis.

A quick review of the steps in testing a statistical hypothesis

1. State the research hypothesis H_1, and the null hypothesis H_0.
2. Examine the research hypothesis H_1, and determine if a one-tail test or a two-tail test is in order:

   If the research hypothesis suggests a difference and a direction (a positive or a negative relationship), then a one-tail test is in order;
   If the research hypothesis suggests only a difference, then a two-tail test is in order.

3. Determine the appropriate statistical test to use.
4. Select a level of significance at which to conduct the test.
5. Identify the **critical value** of the **test statistic**.

6. Compute the value of the test statistic using the data.
7. Compare the observed (computed) value of the test statistic to the critical value of the test statistic, and reach a decision regarding H_0 (See Figure 11.2):

   If the observed value of the test statistic falls on or in the **critical region**, then you REJECT H_0, and ACCEPT the alternative research hypothesis H_1;
   If the observed value of the test statistic fails to fall on or in the critical region, then you FAIL TO REJECT H_0.

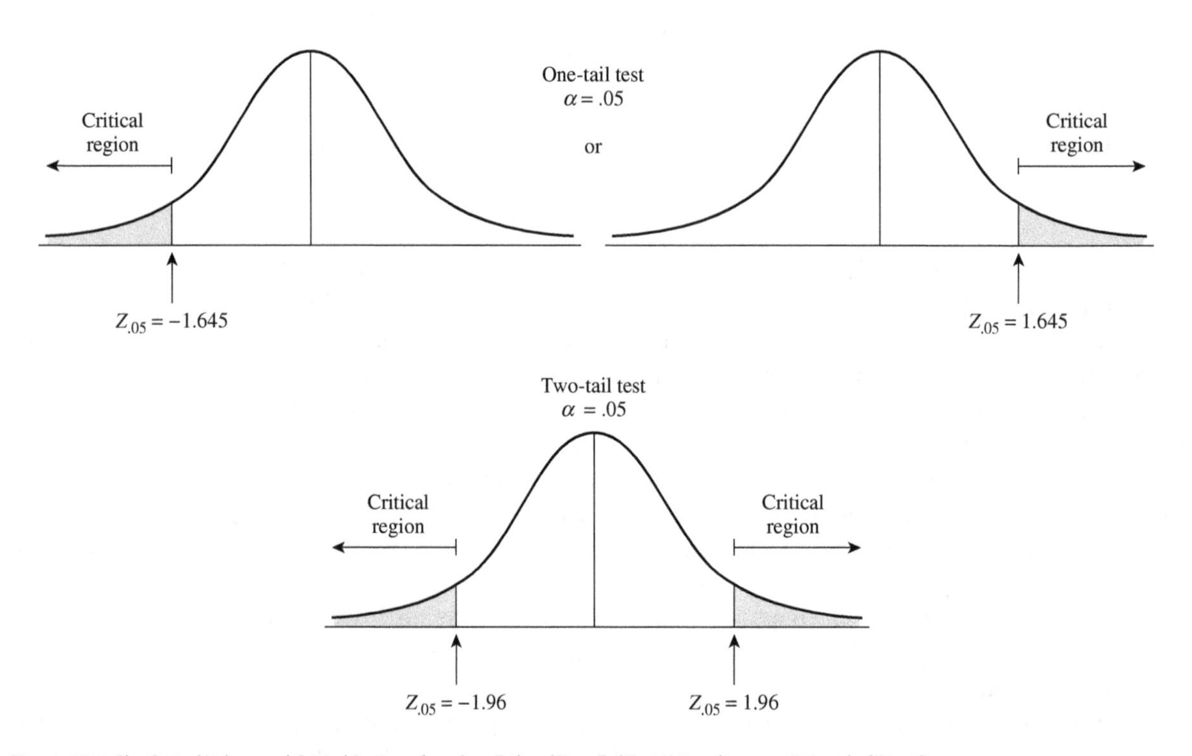

Figure 11.2 The Critical Values and Critical Regions for a One-Tail and Two-Tail Test Using the $\alpha = .05$ Level of Significance

The *Z* Test for Comparing a Sample Mean $\bar{X}$ to a Known Population Mean μ

The *Z* test is commonly used to test for a statistically significant difference between a sample mean $\bar{X}$ and a known population mean μ. The distance between the sample mean and a known population mean can be converted to a *Z* score by the following formula.

Formula to Convert the Difference between a Sample Mean and a Population Mean to a Z Score

$$Z = \frac{\bar{X} - \mu}{\sigma_{\bar{X}}}$$

where

$\bar{X}$ = the sample mean of interest
μ = the population mean (also equal to the mean of the sampling distribution)
$\sigma_{\bar{x}}$ = the standard error of the distribution ($\sigma/\sqrt{n}$).

The resulting observed value for the *Z* test is then compared to the critical value associated with the size of the critical region in the tail of the normal distribution. The size of the critical region is determined by the level of significance, or **alpha (α) level**, which we choose in order to test the null hypothesis. Commonly chosen alpha levels are as follows: .10, .05, .025, .01, and .001. The critical value of the test statistic is the value of the *Z* score that marks the boundary of a given critical region in the tail of the normal curve. For example, a one-tail test conducted at the $\alpha = .01$ level is the *Z* score that marks the boundary of the extreme .01 or 1% of the normal distribution. That *Z* score is either + 2.33 or −2.33 depending on the direction implied in the research hypothesis. Remember that with a two-tail test of the hypothesis, one in which no direction is implied in the research hypothesis, we must split the size of the critical region and place half in one tail and half in the other tail. The critical *Z* values for commonly chosen alpha levels for both one-tail and two-tail tests are presented below. These values can also be found at the end of Table 1 in the Appendix containing areas under the normal curve associated with given *Z* scores.

Critical *Z* Values

Alpha Level	One-tail test + or −	Two-tail test + and −
.10	1.28	1.645
.05	1.645	1.96
.025	1.96	2.24
.01	2.33	2.575
.001	3.08	3.30

The *Z* test is properly applied when the following assumptions are met.

1. The sample has been selected on a random basis.
2. The dependent variable is normally distributed within the population (normality becomes less important as the size of the sample increases).
3. The dependent variable has been measured at the interval or ratio level (allowing us to legitimately compute a mean).
4. The true mean (μ) of the total population to which our sample mean will be compared is known.
5. The population standard deviation (σ) is known, and not estimated from the sample data.

An example of the *Z* test

The mean heart rate of the general population is known to be $\mu = 72$ beats per minute with a standard deviation of $\sigma = 12$ beats per minute. It is hypothesized that individuals who have sedentary occupations and who do not perform any physical exercise on a regular basis will have a significantly higher heart rate. The null hypothesis that we will actually subject to statistical testing is that there is no difference between the mean heart rate of those in sedentary occupations who do not perform any physical exercise on a regular basis compared to the general population. Both the research hypothesis (H_1) and the null hypothesis (H_0) are represented symbolically below.

$H_1 : \mu > 72$

$H_0 : \mu = 72$

Note that μ in this case refers to the mean heart rate of the population of individuals in sedentary occupations who do not perform any physical exercise on a regular basis, and that μ will be estimated by our observation of a probability sample drawn from that population. The value "72" represents the known mean heart rate of the population in general.

Notice that the research hypothesis indicates an expected direction for the difference between the population mean of sedentary individuals estimated by our sample and the known mean of the population in general, so a one-tail test is appropriate in this case. To evaluate the hypothesis, we will obtain a sample of $n = 36$ individuals with sedentary occupations who do not perform any physical exercise on a regular basis, and compute their mean heart rate. We will then test for a statistically significant difference between the population mean estimated by our sample and the known mean of the population in general at the $\alpha = .05$ level using the Z test. The necessary data are presented below:

Population mean $\mu = 72$

Population standard deviation $\sigma = 12$

Sample size $n = 36$

Standard error of the sampling distribution $= 12/\sqrt{36} = 2.00$

Sample mean $\bar{X} = 75.5$

Level of significance $\alpha = .05$

Critical value of the test statistic $Z_{.05} = 1.645$

The computation of the Z test statistic is as follows:

$$Z = \frac{75.5 - 72}{2.00} = \frac{3.5}{2.00} = 1.75$$

The observed value of the test statistic, $Z = 1.75$, is clearly in the critical region that begins at $Z_{.05} = 1.645$, so the proper decision is to **reject the null hypothesis**. The mean heart rate of the sample of $n = 36$ individuals with sedentary occupations who do not perform physical exercise on a regular basis is statistically significantly higher than the mean heart rate of the population in general.

Keep in mind what a finding of statistical significance means in this and other similar examples of hypothesis testing. The finding of a statistically significant difference simply means that the mean heart rate of the sample is different enough from the mean heart rate of the population to suggest that the difference is not due to sampling error. The difference is large enough to suggest that it is a real difference, and that if we had data for the entire population of individuals with sedentary occupations who do not perform physical exercise on a regular basis, we would find that their mean heart rate (μ) would in fact be higher than the mean heart rate of the general population. Of course, we have a chance of

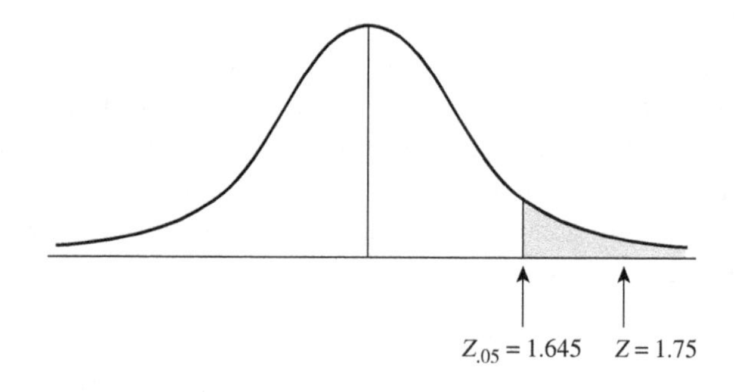

Figure 11.3 Observed and Critical Values of *Z*

being wrong in concluding that there is a real difference based on our observation of the sample, and that chance of being wrong is 5%, corresponding to the level of significance at which we chose to test the null hypothesis.

A finding of statistical significance involves more than the mean

All too often, we focus on the difference between the sample mean and the population mean as the key component in a finding of statistical significance. Granted, there can be no finding of a statistically significant difference unless there is at least a difference between the sample mean and the population mean but all of the components of the test statistic formula play a role in a finding of statistical significance. A finding of statistical significance may also depend on the amount of variation around the mean (as measured by the standard deviation) in the population or on the sample size. The smaller the variation around the mean in the population, the more likely a statistically significant result will be found. If you examine the formula for the Z test, you will see that all other things being equal the smaller the standard deviation, the smaller size of the standard error, and the more likely a significant result will occur. Similarly, the larger the sample size, the smaller the size of the standard error, and the more likely a significant result will occur. Conversely, the more variation around the mean in the population (as evidenced by a larger standard deviation), the less likely a statistically significant result will occur. A larger standard deviation will result in a larger standard error that reduces the chances for a statistically significant result. Likewise, the smaller the sample size, the larger the standard error, and the less likely a statistically significant result will occur.

For example in the case we have just examined, had the population standard deviation been only one point higher ($\sigma = 13$ instead of $\sigma = 12$), the following result would have been obtained:

Population mean $\mu = 72$

Population standard deviation $\sigma = 13$

Sample size $n = 36$

Standard error of the sampling distribution $= 13/\sqrt{36} = 2.17$

Sample mean $\bar{X} = 75.5$

Critical value of the test statistic $Z_{.05} = 1.645$

The computation of the Z test statistic would have been the following:

$$Z = \frac{75.5 - 72}{2.17} = \frac{3.5}{2.17} = 1.613$$

which is a nonsignificant result.

Similarly, had the sample size been reduced only slightly to $n = 30$, from the original $n = 36$, a non-significant result would have been found as well, as illustrated below.

Population mean $\mu = 72$

Population standard deviation $\sigma = 12$

Sample size $n = 30$

Standard error of the sampling distribution $= 12/\sqrt{30} = 2.19$

Sample mean $\bar{X} = 75.5$

Critical value of the test statistic $Z_{.05} = 1.645$

The computation of the Z test statistic would have been the following:

$$Z = \frac{75.5 - 72}{2.19} = \frac{3.5}{2.19} = 1.598$$

which is a nonsignificant result.

The role of the amount and type of variance around the mean can be so important with respect to statistical significance, that in some situations the analysis of variance becomes a first step in hypothesis testing. This is especially true when a research strategy more complicated than the current example of comparing a single sample mean to a known population mean is undertaken. For example, we might have several sample means, and we want to know if there are any significant differences among them. Rather than conducting multiple Z tests, we can perform an initial analysis of variance to see if any of the means are significantly different. Analysis of variance is the subject of the next chapter.

It is also important to understand what the finding of statistical significance does not mean in this case. It does not mean that there is anything medically wrong with the individuals in the sample or the population they represent. It does not mean that their life expectancy will be lower, or that they are at a higher risk of developing heart disease. The finding of statistical significance just means that the heart rate of the sample is higher by 3.5 beats per minute than the population in general, and this difference is large enough to suggest that it is a real difference similar to what would be found in the population from which the sample was drawn. Can a higher heart rate be an indication of a medical problem, and if so, how much higher must a heart rate be to result in a medical problem? These are both important questions, but they are not statistical questions, they are medical questions. Our statistical analysis can only tell us if there is a difference between the two groups, and if so, is that difference large enough to suggest that the difference is real (statistically significant) or that the difference is within the range that can be attributed to sampling variability (not statistically significant).

A second example of the *Z* test

To fulfill the requirements of a class assignment, a sociology student has designed a study to examine the effects of marriage on the grade point average of full-time undergraduate students at her university. The student has been told that the mean grade point average of all full-time undergraduates is $\mu = 2.63$, with a standard deviation of $\sigma = 0.95$ points. The student plans to obtain a sample of married, full-time undergraduate students and to compute their mean grade point average. The research hypothesis of the student is that the mean grade point average of the married students will be different than the grade point average of the full-time undergraduate population. The null hypothesis is that there will be no difference between the mean grade point average of the married students and the grade point average of the full-time undergraduate population. Represented symbolically, the research hypothesis H_1, and the null hypothesis H_0 are as follows:

$H_1 : \mu \neq 2.63$

$H_0 : \mu = 2.63$

The student does not specify a direction for the relationship, because she can make a logical argument that marriage may act to increase academic performance, or that it may act to decrease academic performance. Marriage may act to increase academic performance because marriage often results in a higher level of maturity and sense of responsibility that may carry over into one's academic pursuits. Alternatively, marriage can place great demands on one's time and energy, and that may detract from the time that one would otherwise devote to academic pursuits resulting in a reduction in one's grade point average.

The student has adopted a level of significance of $\alpha = .01$, and has determined that a two-tail test of the hypothesis is in order. She surveys a total of $n = 100$ married, full-time undergraduate students, and observes a mean grade point average of $\bar{X} = 2.87$. Summarizing the relevant data:

$\mu = 2.63$
$\sigma = 0.95$
$\bar{X} = 2.87$
$n = 100$
$\sigma_{\bar{x}} = 0.095$
$\alpha = .01$
$Z_{.01} = \pm 2.575$

Computing the Z test statistic:

$$Z = \frac{2.87 - 2.63}{0.095} = \frac{0.24}{0.095} = 2.53$$

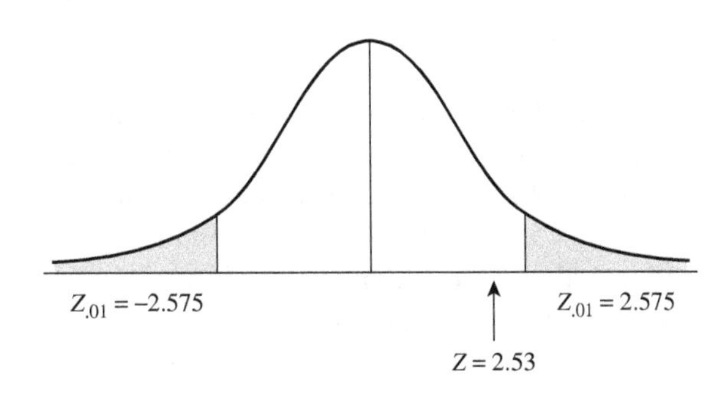

Figure 11.4 Observed and Critical Values of *Z*

The observed value of the test statistic, $Z = 2.53$, is not in either of the critical regions, so the correct interpretation is to Fail to Reject H_0. The mean grade point average of the married students is higher than that of the general student population (2.87 vs. 2.63), but the difference is not statistically significant, and falls into the range of what might be expected as sampling variability.

Note that had the student chosen the $\alpha = .05$ level of significance, the critical values would have been $Z_{.05} = \pm 1.96$, and the observed test statistic would have been in the right-hand critical region. At the $\alpha = .05$ level, we would have rejected the null hypothesis, and concluded that the mean grade point average of the married students was statistically significantly higher than that of the general student population! Both the $\alpha = .01$, and $\alpha = .05$ levels of significance are commonly used, and both are legitimate levels at which to conduct statistical tests. In this example, one would lead to a finding of non-significance, and the other would lead to a finding of statistical significance. There are two important points to make here. First, anytime a statistically significant result is reported, one of the first questions you want to ask is "What level of significance was used to test the hypothesis?" Second, the level of significance at which you will conduct your statistical tests should be chosen prior to examining the data. It is not ethical to obtain a test statistic, and then choose a level of significance that supports your desired result.

It is particularly important to select a level of significance, which you will use in your research if you will be using a computer package to analyze the data. Many of the leading computer packages used in behavioral science research will report the value of the observed test statistic and the probability level at which the test statistic would be considered significant. You will then need to compare the reported probability level with the level of significance you have chosen. For example, a typical statistical analysis computer package would report the results for the grade point average analysis as follows:

$$Z = 2.53 \; (p = 0.0114).$$

The results indicate that the observed value of the test statistic is $Z = 2.53$ and the value in parenthesis represents the exact probability associated with the test statistic. In this case, the test statistic would be considered statistically significant at the $\alpha = .0114$ level. We would reject the null hypothesis at any level of significance we had chosen that is 0.0114 or greater; we would fail to reject the null hypothesis at any level of significance less than 0.0114. This is consistent with our findings above when we failed to reject H_0 at the $\alpha = .01$ level of significance, but we would have rejected H_0 had we chosen the $\alpha = .05$ level of significance. An **alpha level** of .05 is greater than the 0.0114 probability level, and an alpha level of .01 is slightly less than the 0.0114 probability level.

The t Distribution

There are times when it is not appropriate to use the *Z* test for comparing a sample mean to a known population mean, and one should instead use a similar statistical test called the Student's t-test (or just the t-test for short). Actually, it is something of an understatement to refer to the t-test as being similar to the *Z* test. As you will see below when the formula is presented, the tests are actually identical. What will be different between the *Z* test and the t-test are the critical values used to interpret the observed value of the test statistic. The critical values for the *Z* test always come from the normal distribution, and as a result the critical values are always the same for any *Z* test conducted at a given level of significance. For example, a one-tail test conducted at the $\alpha = .05$ level is always +1.645 or −1.645 depending on the expected direction stated in the research hypothesis, and the critical values for a two-tail test conducted at the $\alpha = .05$ level are always ±1.96. However, the t-test does not use critical values from the normal distribution. Critical values for the t-test come from the t-distribution, or more accurately, the set of t-distributions, because the t-test is based on a number of related sampling distributions that resemble a normal distribution, but are not quite normal in form.

You may recall an earlier brief discussion of the "t" distribution in Chapter 6 on the normal curve. A statistician named W.S. Gossett found that the sampling distribution of the mean was not quite normal in form when the value of the population standard deviation "σ" had to be estimated by the sample standard deviation "*s*," and when based on small sample sizes. Gossett chose to publish his findings under the pseudonym of "Student," because his employer did not allow the publication of company sponsored research, hence the name "Student's" t-test.

In many respects, the t distribution is similar to the *Z* or normal distribution, but an important difference between the two is that the t distribution has an added degree of uncertainty resulting from the fact that the true value of the population standard deviation "σ" is unknown, and must be estimated by the sample standard deviation "*s*." Just as sample means ($\bar{X}$) based on the same size sample "*n*" drawn from a given population can yield different estimates of the true population mean μ, so too will different sample standard deviations (s) based on the same size sample "*n*" yield different estimates of the true population standard deviation σ. However, as the sample size "*n*" increases, the accuracy of the estimate of population standard deviation "σ" based on the sample standard deviation "*s*" also increases. In the case of large sample sizes, the difference between the "t" distribution and the normal distribution becomes insignificant, and one can use the *Z* test based on the normal distribution even without knowing the true value of the population standard deviation σ. **Definitions of "large sample size" vary, but as a rule a sample size of $n = 30$ or greater is sufficient to justify the use of the *Z* test and normal distribution**. It is important to note that even with small sample sizes of $n < 30$, the *Z* test and normal distribution may be used when the true value of the population standard deviation σ is known.

When the population standard deviation σ is unknown, the primary problem caused by with the small sample size is that our estimate of the population standard deviation σ tends to be biased. Again, from an earlier chapter (Chapter 5 on measures of variation), you should recall that we tried to adjust for this fact by adjusting the formula for a sample standard deviation that was to serve as the basis for an estimate of the population standard deviation. When estimating a population standard deviation using sample data we use the following formula for a standard deviation.

Computational Formula for Sample Standard Deviation Used to Estimate the Population Standard Deviation σ

$$s = \sqrt{\frac{\sum X^2 - \frac{(\sum X)^2}{n}}{n-1}}$$

We saw in Chapter 5 that "*n* – 1" is sometimes called "degrees of freedom," and that it refers to the number of elements in the sample that are free to take on any value, and still allow us to preserve the characteristic that the sum of the deviations of each score from the mean in a population will always sum to zero. I mentioned earlier that there is a set of t distributions, and each is slightly different depending on the size of the sample on which the distribution is based. We distinguish each t distribution not by the sample size "*n*" upon which it is based, but by its associated degrees of freedom (*n* – 1).

Given the diversity of multiple t distributions, it is not possible to present a single distribution and have access to all of the critical values necessary to conduct t-tests the way we can with the normal distribution and the Z test. Table 2 in the Appendix contains critical t values at the α = .05, and α = .01 levels of significance for selected degrees of freedom under the assumption of both a one-tail and a two-tail test of the null hypothesis. A portion of the table containing critical values for a one-tail test at the α = .05 level of significance is presented below.

Degrees of Freedom (*n* – 1)	α = .05
1	6.314
2	2.920
3	2.353
15	1.753
25	1.725
30	1.697
75	1.665

Notice that the critical t value at the α = .05 level (the t value that marks the boundary of the extreme 5% of the t distribution just as the *Z* value 1.645 marks the boundary of the extreme 5% of the normal distribution) at degrees of freedom (df) = 1 is quite high at 6.314. Notice also that as the degrees of freedom increase, the critical values of their associated t distribution become smaller. By the time we have df = 30, the critical t value at the α = .05 level is 1.697, which is very close in size to the *Z* value of 1.645. This is why as a general rule, it is permissible to conduct a *Z* test when there is a large sample size even in the absence of knowledge of the value of the population standard deviation σ. However, without knowledge of the population deviation σ, and a small sample size, one should not conduct a *Z* test, but should instead conduct a t-test as described in the following section.

The t-Test for Comparing a Sample Mean $\bar{X}$ to a Known Population Mean μ

The logic of the t-test for comparing a sample mean to a known population mean is identical to that of the *Z* test for comparing a sample mean to a known population mean that we have already examined. The formula is as follows:

$$t = \frac{\bar{X} - \mu}{s_{\bar{X}}}$$

where

$\bar{X}$ = the sample mean of interest
μ = the population mean (also equal to the mean of the sampling distribution)
$s_{\bar{X}}$ = the estimated standard error of the sampling distribution $(s/\sqrt{n})$.

The t-test can be conducted as a one-tail test or a two-tail test of the hypothesis, and just as before with the *Z* test, the null hypothesis is a statement of no difference between the sample mean and the population mean. The assumptions required for the *Z* test also apply for the t-test with the exception that the population standard deviation (σ) is unknown, and must be estimated from the sample data. A typical research application of the t-test is as follows.

State funding to a local high school is in part based on mean student attendance with funding reduced proportionally as student attendance declines. The school principal retains a sociologist as a consultant in an effort to improve attendance and increase the level of state funding. The sociologist designs an incentive program where students most prone to cut class will be rewarded for every day they attend school. The research hypothesis is that the mean attendance of the experimental group will be higher than the mean of the population, or stated symbolically:

$H_1: \mu > .85$

The null hypothesis is a statement of no difference, or stated symbolically:

$H_0: \mu = .85$

The program is tried experimentally on a group of $n = 25$ students for a period of 3 months. The results are as follows:

Mean attendance level of at risk students: $\mu = .85$ (indicating proportion of days in attendance)

Mean attendance level of experimental group of $n = 25$ at risk students: $\bar{X} = .91$; with a standard deviation of $s = .15$

In this case, we have a one-tail test because the research hypothesis provides an expected direction, we expect the experimental group to have a higher level of attendance than the population in general. We will be comparing a sample mean to a known population mean for a sample of $n = 25$, and because the population standard deviation "σ" is unknown, we will be estimating the standard error of the sampling distribution using the sample standard deviation "*s*," so a t-test is the appropriate choice, and we will test the null hypothesis at the $\alpha = .05$ level.

The critical t value may be obtained from the appropriate section of Table 2 (Critical Values of *t* in Appendix A) for a one-tail test at $\alpha = .05$, and 24 degrees of freedom. The critical t value is as follows:

$t_{.05} = 1.711$

Summarizing the data for the t-test:

$\mu = .85$
$\bar{X} = .91$

$n = 25$
$s = .15$
$s_{\bar{X}} = .15/\sqrt{25} = .03$

$$t = \frac{.91 - .85}{.03} = \frac{.06}{.03} = 2.00$$

The observed t value of 2.00 is larger than the critical value of 1.711, so our decision is to REJECT H_0. The experimental program resulted in a statistically significant improvement in the attendance level of the experimental group (See Figure 11.5).

Keep in mind that statistical testing is a useful tool in decision making, but a result of statistical significance should seldom be the sole determinant of a decision. In this case for example, we have seen that the experimental program produced statistically significant results, but that does not necessarily mean that the program should be adopted and applied school-wide. What other questions or issues should be considered before a final decision is made? Certainly one question should be the cost of the program. Higher attendance results in a higher level of funding from the state, but how much does the incentive program cost to achieve the higher attendance? It is possible that the cost of the incentive program would be greater than the additional funding received from the state, and as a result the school would be in worse financial shape even in the face of improved attendance. A second question that should be considered is whether the improved attendance resulting from the incentive program is permanent, or is it just a temporary phenomenon that will disappear after the novelty of the incentive program wears off? Of course this question cannot be answered by this study, and a second study of a longer duration would be required to adequately answer the question. As you may have already seen in some of the content courses of your major, it is common in behavioral science research for a research study to create as many questions as it answers. It is by pursuing answers to these questions through the use of a scientific method and the application of statistical analysis that we build the body of knowledge associated with our respective disciplines.

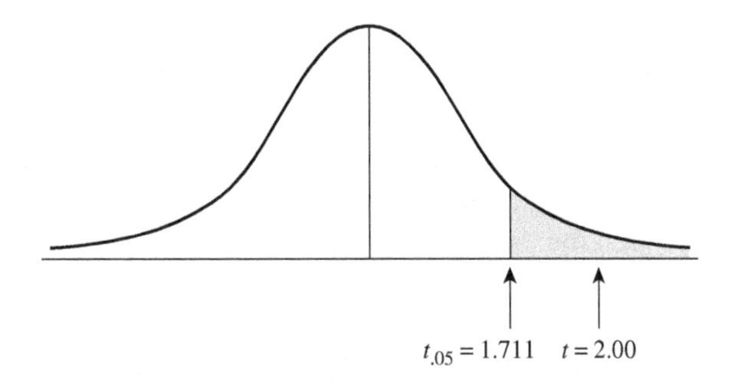

Figure 11.5 Observed and Critical Values of *t*

A second example of the t-test

The 1-year recidivism rate for juvenile offenders in a particular court jurisdiction is known to be $\mu = .47$, indicating that 47% of all juvenile offenders convicted of a crime are arrested for a second offense within 1 year. A new shock probation sentencing program where juveniles are sent to a 6-week "boot camp" facility featuring strict military style discipline is implemented on a trial basis on a group of $n = 16$ juvenile offenders. Expectations among the administrators of the program are mixed concerning the results. Some think that the shock probation program will be highly effective, and as a result expect the recidivism rate to decline. Others think that the short duration of the program may send the wrong signal to the juveniles, who will view it as a "slap on the wrist," and the recidivism rate may increase. The court has retained you to evaluate the results using a two-tailed test of the hypothesis at the $\alpha = .01$ level.

Table 2 in the Appendix indicates that the appropriate critical values of the test statistic for a two-tail test with 15 degrees of freedom at the $\alpha = .01$ level are $t_{.01} = \pm 2.947$. The research and null hypotheses are stated symbolically below:

$H_1: \mu \neq .47$

$H_0: \mu = .47$

The data are as follows:

$\mu = .47$
$\bar{X} = .42$
$s = .14$
$n = 16$
$s_{\bar{X}} = .14/\sqrt{16} = .035$

$$t = \frac{.42 - .47}{.035} = \frac{-.05}{.035} = -1.43$$

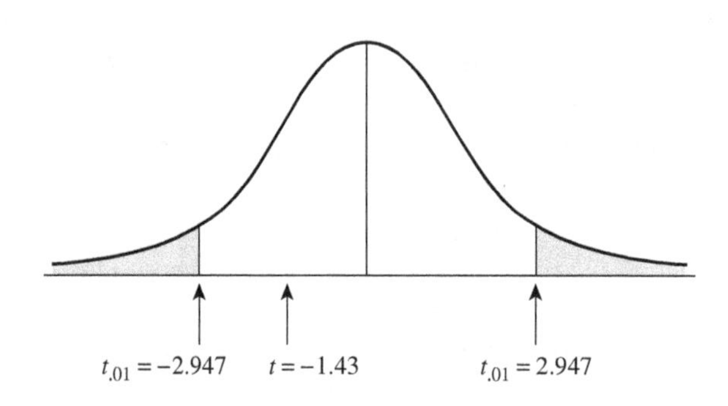

Figure 11.6 Observed and Critical Values of *t*

The observed value of the test statistic of t = -1.43 is not in either of the critical regions (See Figure 11.6), so the appropriate decision is to FAIL TO REJECT H_0. The shock probation program is associated with a lower recidivism rate, but the difference between the recidivism rate observed among the sample of n = 16 juveniles is not statistically significantly different than the recidivism rate for the population. The difference between the two recidivism rates is within the range that could reasonably be expected as a result of sampling error.

Statistical Tests for Two Independent Sample Means

The two statistical tests that we have examined thus far for comparing a sample mean to a known population mean, the Z test and the t-test, are useful statistical tests, but in reality, they are not frequently used in the practice of behavioral science research. The reason the tests are not used that often, is that in the majority of cases, we either know the value of the population mean for both groups we wish to compare, or we do not know the value of the population mean for either of the comparison groups. In the first case where we know the value of both population means, we do not need any statistical test, because as you should recall, you can tell if two population means are different by simply looking at their respective values. In the second case where we do not know the value of either population mean, we cannot use the Z test or t-test that we have just examined.

A much more common research situation is one in which we have data for two sample means, and we wish to compare the sample means for a statistically significant difference. The fact that we have two independent sample means does not necessarily mean that we have gone through a sampling process twice, although it can mean that. The most common situation resulting in the comparison of two independent sample means is when a single sample is drawn from a population, and then relevant subgroups from the sample are examined for statistically significant differences. For example, we might select a single sample and then compare the mean value on a variable of interest for the males in the sample with the mean value on the same variable for the females. To say that the sample means are "independent" indicates that the groups are mutually exclusive. It is possible to have two independent sample means that result from two separate sampling efforts. For example, we might want to compare college students enrolled at private colleges or universities with college students enrolled at state-supported colleges or universities. In this case, we would actually be taking a separate sample from each population of interest. A second situation that might involve two separate sampling efforts would be when we want to examine the same population over time. We might select and analyze a sample, and then return to select a second sample 5 years later to assess changes over time.

In each of the scenarios described above, we would have two independent sample means that we wish to compare for evidence of a statistically significant difference. Just as before, our research might be guided by a research hypothesis that calls for a one-tail test if a direction is specified, or a two-tail test if we are only looking for a difference between the two sample means. Also as before, the null hypothesis is actually subjected to the statistical test at a specified level of significance. There are two statistical tests that are commonly used to test two independent sample means for a statistically significant difference: a t-test and a *Z* test. Unlike before, the choice of the appropriate test is not based on knowledge of the population standard deviation or the sample size. With two independent sample means, we do not know the population standard deviations (or the population variances), and must estimate them by using the respective sample standard deviations.

The difference between the t-test and the *Z* test is the way in which we compute the standard error of the sampling distribution. Both tests assume that the respective sample variances (s^2) are accurate estimates of the true population variances (σ^2), although each sample variance will be subject to sampling variation. The t-test is the more appropriate choice under the assumption that the two sample variances are equal. (There are several procedures available to test for a statistically significant difference between two variances, but we have not included them here.) Under the assumption of equivalent sample variances, we will obtain our best estimate of the true value of the population variance by using a weighted average of the two sample variances, a process often referred to as "pooling the variance." Each sample variance is weighted by its respective degrees of freedom, and although this process may be a little cumbersome it will yield a better estimate of the true population variance that we will then use to compute the standard error of the difference (the equivalent of the standard error in a single sample t-test). The t-test for two independent sample means is described below.

The t-test for two independent sample means

The t-test for two independent sample means is an inferential statistical test used to determine if two population means are different based on evidence from the two observed sample means. The t-test is properly applied under the following assumptions:

1. The mean of the variable of interest is measured in two independent random samples;
2. The variable of interest has been measured at the interval or ratio level (allowing us to legitimately compute a mean);
3. The population of raw scores for the variable of interest is normally distributed (this assumption is less important for large sample sizes of $n > 30$);
4. The population variance for the variable of interest must be estimated from the sample data; and,
5. The two population variances are equivalent (allowing us to use a pooled estimate of the population variance).

The null hypothesis, H_0, is usually a statement of no difference between the two population means, and is written as follows:

$$H_0 : \mu_1 - \mu_2 = 0$$

The null hypothesis is written in the form stating that the difference between the two population means will equal zero, rather than just stating that the two means equal one another because there are times when the t-test for two independent samples is used under the assumption of a nonzero difference between the two population means. We will examine an example of a nonzero difference between the two population means after first presenting the formula for the t-test, and an illustration of it using the more typical situation of a hypothesized zero difference.

The Formula for the t-Test for Two Independent Sample Means

$$t = \frac{(\overline{X}_1 - \overline{X}_2) - (\mu_1 - \mu_2)}{s_{\overline{X}_1 - \overline{X}_2}}$$

where

$(\overline{X}_1 - \overline{X}_2)$ = the difference between the two observed sample means

$(\mu_1 - \mu_2)$ = the hypothesized difference between the two population means (usually assumed to be zero)

$s_{\overline{X}_1 - \overline{X}_2}$ = the standard error of the difference (analogous to the standard error in the single sample t-test)

The standard error of the difference is based on a pooled estimate of the population variance using sample data from each of the two samples weighted by their respective degrees of freedom. The pooled estimate of the population variance is obtained as follows:

Computation for the Pooled Estimate of the Variance:

$$s^2\text{pool} = \frac{(n_1 - 1)s_1^2 + (n_2 - 1)s_2^2}{(n_1 - 1) + (n_2 - 1)}$$

where

s_1^2 = the estimate of the population variance for sample one

s_2^2 = the estimate of the population variance for sample two

$(n_1 - 1)$ = the degrees of freedom for sample one

$(n_2 - 1)$ = the degrees of freedom for sample two

The pooled estimate of the population variance is then used as the basis for the standard error of the difference as follows:

Computation of the Standard Error of the Difference:

$$s_{\overline{X}_1 - \overline{X}_2} = \sqrt{s^2\text{pool} + \left(\frac{1}{n_1} - \frac{1}{n_2}\right)}$$

The formula for the standard error of the difference may seem to be foreign to you, but the concept is very similar to the standard error used in the single sample t-test. As you recall from earlier in the chapter, the standard error in the single sample t-test is computed by dividing the estimate of the population standard deviation "$s_{\overline{X}}$" by the square root of the sample size "n." In the case of the standard error of the estimate, we are working with a pooled estimate of the population variance, and when it is multiplied by the term $(1/n_1 + 1/n_2)$, we have the equivalent of the variance divided by the sample size. When we take the square root of that term we have the standard error of the difference used in the two sample case that is equivalent to the standard error for the one sample case.

Computational Formula for the Two Sample t-Test

$$t = \frac{(\overline{X}_1 - \overline{X}_2) - (\mu_1 - \mu_2)}{\sqrt{\frac{(n_1 - 1)s_1^2 + (n_2 - 1)s_2^2}{(n_1 - 1) + (n_2 - 1)}\left(\frac{1}{n_1} + \frac{1}{n_2}\right)}}$$

As you can see, the computation for the t-test for two independent sample means can be quite cumbersome, but the logic of the test follows directly from what we have seen previously with the single sample t-test. The two sample t-test may be used for either a one-tail test of the hypothesis or a two-tail test of the hypothesis. Critical values for the test statistic are obtained from the appropriate section of Table 2 in the Appendix using $(n_1 - 1) + (n_2 - 1)$ degrees of freedom.

An example of the t-test for two independent samples

A researcher is provided with the following income data for a sample of males, and a sample of females, and is asked to evaluate the data for a statistically significant difference between the two income levels. The research hypothesis and null hypothesis are stated symbolically below, and the null hypothesis is to be tested at the $\alpha = .05$ level for a two-tail test.

H_1: $\mu_{male} - \mu_{female} \neq 0$
H_0: $\mu_{male} - \mu_{female} = 0$

Data:	Male	Female
Mean, $\overline{X}$	\$32,500	\$32,400
Sample size n	20	24
Standard deviation s	\$155	\$160
Variance s^2	\$24,025	\$25,600

$df = (20 - 1) + (24 - 1) = 42$
$t_{.05} = \pm 2.021$

Computing the observed t-test value:

$$t = \frac{(32{,}500 - 32{,}400) - (\mu_{male} - \mu_{female})}{\sqrt{\frac{(20-1)24{,}025 + (24-1)25{,}600}{(20-1) + (24-1)}\left(\frac{1}{20} + \frac{1}{24}\right)}}$$

$$= \frac{(100) - (0)}{\sqrt{\frac{(19)24{,}025 + (23)25{,}600}{(19) + (23)}\left(\frac{1}{20} + \frac{1}{24}\right)}}$$

$$= \frac{100}{\sqrt{\frac{456{,}475 + 588{,}800}{42} \times (.05 + .042)}}$$

$$= \frac{100}{\sqrt{\frac{1,045,275}{42} \times .092}}$$

$$= \frac{100}{\sqrt{24,887.5 \times .092}}$$

$$= \frac{100}{\sqrt{2,289.650}}$$

$$= \frac{100}{47.85} = 2.09$$

The observed value of the test statistic, $t = 2.09$, is larger than the critical value, $t_{.05} = \pm 2.021$ (See Figure 11.7), so the correct decision is to REJECT H_0, there is a statistically significant difference between the two means. Note that the size of the difference between the two means is only \$100 per year. Once again, statistical significance should not be equated with any sense of theoretical importance. Our result simply indicates that there is evidence from the two samples to indicate that the population mean incomes for males and females differ by \$100, subject to a 5% risk of being wrong (the size of our chosen alpha level).

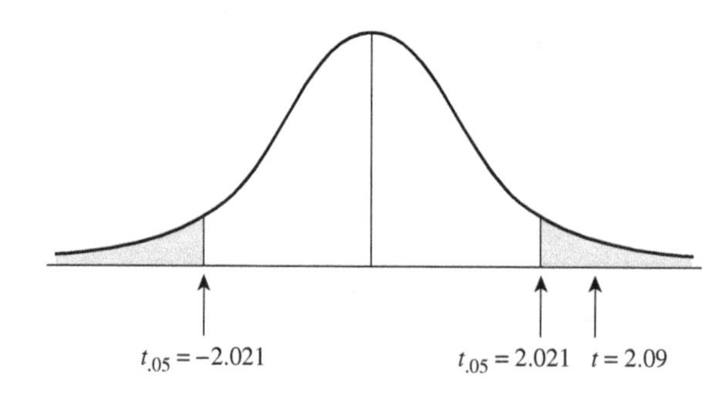

Figure 11.7 Observed and Critical Values of *t*

The t-test for two independent samples when $\mu_1 \neq \mu_2$

The t-test for two independent samples may also be computed under the assumption that the two population means are not equal. In most cases under the assumption of nonequal population means, one is not expecting evidence of a significant difference between the two population means because a difference is already assumed to exist. Instead of expecting a difference, one is expecting to find evidence of a change in the amount of the difference. For example, the reading level of mainstream (ms) fifth grade students may be known to be three grades higher than fifth grade students enrolled in the special education (se) program ($\mu_{ms} = 5$, $\mu_{se} = 2$). The students in the special education program might then be subjected to an experimental reading program designed to improve their reading ability. After a suitable period of time, an independent sample of $n = 25$ children from the mainstream population is observed to have a reading level of $\overline{X}_{ms} = 5.2$ years, and an independent sample of $n = 15$ children from the special education population is observed to have a reading level of $\overline{X}_{se} = 3.0$ years. Because the experimental reading program is designed to improve the reading level of the special education students, the research hypothesis is that the difference between the two groups is less than the historic 3-year difference. A one-tail test of the hypothesis is in order, and will be evaluated at the $\alpha = .05$ level. The research and null hypotheses are stated below symbolically, along with a summary of the experimental data.

$H_1: \mu_{ms} - \mu_{se} < 3.0$
$H_0: \mu_{ms} - \mu_{se} = 3.0$

	Mainstream Class	Special Education Class
Mean $\bar{X}$	5.2	3.0
Standard deviation s	1.50	1.80
Variance s^2	2.25	3.24
Sample size n	25	15

Degrees of freedom = (25 – 1) + (15 – 1) = 24 + 14 = 38
Critical t value at $\alpha = .05 = t_{.05} = -1.697$

Note that Table 2 in Appendix A with critical t values does not contain a value for 38 degrees of freedom. Our choice is to use either the value for df = 30 or df = 40. Even though 38 is closer to 40 than it is to 30, the proper choice is to use the critical value for df = 30. The critical value associated with df = 40 will be slightly smaller than the appropriate value associated with df = 38, and the critical value associated with df = 30 will be slightly larger than the appropriate value associated with df = 38. In the absence of the exact value required, the appropriate choice is the more conservative value, which is this case is the larger critical value making it harder to reject H_0.
Computing the value of the t-test statistic yields the following:

$$t = \frac{(5.2 - 3.0) \; - \; (5.2 - 2.0)}{\sqrt{\dfrac{(25-1)2.25 + (15-1)3.24}{(25-1) + (15\;-1)}\left(\dfrac{1}{25} + \dfrac{1}{15}\right)}}$$

$$= \frac{(2.2) \; - \; (3.0)}{\sqrt{\dfrac{(24)2.25 + (14)3.24}{(24) + (14)}\left(\dfrac{1}{25} + \dfrac{1}{15}\right)}}$$

$$= \frac{-0.8}{\sqrt{\dfrac{54 \; + \; 45.36}{38} \times (.04 + .067)}}$$

$$= \frac{-0.8}{\sqrt{\dfrac{99.36}{38} \times .107}}$$

$$= \frac{-0.8}{\sqrt{2.615 \times .107}}$$

$$= \frac{-0.8}{\sqrt{.2798}}$$

$$= \frac{-0.8}{.529} = -1.512$$

The observed value of the test statistic, $t = -1.512$ does not reach the critical region of the curve beginning at $t_{.05} = -1.697$ (See Figure 11.8), so the proper decision is to FAIL TO REJECT H_0. The

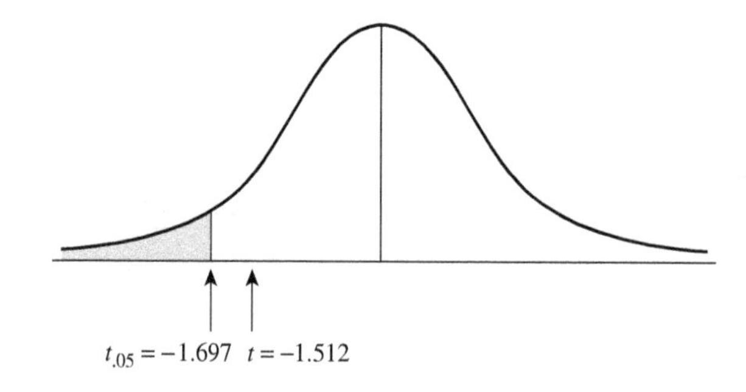

Figure 11.8 Observed and Critical Values of t

sample data indicate a slight reduction in the difference between the reading levels of the mainstream students and the special education students (2.2 years difference compared to the historical 3.0 years difference), but the difference is not statistically significant, and can be attributed to sampling variation.

The t-test for two independent samples that we have examined can be rather cumbersome when computed by hand, especially when computing the pooled estimate of the variance. The practice of using a pooled estimate of the population variance is only legitimate if the two sample variances are roughly equivalent. The Z test for two independent samples is a more appropriate alternative to the t-test for two independent samples when the sample variances are not equivalent. The Z test for two independent samples is presented below. One should note that the test for two independent samples where the sample variances are not equivalent may also be applied as a version of the t-test using the same formula as described below for the Z test, and using $N-2$ degrees of freedom for obtaining the critical values (where $N =$ the total sample size of both groups).

The *Z* test for Two Independent Samples

The Z test for two independent samples is similar to the t-test for two independent samples with three notable exceptions. First, the Z test requires all of the assumptions associated with the t-test, with the exception of the assumption that the two sample variances are equivalent. The Z test assumes that the two sample variances are not equivalent. Second, as is the case with other Z tests, the critical values for the test statistic are based on the normal distribution, and can be found at the end of Table 1 in Appendix A. Third, the Z test is a much simpler test to apply than the t-test. The numerator is computed in the same manner for each test, but the standard error for the Z test is simply the square root of the quantity of the variance of the first sample divided by its sample size plus the variance of the second sample divided by its sample size. The formula is as follows.

Computation Formula for the Z test for Two Independent Samples

$$Z = \frac{(\bar{X}_1 - \bar{X}_2) - (\mu_1 - \mu_2)}{\sqrt{\frac{s_1^2}{n_1} + \frac{s_2^2}{n_2}}}$$

where

$s_1^2 =$ the estimate of the population variance for sample one

$s_2^2 =$ the estimate of the population variance for sample two

$n_1 =$ the sample size for sample one

$n_2 =$ the sample size for sample two

Unless the two sample variances are radically different, the results from the t-test and the Z test for two independent samples will be very similar, and in some cases will be identical. For example, consider the following income data for which we previously computed a t-test value of $t = 2.09$.

Data:

	Male	Female
Mean X	\$32,500	\$32,400
Sample size n	20	24
Standard deviation s	\$155	\$160
Variance s^2	\$24,025	\$25,600

Performing a Z test for two independent samples yields the following result.

$$Z = \frac{32,500 - 32,400 - (0)}{\sqrt{\frac{24,025}{20} + \frac{25,600}{24}}}$$

$$= \frac{100}{\sqrt{1201.25 + 1066.67}}$$

$$= \frac{100}{\sqrt{2267.92}}$$

$$= \frac{100}{47.62} = 2.099$$

The value of the observed Z test statistic of $Z = 2.099$ is almost identical to the previously observed t value of $t = 2.09$. The critical value for the Z test for a two-tail test at $\alpha = .05$ is 1.96, and our decision with respect to the null hypothesis would be to REJECT H_0, just as before when the t-test was conducted (See Figure 11.9).

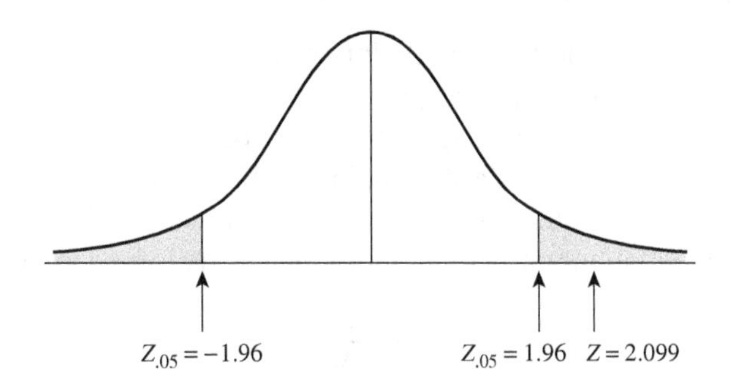

Figure 11.9 Observed and Critical Values of *Z*

The *Z* test for Significant Differences between Two Proportions

One final application of a Z test that we will examine in this chapter is a Z test used to evaluate two proportions for a statistically significant difference. The logic of the Z test for a significant difference between two proportions follows the same pattern that we have seen earlier when examining differences between means. In fact, you may recognize it as exactly the same type of test, if you recall the discussion of the binomial distribution from Chapter 7 on probability.

We saw in Chapter 7 that a binomial variable was one that could take on only one of two possible outcomes. For example, the proportion of individuals who support political candidate "A" and the proportion of individuals who do not support political candidate "A," represent a binomial variable. We also saw that a binomial distribution had a mean that could be computed as, $\bar{X} = n \times p$ (where "n" is the sample size, and "p" is the proportion of the sample favoring candidate "A"), and a standard error, σ_p that could be computed as follows

$$\sqrt{\frac{(p \times q)}{n}}$$

where

"p" = the proportion of the sample favoring candidate "A"
"q" = the proportion of the sample not favoring candidate "A"

Given any two proportions, we can compute a mean for each, and we can compute a standard error for each. The sample statistics of the mean and the standard error have been the basis of each of the tests of significance that we have examined thus far in this chapter, and they are the basis of the Z test used to determine if a statistically significant difference exists between two proportions as we will see below. The Z test for two proportions can be conducted as a one-tail test or a two-tail test just as the other statistical tests we have examined, and the population proportions about which we are making inferences may be assumed to be equal or not equal similar to the assumptions that could be made about two population means in the t-test or the Z test for two independent means presented previously. For example, in most cases, the null hypothesis will assume that the two population proportions are equal (or have a difference equal to 0), but in some cases the null hypothesis will assume that the two populations proportions are different by a given amount, and we will be testing to see if there has been a change in the degree of difference. The latter case is analogous to the t-test example where the mean reading levels of the mainstream students and the special education students were hypothesized to be different by 3 grade levels, and we tested to see if that had changed as a result of the application of an experimental reading program for the special education students.

Computing proportions and the standard error of the difference

Just as we have two sample means and two population means in the situation for the Z test for two independent sample means, we have two sample proportions and two population proportions in the Z test for significance for two proportions. We can designate the proportion of a population in a given condition (for example, those in the population who favor candidate "A") as "p," and we can designate the proportion of the sample taken from the population in a given condition as "p'" (read as: p prime).

The proportion of the population in a given condition can easily be computed as follows:

$$p = \frac{X}{N}$$

where

X = the number of individuals in the population in the given condition
N = the population size

Similarly, the proportion of the sample in a given condition can be computed in the same fashion:

$$p' = \frac{X}{n}$$

where

X = the number of individuals in the sample in the given condition
n = the sample size.

In the cases where the Z test for two proportions is applied, we will have two sample proportions which we can designate as "p'_1," and p'_2" respectively. Each population proportion will have an associated variance σ^2, which can be computed as:

$$\sigma^2 = \frac{p \times q}{N}$$

where

p = the proportion of the population in the given condition of interest
q = the proportion of the population not in the given condition of interest (Note: $q = 1 - p$)
N = the population size.

The standard error of the difference $\sigma_{(p'_1 - p'_2)}$ can then be computed as follows:

$$\sigma_{(p'_1 - p'_2)} = \sqrt{\sigma^2_{p_1} + \sigma^2_{p_2}} = \sqrt{\frac{p_1 q_1}{n_1} + \frac{p_2 q_2}{n_2}}$$

Unfortunately, we usually have a problem in computing the standard error when working with two sample proportions because we seldom know the true value of the population variance. This is a problem similar to what we encountered earlier when working with two independent sample means, and we will solve it in much the same fashion by using a pooled estimate of the population variance based on the sample data.

We can create a weighted estimate $(\bar{p})$ of the population proportion p as follows:

$$\bar{p} = \frac{n_1 p'_1 + n_2 p'_2}{n_1 + n_2}$$

Note: because $p' = X/n$, we can simplify the computation of the weighted estimate of p as follows:

$$\bar{p} = \frac{X_1 + X_2}{n_1 + n_2}$$

Our weighted estimate of $\bar{q}$, is easily obtained as follows:

$$\bar{q} = 1 - \bar{p}$$

Finally, our weighted estimate of the standard error of the difference is obtained by the following equation:

$$\sigma_{(p'_1 - p'_2)} = \sqrt{\bar{p}\,\bar{q}\left(\frac{1}{n_1} + \frac{1}{n_2}\right)}$$

Formula for the *Z* test for a significant difference between two proportions

The formula for the Z test for a significant difference between two proportions is very similar to the t-test or Z test for a significant difference between two independent means. The numerator consists of the difference between the two observed sample proportions minus the expected difference between the two population proportions. Just as the case in the previous test, the expected difference between the two population proportions is usually assumed to be zero, but that is not always the case. The denominator of the test statistic consists of the standard error of the difference, which is based on the weighted estimate of the population variance. The formula is as follows:

$$Z = \frac{(p_1' - p_2') - (p_1 - p_2)}{\sqrt{\overline{p}\,\overline{q}\left(\frac{1}{n_1} + \frac{1}{n_2}\right)}}$$

where

p' = the sample proportion, $p' = X/n$

p = the population proportion, $p = X/N$

$\overline{p}$ = the pooled estimate of the population proportion based on the sample data:

$$\overline{p} = \frac{X_1 + X_2}{n_1 + n_2}$$

$\overline{q}$ = the pooled estimate of the proportion of the population not in the given condition: $\overline{q} = 1 - \overline{p}$

Appropriate application of the Z test assumes independence of the samples, and that the value of $n \times p$ and $n \times q$ for both samples not be exceedingly small. As a general rule, $n \times p$ and $n \times q$ should be equal to or greater than 5. The test may be applied as a one-tail or a two-tail test depending on the nature of the research hypothesis, and critical values may be obtained from Table 1 in Appendix A.

An example of the *Z* test for proportions

A total of $X_1 = 42$ convicted felons from a sample of $n_1 = 100$ men released on parole is found to be arrested again within a 5-year period. A total of $n_2 = 50$ convicted felons is given a 6-month prerelease orientation program aimed at reducing the recidivism rate. After a 5-year period, a total of $X_2 = 19$ of the felons completing the orientation program have been arrested within a 5-year period. We are interested in evaluating the effectiveness of the prerelease orientation program that was expected to reduce the recidivism rate. The null hypothesis is to be evaluated at the $\alpha = .05$ level.

$H_1: p_1 > p_2$

$H_0: p_1 = p_2$

$\alpha = .05$

$Z_{.05} = 1.645$ for a one-tail test

Data:

$X_1 = 42;\ n_1 = 100;\ p'_1 = .42$

$X_2 = 19;\ n_2 = 50;\ p'_2 = .38$

$$\overline{p} = \frac{42 + 19}{100 + 50} = \frac{61}{150} = .407$$

$\overline{q} = 1 - .407 = .593$

Computing the value of the test statistic yields:

$$Z = \frac{(.42 - .38) - (0)}{\sqrt{.407 \times .593\left(\frac{1}{100} + \frac{1}{50}\right)}}$$

$$= \frac{.04}{\sqrt{.241 \times (.01 + .02)}}$$

$$= \frac{.04}{\sqrt{.00723}}$$

$$= \frac{.04}{.085} = .471$$

The observed value of the test statistic is $Z = .471$, the critical region begins at $Z_{.05} = 1.645$ (See Figure 11.10), so the correct decision is to FAIL TO REJECT H_0. There is no evidence that the prerelease orientation program reduced the 5-year recidivism rate of the convicted felons.

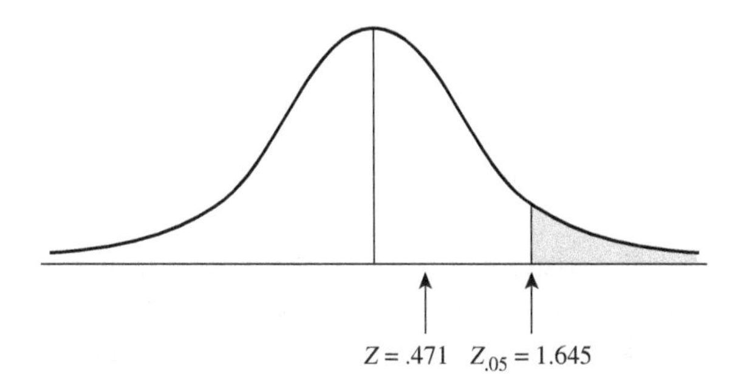

Figure 11.10 Observed and Critical Values of *Z*

A second example of the *Z* test for proportions

A small college admitted a freshman class of $n_1 = 200$, full-time students and found that a total of $X_1 = 140$ had been retained as of the beginning of the sophomore year, resulting in a retention proportion of $p'_1 = .70$. The college created a program aimed at improving the retention rate of its first year students, and found that in the next freshman class of $n_2 = 220$ students, a total of $X_2 = 176$ were retained as of the beginning of the sophomore year resulting in a retention proportion of $p'_2 = .80$. We are to conduct a statistical analysis of the data to see if there is a significant difference between the two retention proportions at the $\alpha = .01$ level. Because the retention program was aimed at improving the retention rate, we expect the retention rate of the second class to be higher, so a one-tail test of the hypothesis is in order. We do not expect any difference between the two population proportions.

$H_1 : p_2 > p_1$

$H_0 : p_2 = p_1$

$\alpha = .01; Z_{.01} = -1.96$ for a one-tail test

(Note: the critical value is negative because we expect the retention rate of the second group to be higher, resulting in a negative value for the *Z* test statistic.)

Data:

$X_1 = 140; n_1 = 200; p'_1 = .70$

$X_2 = 176; n_2 = 220; p'_2 = .80$

$$\bar{p} = \frac{140+176}{200+220} = \frac{316}{420} = .752;$$

$\bar{q} = 1 - .752 = .248$

Computing the value of the test statistic yields the following:

$$Z = \frac{(.70 - .80) - (0)}{\sqrt{.752 \times .248\left(\frac{1}{200} + \frac{1}{220}\right)}}$$
$$= \frac{-.10}{\sqrt{.1865 \times (.005 + .0045)}}$$
$$= \frac{-.10}{\sqrt{.00177}}$$
$$= \frac{-.10}{.042} = -2.38$$

The observed value of the test statistic is $Z = -2.38$; the critical region begins at $Z.01 = -1.96$, so the correct decision is to REJECT H_0. There is a statistically significant difference in the proportion of students retained in the second class of students compared with the first class of students (See Figure 11.11.)

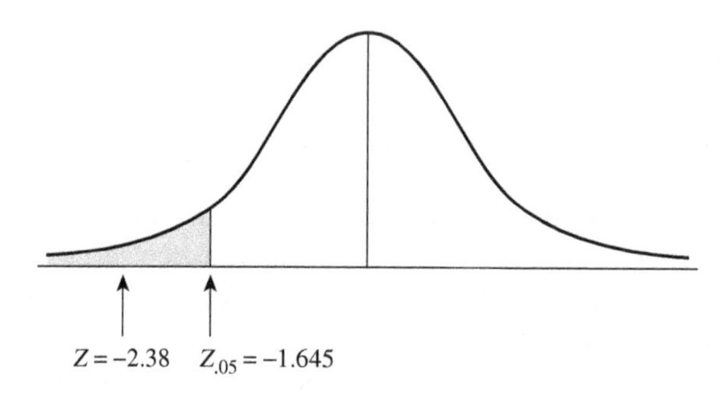

Figure 11.11 Observed and Critical Values of *Z*

The t-Test for Two Related Samples

The final statistical test we will examine in this chapter is a version of the t-test used to compare two related samples. To call the samples related is another of the understatements sometimes made in statistical terminology. In most cases, what we refer to as related samples are actually the same sample observed at two different points in time. Usually, we are examining data resulting from some variation of the classic experimental design, which is outlined briefly below.

Essentially, an experimental design requires the observation or measurement of a sample at two points in time. At time one (t_1) the level of the dependent variable (Y) is measured. The time one measurement is referred to as the pretest. The level of the dependent variable (Y) is measured again at time three (t_3).

The time three measurement is referred to as the posttest. In between t_1 and t_3, at time two (t_2), the sample is exposed to the independent variable (X). The independent variable (X) is often referred to as the stimulus in an experimental design.

Typically, the researcher has a hypothesis that exposure to the independent variable (X) will bring about some change in the dependent variable (Y). In the experimental design, we pretest to find the level of the dependent variable (Y), expose the subjects to the stimulus (the independent variable X), and then posttest the subjects. Comparing the pretest scores with the posttest scores will allow us to determine the effect of "X" on the subjects.

The experimental design may be conducted by using the same group of subjects for the pretest, stimulus, and posttest, or by using the technique of matched pairs. With the matched pairs approach, two groups of equal size are formed with the goal of making them as identical or "matched" as possible. A pretest is conducted to determine if a sufficient match has been obtained. One group is then exposed to the stimulus, and the other group is not. Following exposure to the stimulus, the level of the dependent variable is measured in both groups, and the comparison that follows can be used to determine the effects of the independent variable (X). In actual practice, even the simplest version of the classic experimental design is much more involved than I have presented here, but a more detailed discussion of the classic experimental design is more appropriate for a text in research methods. The brief overview presented here should serve as sufficient background to allow you to make sense of the statistical analysis often used with experimental designs. The t-test for related samples is often the statistic of choice when evaluating the results from a classical experimental design using a single group pretest–posttest design, or when using a matched pairs design.

The formula for the t-test for related samples

The t-test for related samples is one of the more cumbersome tests to compute, and at first glance, it will appear quite different from the other t-tests and Z tests we have examined up to this point. As we examine the formula in more detail, you will probably see that the t-test for related samples is actually quite similar to the t-tests we have examined previously. The key to understanding the t-test for related samples is to realize that the computations are based on the difference (D) between the posttest scores and the pretest scores on an individual by individual basis. In the case of a single group design, D is based on the difference between the posttest scores (t_3) for the experimental group and the pretest scores (t_1) for the experimental group. In the case of a matched pairs design, D is based on the difference between the posttest scores (t_3) for the experimental group and the posttest scores (t_3) for the matched comparison or control group. The formula is as follows:

$$t = \frac{\sum D / n}{\dfrac{\sqrt{\dfrac{\sum D^2 - \dfrac{(\sum D)^2}{n}}{n-1}}}{\sqrt{n}}}$$

where

D = the difference between the pretest and posttest scores for each individual in the sample
$\sum D$ = the sum of the difference between the observed scores for each individual in the sample
n = the sample size
$\sum D^2$ = each difference between the observed scores squared on a case-by-case basis, and then summed
$(\sum D)^2$ = the sum of the difference between the observed scores for each individual in the sample, quantity squared

Making sense of the formula

The formula may seem very different than the others we have examined, but it is really quite similar. The numerator consists of the sum of the differences divided by the sample size "n." Typically, in the t-tests and Z tests examined previously we have seen the numerator as a "difference between two means." While it may not seem so, that is exactly what we have in the t-test for related samples. The sum of the differences divided by "n" can be thought of as a "mean difference" between the two sets of scores. In fact, if you were to compute the mean for each set of scores, and then subtract one from the other, you would obtain the same result as the sum of the differences divided by "n." So the numerator is actually the same; it is a difference between means, but it is computed in a slightly different manner.

The denominator consists of two parts. The portion under the radical or square root sign may seem more familiar to you if you replace the "D" with an "X." You might then recognize that part of the formula as the computational formula for a standard deviation. And how do we convert a standard deviation into a standard error? By dividing it by the square root of the sample size. So the denominator is simply a standard deviation based on the difference between the two sets of scores, which is then converted to a standard error when it is divided by the square root of the sample size.

Looking at the entire formula for the t-test for related samples, we have a numerator consisting of a "mean difference," and a denominator consisting of a standard error. In reality, the formula is practically identical to all of the t-tests or Z tests we have worked with up to this point.

The t-test for related samples can be applied in a one-tail test or a two-tail test situation depending on the nature of the research hypothesis. Critical values may be obtained from Table 2 in Appendix A, using $n - 1$ degrees of freedom. The t-test for related samples is a parametric test, and the same assumptions apply for this t-test as the others we have examined.

An example of the t-test for related samples

A researcher hypothesizes that viewing violent content programming will increase the level of aggressive behavior in preschool-aged children. A group of $n = 9$ preschool-aged children is observed at play in an observation room with a variety of toys for a period of 1 hour. The number of aggressive actions of each child is recorded as a pretest. The children are later shown several violent content cartoons (exposure to the stimulus), and are then allowed back into the observation room for another 1 hour period of play. The number of aggressive actions of each child is again recorded for the posttest. The data are presented below and are to be evaluated at an alpha level of $\alpha = .01$. Direction is implied in the research hypothesis, so a one-tail test is in order. The research hypothesis and null hypothesis are stated below symbolically along with the critical value of the test statistic.

H_1: Posttest scores > Pretest scores
H_0: Posttest scores = Pretest scores
$df = 8$
$t_{.01} = 2.896$

Note that the hypotheses are referring to the posttest scores and pretest scores as being representative of population values estimated by the sample results. If we wanted to know if the sample posttest scores were greater than the sample pretest scores we would simply look at the values. The statistical test we are applying is intended to tell us if there is enough difference between the posttest scores and the pretest scores to suggest a similar difference in the populations that they represent. The data are as follows:

Individual	Pretest	Posttest	D	D^2
1	3	6	3	9
2	5	8	3	9
3	2	3	1	1
4	8	7	−1	1
5	6	8	2	4
6	0	2	2	4
7	5	7	2	4
8	2	4	2	4
9	3	8	5	25
			$\Sigma D = 19$	$\Sigma D^2 = 61$

Note that the difference (D) is obtained above by subtracting the pretest score from the posttest score on an individual by individual basis. It is equally permissible to subtract the posttest score from the pretest score to obtain the value of "D," but it is important that you subtract in a consistent manner down the column. Further, the sign of the sum of the difference is important, because it indicates the nature of the direction of the relationship. The research hypothesis assumes that the posttest scores for aggressive behavior will be higher than the pretest scores, so a positive difference is expected given the way "D" has been obtained above. Had we subtracted the posttest scores from the pretest scores, a negative sum of the differences would be consistent with the expectation stated in the research hypothesis.

Computing the value of the test statistic yields the following:

$$t = \frac{19/9}{\dfrac{\sqrt{\dfrac{61 - \dfrac{(19)^2}{9}}{8}}}{\sqrt{9}}}$$

$$= \frac{2.11}{\dfrac{\sqrt{\dfrac{61 - \dfrac{361}{9}}{8}}}{3}}$$

$$= \frac{2.11}{\dfrac{\sqrt{\dfrac{61 - 40.11}{8}}}{3}}$$

$$= \frac{2.11}{\dfrac{1.62}{3}}$$

$$= \frac{2.11}{.54} = 3.91$$

The observed value of the test statistic is $t = 3.91$, the critical value of the test statistic was $t_{.01} = 2.896$ (See Figure 11.12), so the correct decision is to REJECT H_0. The posttest aggression scores of the sample were statistically significantly higher than the pretest scores at the alpha = .01 level. The increase in the aggression level of the play of the children is beyond that which could be attributed to sampling error or random chance. Note that the same limitations on the interpretation of "statistical significance" apply here as they have in other cases. The statistically significant result does not necessarily mean that the children have become "dangerous" in any sense, or that they are a threat to themselves or others. The results simply indicate that after viewing the violent content material, the children now represent a different population, a population with a higher level of aggressive behavior. Whether or not that is a problem is not a statistical question.

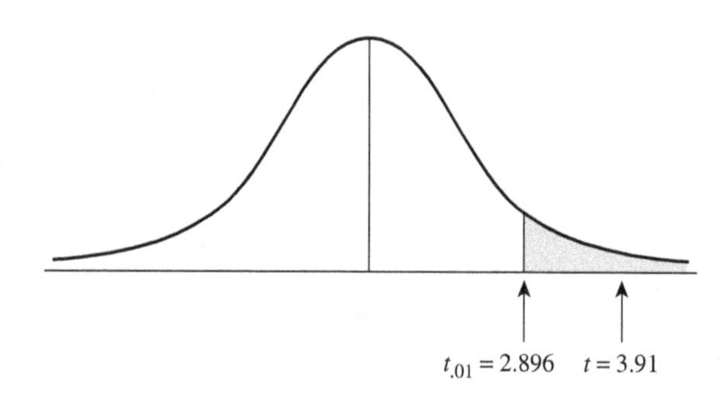

Figure 11.12 Observed and Critical Values of *t*

A second example of the t-test for related samples

A researcher is interested in the effects of playing upbeat background music on worker productivity among a sample of clerical workers. The research hypothesis is that the music will have an effect on the level of worker productivity, as measured by the number of key strokes per minute among the members of the sample. The researcher does not predict a direction in the effect, because the upbeat music may serve to improve productivity by helping the workers adopt a faster pace of work, but it also might reduce productivity by serving as a distraction. The researcher obtains a matched pairs sample of $n = 10$ clerical workers who are employed at a data entry firm, who are matched with another sample of $n = 10$ employees who will serve as a comparison group. The employees are matched in terms of experience, average key strokes per minute, sex, and age. The experimental group is exposed to upbeat background music during their shift for a period of 1 month. The comparison group is not exposed to any background music at all. At the end of the 1-month observation period, the following results are obtained. The results are to be evaluated at the $\alpha = .05$ level on a two-tail test of the hypothesis.

H_1: Posttest experimental ≠ Posttest control
H_0: Posttest experimental = Posttest control

$\alpha = .05$
df = 9
$t_{.05} = \pm 2.262$

Data: Keystrokes per Minute for the Control Group and the Experimental Group

Posttest Control	Posttest Experimental	D	D^2
120	125	5	25
95	110	15	225
110	90	–20	400
85	92	7	49
80	78	–2	4
92	88	–4	16
100	120	20	400
98	99	1	1
84	79	–5	25
80	94	14	196
		$\Sigma D = 31$	$\Sigma D^2 = 1341$

$$t = \frac{31/10}{\dfrac{\sqrt{\dfrac{1341 - \dfrac{(31)^2}{10}}{9}}}{\sqrt{10}}}$$

$$= \frac{3.10}{\dfrac{\sqrt{\dfrac{1341 - \dfrac{961}{10}}{9}}}{3.16}}$$

$$= \frac{3.10}{\dfrac{\sqrt{\dfrac{1341 - 96.1}{9}}}{3.16}}$$

$$= \frac{3.10}{\dfrac{117.6}{3.16}}$$

$$= \frac{3.10}{3.72} = 0.833$$

The observed value of the test statistic is $t = 0.833$, the critical values of the test statistic were $t_{.01} = \pm 2.262$ (See Figure 11.13), so the correct decision is to FAIL TO REJECT H_0. The posttest productivity scores of the experimental group were higher than the posttest productivity scores of the comparison group, but the difference is not statistically significant. The results fall within the level of change that could be attributed to sampling error or random chance.

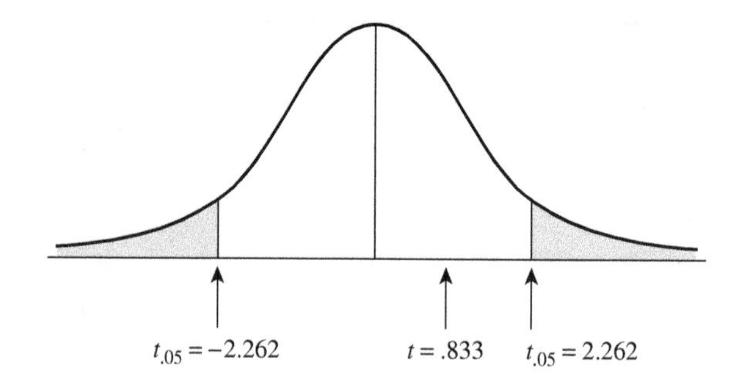

Figure 11.13 Observed and Critical Values of *t*

Computer Applications

Note: SPSS uses the t-test for tests of significance for means rather than the optional *Z* test discussed for some applications in the text.

1. Load the GSS data set and conduct a single sample t-test for the variables "Age" and "CONRINC." State the proper null hypothesis based on the following research hypotheses:

   $H_1 : \mu_{age} > 45$

   $H_1 : \mu_{CONRINC} > 15$

2. Using the GSS data set, conduct an independent samples t-test for a significant difference in "**CONRINC**" for males and females. State the proper null hypothesis based on the following research hypothesis:

   $H_1 : \mu_{male} > \mu_{female}$

3. Using data from one of the examples in the text, conduct a related samples t-test and interpret the results.

How to do it

Load the GSS data set. Click on "Analyze" "Descriptive" "Compare Means" and "One Sample T Test" to open the single sample t-test dialog box. Highlight "Age" and move it to the Test Variables selected box. Enter the value 45 (for 45 years of age) as the hypothesized value in the Test Value box. Click on "OK" to compete the procedure. Interpret the results.

Click on "Statistics" "Compare Means" and "One Sample T Test" to open the single sample t-test dialog box. Highlight "CONRINC" and move it to the Test Variables selected box. Enter the value 15 (for 15 thousand dollars) as the hypothesized value in the Test Value box. Click on "OK" to compete the procedure. Interpret the results.

Click on "Statistics" "Compare Means" and "Independent Samples T Test" to open the independent sample t-test dialog box. Highlight "CONRINC" and move it to the Test Variables selected box. Highlight "Sex" and move it to the Grouping Variable selected box. Click on "Define Groups" to enter the values associated with the two groups to be compared (enter a "1" for Group 1 which will represent males, and enter a "2" for Group 2 that will represent females). Click on "OK" to compete the procedure. Interpret the results.

Use the "File" "New" "Data" command to clear the GSS data set and obtain a clear data editor screen. Input desired data from the text and conduct a related samples t-test. Click on "Statistics" "Compare Means" and "Paired Samples T Test" to open the paired samples t-test dialog box. Select the two variables to be compared by highlighting them and clicking on the appropriate direction arrow. Click on "OK" to complete the procedure. Interpret the results.

Summary of Key Points

This chapter has covered a variety of statistical tests, and sometimes choosing the appropriate test can be confusing. The process is made easier by realizing that each test is appropriate for a specific type of statistical situation. The choice is made easier if you approach it in a systematic fashion. First, examine the characteristics of the statistical question, and then match those characteristics with the appropriate test. For example, we can briefly review the tests examined in this chapter in a type of flow chart method focusing on the characteristics of each situation we have examined, and then matching the situation to the appropriate test.

The first situation in behavioral science research we examined is when we want to **compare a sample mean to a known population mean**. We have two possible tests: a *Z* test and a t-test. The tests are identical; the difference is in the choice of critical values, with the *Z* test using critical values from the *Z* table, and the t-test using critical values from the t table with $(n - 1)$ degrees of freedom. Which test is the more appropriate choice? Use the *Z* test when the population standard deviation σ is known or when the population standard deviation σ is unknown and the sample size is large $(n > 30)$. When the population standard deviation σ is unknown and the sample size is small $(n < 30)$, the t-test is the more appropriate choice.

A more common research situation is **comparing two independent sample means**. We examined two statistical tests: the t-test and the *Z* test. The t-test is the appropriate choice when the sample variances are equivalent allowing us to use a pooled estimate of the population variance by weighting each sample variance by its associated degrees of freedom. Critical values come from the t table with degrees of freedom equal to $[(n_1 - 1) + (n_2 - 1)]$. The *Z* test for two independent sample means is the appropriate choice when the two sample variances are not equivalent. Critical values come from the *Z* table.

Analyzing **two sample proportions for a statistically significant difference** is accomplished by the *Z* test we examined. Critical values are taken from the *Z* table.

The final situation we examined involved testing **two related samples for a significant difference**. Related samples may be the result of a pretest–posttest comparison for an experimental group only design or the result of a matched pairs design. The appropriate test is the t-test for related samples. Critical values come from the t table with $(n - 1)$ degrees of freedom.

Research Hypothesis (H_1)—A hypothesis stating the researcher's expectation regarding the relationship between variables. When the direction of the relationship is stated in the research hypothesis, a one-tail test should be conducted.

Null Hypothesis (H_0)—A statement of no relationship between variables crafted to state the opposite of what the researcher expects to find. The null hypothesis is subjected to the statistical test.

Alpha Level—The amount of error one is prepared to accept in hypothesis testing when reaching a decision on the null hypothesis.

Critical Region—The area of the sampling distribution associated with a test statistic corresponding to the level of significance (alpha level) chosen for a statistical test.

Critical Value—The value of the test statistic marking the beginning of the critical region. The observed value of the test statistic must equal or exceed the critical value in order to reject the null hypothesis.

Test Statistic—The value from a statistical test used to test the null hypothesis.
Statistical Significance—A result indicating that a difference observed in sample data is sufficiently large to suggest that a similar difference exists in the respective population parameters.

Questions and Problems for Review

1. Identify the most appropriate statistical test in the following situations, and indicate if a one-tail or a two-tail test of the hypothesis is in order:
   A. H_1: $\mu_1 \neq \mu_2$; sample variances are unequal
   B. H_1: $\mu < 20$; σ is unknown; $n = 16$
   C. H_1: $\mu \neq 20$; σ is known
   D. H_1: Posttest scores ≠ Pretest scores
   E. H_1: $\mu_1 < \mu_2$; sample variances are equal

2. Identify the assumptions that must be met to correctly apply a t-test for independent sample means. How do these assumptions differ from those required for a t-test to compare a sample mean to a known population mean?

3. Test the null hypothesis below at the $\alpha = .01$ level. Compute the appropriate test statistic, show the critical value, and interpret the results.

   $H_1 : \mu \neq 51$

   $H_0 : \mu = 51$

   $\bar{X} = 45$
   $s = 12$
   $n = 16$

4. A researcher expects the posttest scores measuring systolic blood pressure levels for a group of $n = 8$ subjects to be lower than the pretest scores of the same individuals following a series of stress reduction counseling sessions. Evaluate the following results using the appropriate statistical test. Compute the value of the test statistic, indicate the critical value of the test statistic at $\alpha = .05$, and reach a decision regarding the null hypothesis.

Individual	Pretest	Posttest
1	140	135
2	120	125
3	147	145
4	156	125
5	187	160
6	146	152
7	155	145
8	140	120

5. Test for a significant difference in the starting hourly wage of a sample of government clerical workers compared with a sample of private sector clerical workers. Assume that the sample variances are equivalent. Use an alpha level of $\alpha = .05$ to test the null hypothesis. Compute the test statistic, indicate the critical value and reach a decision regarding the null hypothesis.

	Group	
	Government	Private Sector
Mean	\$12.45	\$11.85
Sample Size	125	100
Standard Deviation	\$2.08	\$1.98

6. Evaluate the data from problem 4 above assuming that the sample variances are not equivalent. Compare the results with those obtained previously, and discuss how they are similar or different.

7. Compute the appropriate test statistic, identify the critical value, and reach a conclusion with respect to H_0 using the information provided below:

H_1: $\mu > \$45{,}405$
H_0: $\mu = \$45{,}405$
$\alpha = .01$
$\bar{X} = \$46,280$
$n = 25$
$\sigma = \$1{,}950$

8. Read a journal article appropriate to your discipline that uses one of the statistical tests discussed in this chapter. Identify the test used, the value of the test statistic, the critical value, and the decision reached regarding H_0. Was the most appropriate test used, or would you have suggested another test?

9. In 2012, a Midwestern college had $N = 200$ graduates apply for admission to various law schools with $X = 110$ being accepted. The college created a prelaw advising program intended to improve the rate of acceptance of its students. In 2014, the college had $N = 230$ graduates apply to various law schools with a total of $X = 130$ accepted. Conduct the appropriate test to determine if the proportion of students accepted to law school improved as expected as a result of the prelaw advising program. Compute the test statistic, report the critical value assuming an $\alpha = .05$ level of significance, and reach a decision regarding the null hypothesis.

10. Which statistical test would be appropriate to determine if the population mean income of recent college graduates in the United States is significantly different from the population mean income of recent college graduates in Canada?

11. A random sample of $n = 9$ counties in a western state were selected for participation in a crime intervention program intended to reduce the homicide rate. Pretest data reflect the homicide rate of each county prior to the implementation of the intervention program, and the posttest data reflect the homicide rate at the conclusion of the trial period. Evaluate the results using the appropriate test statistic at the $\alpha = .01$ level. Compute the test statistic, report the critical value, and reach a decision regarding the null hypothesis.

County	Pretest	Posttest
1	9.9	8.5
2	10.0	9.0
3	25.5	25.5
4	6.5	8.2
5	12.6	8.4
6	33.5	20.0
7	8.9	8.4
8	9.0	8.2
9	8.8	8.9

CHAPTER 12

Analysis of Variance

Key Concepts

ANOVA	Mean Square	Tukey's HSD Test
Factor	Variance within Groups	Fisher's Protected t-Test
Treatment	Variance between Groups	Eta Squared
F Distribution	Post Hoc Test	Omega Squared

Introduction

Analysis of Variance is an important tool in statistical analysis, and one that follows directly from the tests of significance for means that we have just examined in the previous chapter. Analysis of Variance, or **ANOVA** as it is often abbreviated, is a parametric statistical technique used to evaluate a set of sample means for statistically significant differences. In that respect, it is similar in function to the t-test or *Z* test that we have already examined. The difference is that ANOVA is used when we need to compare several group means for a statistically significant difference instead of only two groups as is the case with a t-test or *Z* test.

Consider an example where we might want to examine mean income by educational attainment. We might want to examine the mean incomes of individuals with (A) less than a high school education, (B) a high school education, (C) some college, and (D) college graduates for statistically significant differences. We could conduct a t-test to determine if a statistically significant difference exists between each pair of means, but we would have to conduct a total of six different t-tests to examine every possible combination of pairs of means (A with B, A with C, A with D, B with C, B with D, and C with D). Even though we could conduct these multiple t-tests to search for statistically significant differences in income levels associated with different levels of educational attainment, there are two reasons not to do so. First, conducting a t-test for every possible combination of means would be a rather cumbersome and time-consuming procedure. Second, and more importantly, is the fact that hypothesis testing with statistical tests always involves a risk of being wrong. We have referred to the risk of rejecting the null hypothesis when the correct decision should be to fail to reject the null hypothesis as Type I Error, which is equal to the alpha level or level of significance associated with the statistical test. As we have seen in previous chapters, an alpha level of $\alpha = .05$ means that we have a 5% chance of being wrong in our conclusion regarding the null hypothesis. There is another way to interpret the size of the alpha level that illustrates the problem of performing multiple t-tests. An alpha level of $\alpha = .05$ means that we will probably reach the incorrect decision regarding the null hypothesis 5 times out of every 100 t-tests that are performed. When we perform one t-test, the odds are in our favor that we will reach the proper decision, but every additional t-test brings us that much closer to a wrong decision. In our simple example above, we would have to perform a total of six t-tests, and the probability of reaching the proper decision each time when conducting six t-tests at the .05 level is much smaller ($p = .95^6 = .735$), so an alternative method of examining the data would be preferable. ANOVA represents that alternative method.

One-Way Analysis of Variance

Analysis of Variance is exactly what the title suggests that it is; we are examining groups for statistically significant differences by analyzing the variance present in the data. Certainly one key to understanding the process is to recall the definition of variance as we discussed the concept in Chapter 5. The statistic "variance" is the mean squared deviation of raw scores about their mean. In other words, variance tells us what the mean or average squared distance is of each score from the mean in a set of data. The greater the value of the variance, the more the scores are spread out or vary from their mean. Conversely, the smaller the value of the variance, the closer the scores are to their mean. You should recall from the last chapter that the size of the difference between two means is not the only **factor** that accounts for a statistically significant result. The respective sample sizes have an impact, and so do the sizes of the variances. In the case of ANOVA, we are simply focusing on the "variance" aspect of the situation, or as you will see below, we are focusing on a part of the variance to decide if a group of means contains some statistically significant differences.

To better understand the ANOVA procedure, it will be worthwhile to review the logical and computational formulas for the variance. The logical formula for the variance, s^2, is as follows:

$$s^2 = \frac{\Sigma\,(X - \overline{X})^2}{n}$$

In actual computations of variance, we used the alternative computational formula as follows:

$$s^2 = \frac{\Sigma X^2 - \frac{(\Sigma X)^2}{n}}{n}$$

The ANOVA procedure does not actually involve the direct computation of the variance statistic as represented above. Instead we begin with the component of the variance computation represented by the numerator. Variance represents the mean squared deviation of the scores, which is obtained by computing each score's squared distance from the mean, and then dividing by "*n*." In ANOVA, we are initially interested in the "squared distance from the mean" component represented by the numerator. In ANOVA, the "squared distance from the mean" component for the total sample is referred to as the **sum of squares total** (SS_{tot}). ANOVA allows us to partition the total sum of squares present in the data into two parts: that which is attributed to the **sum of squares between groups** (SS_{bn}); and that which is attributed to the **sum of squares within groups** (SS_{wn}).

To illustrate the difference between these two concepts, let's consider a graphic example based on the situation we have introduced above dealing with educational attainment and income. Suppose, we were to take a sample of individuals from each of the four groups identified earlier: those with less than a high school education (LTHS), those with a high school education (HS), those with some college (SC), and those with a college degree (CD). We can think about these individuals in two ways. We can think about them as members of one large group in which case all members of the sample would be expected to be fairly similar to each other no matter to which educational group they belonged. Or, we can think about the sample as representing members of four distinct groups in which case we would expect the members of each group to be fairly similar to each other, but to be different from the members of the other groups. Figures 12.1 and 12.2 provide a graphic depiction of these two different situations with each group's income level represented along the horizontal axis, and the frequency represented along the vertical axis.

In Figure 12.1, the sample represents individuals from each of the four income groups, but, as you can see, group membership does not seem to matter very much. The large amount of overlap from one group to the next suggests that the membership in any particular educational group does not have very much to do with one's income level. In the terminology of ANOVA, we would say that there is a great deal of variation **within** groups. The fact that the four educational level groups overlap a great deal also indicates that there is not very much variation **between** groups. Knowing an individual's educational level would not help us predict their income very well, due to the large amount of variation within each group and the lack of variation between the groups. Group membership would not be strongly related to income level, and as a result the mean income of one group is not likely to be significantly different from any other group.

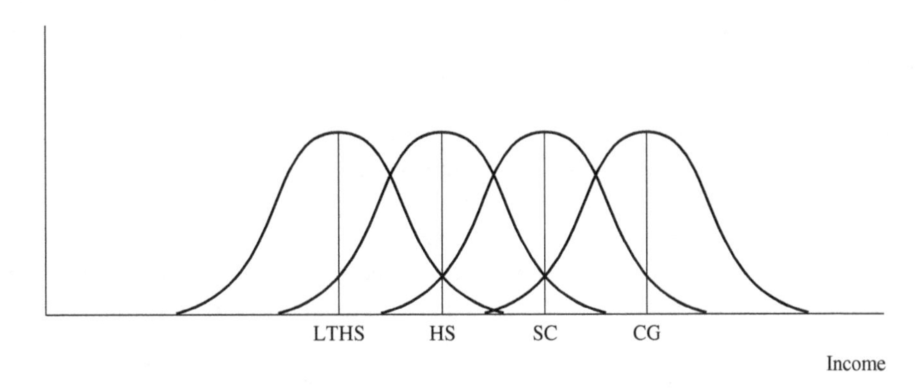

Figure 12.1 The Sample Represents a Single Large Group

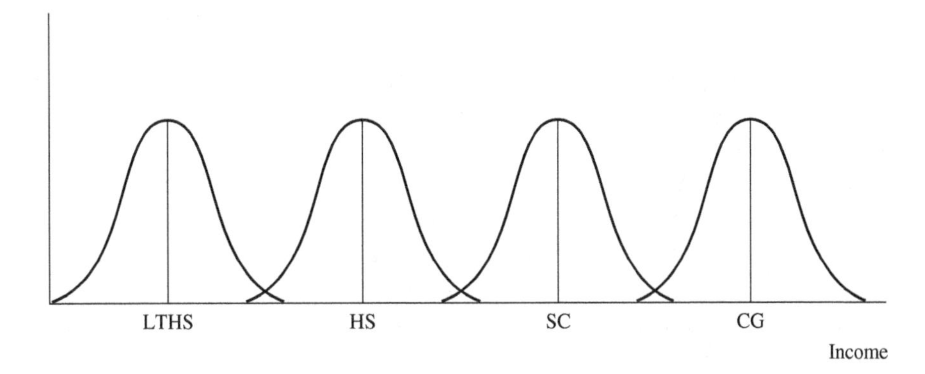

Figure 12.2 The Sample Represents Four Distinct Groups

In Figure 12.2, the individuals display a very different pattern. There is very little variation **within** groups. Each educational group is tightly clustered around its mean income level, so the amount of variance within each group would be small. At the same time, there is a great deal of variation **between** groups. Each group stands out as a distinct entity, and can easily be distinguished from the others. Just as one group varies from the next, we would likely find that each group's mean income would be significantly different from the others.

ANOVA is a statistical technique that allows us to compute, and then analyze the amount of **variance within groups** and **between groups**. The result is a ratio of the variance between groups to the variation within groups. The ratio is interpreted as an "*F*" statistic, and the higher the level of the "***F* ratio**" the more likely we have a situation as depicted in Figure 12.2 indicating that group membership is a statistically significant determinant of the dependent variable. The smaller the level of the "*F* ratio" the more likely we have a situation as depicted in Figure 12.1 indicating that group membership is not a significant factor in determining the level of the dependent variable.

Basic Terms and Assumptions for Analysis of Variance

The ANOVA model that we will examine in this chapter is called a one-way ANOVA. One-way ANOVA means that we are examining the effects of only one independent variable on the dependent variable. The independent variable is referred to as a **factor**, and the different levels of the independent variable are referred to as **treatments**. In the case of our example of education and income, the level of education is the independent variable, or factor, and we are interested in its effect on the dependent variable income. Each educational level group we consider represents a different treatment level. In our example, we are considering four levels or treatments. The number of treatment levels corresponds with the number of group means that are being examined, and we conventionally use the letter "*k*" to represent the number of groups.

The ANOVA process allows us to determine if any of the group means are significantly different by examining the variance present in the data. Our measure of variance is called the **mean square**. When computing the traditional variance, we divide the "sum of squared deviations" by the sample size. We perform a similar procedure in ANOVA. The mean square value is computed by dividing the sum of squares by its associated degrees of freedom. The **total degrees of freedom** is obtained by (N–1) as we have done in previous chapters, where "*N*" is the total sample size. The **degrees of freedom between groups** is obtained by (k–1), where "*k*" is the number of group means being evaluated. The **degrees of freedom within groups** is obtained by (N–k), where "*N*" is the total sample size, and "*k*" is the number of group means being evaluated.

The sum of squares and the degrees of freedom allow us to compute the value of the mean square between groups (MS_{bn}), which is equal to the sum of squares between groups divided by the degrees of freedom between groups ($MS_{bn} = SS_{bn}/df_{bn}$); and the mean square within groups (MS_{wn}), which is equal to the sum of squares within groups divided by the degrees of freedom within groups ($MS_{wn} = SS_{wn}/df_{wn}$). Note that by dividing the appropriate sum of squares by its associated degrees of freedom is analogous to creating a measure of the variance between groups and the variance within groups. We then are able to determine if any of the group means are statistically significantly different by looking at the ratio of the mean square between groups (MS_{bn}) to the mean square within groups (MS_{wn}). This ratio of MS_{bn}/MS_{wn} is referred to as the ***F* statistic**.

The research hypothesis in ANOVA is that a statistically significant difference exists between one or more pairs of the possible comparisons of the group means. The null hypothesis is that the means are equal across all treatment levels. If the *F* statistic is significant, and we reject the null hypothesis, it indicates that at least some of the possible mean differences are statistically significant. Logically we can be sure that there is a statistically significant difference between the group with the lowest mean and the group with the highest mean. Of course there are several other comparisons that could be made resulting in statistically significant differences, but we must use a **post hoc procedure** to identify any additional statistically significant results. Post Hoc procedures consist of several strategies that can be used to identify all of the possible statistically significant results that may be present among the treatment levels associated with a particular factor. We will examine two post hoc procedures: **Tukey's HSD** used when the sample size is equal for all treatment levels, and **Fisher's protected t-test** which can be used for either equal or unequal sample sizes among the treatment levels. First, we will examine the assumptions required for the ANOVA procedure, and then conduct an ANOVA computation based on our example of level of educational attainment and income.

The assumptions required to legitimately perform ANOVA are comparable to those associated with the t-test for two independent means seen in the previous chapter. The procedure requires the following:

1. Random (or probability based) sampling;
2. An interval level or ratio level of measurement for the dependent variable allowing us to legitimately compute a mean;
3. The dependent variable is normally distributed in the population of interest;
4. The variances associated with the various treatment levels are equivalent (this assumption is less important as the sample size increases); and,
5. The sample size of each treatment level does not have to be equal, but equal sample sizes are preferable. (The computations for the ANOVA procedure and the post hoc procedure are easier with equal sample sizes.)

In addition, the null hypothesis assuming that the sample means are equal across all treatment levels is tested assuming a two-tailed test using the F statistic.

Computing the Sum of Squares

The first step in conducting the ANOVA procedure is to compute the **sum of squares** components. We first compute the sum of squares total (SS_{tot}), which represents a measure of the total variation in the entire sample across all of the treatment levels. We then compute the sum of squares between groups (SS_{bn}), which represents the amount of the total variance due to group membership. The final component, sum of squares within groups (SS_{wn}), can then be obtained by subtraction ($SS_{wn} = SS_{tot} - SS_{bn}$). The sum of squares values are then used to compute the mean square values, which form the basis of the F statistic.

The computational formula for the sum of squares components should appear very familiar to you, because they are based on the computational formula for the variance.

Computational Formula for the Sum of Squares Total

$$SS_{tot} = \Sigma X_{tot}^2 - \left(\frac{(\Sigma X_{tot})^2}{N} \right)$$

where

ΣX^2_{tot} = the sum of the Xs squared (each X value squared and then summed) for the entire sample across all groups

$(\Sigma X_{tot})^2$ = the sum of the Xs quantity squared (the sum of the X values and then that quantity squared)

N = the total sample size (all individuals in all groups).

Computational Formula for the Sum of Squares between Groups

$$SS_{bn} = \Sigma \left(\frac{(\Sigma X_i)^2}{n_i} \right) - \frac{(\Sigma X_{tot})^2}{N}$$

where

$(\Sigma X_i)^2$ = the sum of the Xs quantity squared (the sum of the X values within each group, and then that quantity squared, and then summed across all groups)

n_i = the sample size of each group

$(\Sigma X_{tot})^2$ = the sum of the Xs quantity squared (the sum of the X values and then that quantity squared)

N = the total sample size (all individuals in all groups).

Computational Formula for the Sum of Squares within Groups

$$SS_{wn} = SS_{tot} - SS_{bn}$$

where

SS_{tot} = the sum of squares total

SS_{bn} = the sum of squares between groups.

A Computational Example of Analysis of Variance

This first example of an ANOVA assumes that we have taken a total sample of $N = 20$ individuals who represent four different levels of educational attainment: less than a high school education (LTHS), a high school education (HS), some college (SC), and college graduates (CG), with a sample size of $n = 5$ in each group. Income data in thousands of dollars for the four groups are presented below.

Data

Income * $1,000				
LTHS	HS	SC	CG	
10	15	15	20	
15	25	25	50	
20	29	30	60	
25	38	53	65	
30	45	60	80	
$\Sigma X = 100$	$\Sigma X = 152$	$\Sigma X = 183$	$\Sigma X = 275$	$\Sigma X_{tot} = 710$
$n = 5$	$n = 5$	$n = 5$	$n = 5$	$n = 20$
$\bar{X} = 20$	$\bar{X} = 30.4$	$\bar{X} = 36.6$	$\bar{X} = 55$	$\bar{X}_{tot} = 35.5$
$\Sigma X^2 = 2{,}250$	$\Sigma X^2 = 5{,}160$	$\Sigma X^2 = 8{,}195$	$\Sigma X^2 = 17{,}125$	$\Sigma X^2_{tot} = 32{,}694$

Computing the Sum of Squares Total

$$SS_{tot} = \Sigma X^2_{tot} - \left(\frac{(\Sigma X_{tot})^2}{N}\right)$$

$$= 32{,}694 - \left(\frac{(710)^2}{20}\right)$$

$$= 32{,}694 - \left(\frac{504{,}100}{20}\right)$$

$$= 32{,}694 - 25{,}205 = 7{,}489$$

Computing the Sum of Squares between Groups

$$SS_{bn} = \Sigma\left(\frac{(\Sigma X_i)^2}{n_i}\right) - \frac{(\Sigma X_{tot})^2}{N}$$

$$SS_{bn} = \left(\frac{(100)^2}{5} + \frac{(152)^2}{5} + \frac{(183)^2}{5} + \frac{(275)^2}{5}\right) - \frac{(710)^2}{20}$$

$$= \left(\frac{10,000}{5} + \frac{23,104}{5} + \frac{33,489}{5} + \frac{75,625}{5}\right) - \frac{504,100}{20}$$

$$= (2,000 + 4,620.8 + 6697.8 + 15,125) - 25,205$$

$$= 28,443.6 - 25,205 = 3,238.6$$

Computing the Sum of Squares within Groups

$$SS_{wn} = SS_{tot} - SS_{bn} = 7,489 - 3,238.6 = 4,250.4$$

Computing the Degrees of Freedom

Degrees of Freedom Total
$df_{tot} = N - 1 = 20 - 1 = 19$
Degrees of Freedom between Groups
$df_{bn} = k - 1 = 4 - 1 = 3$
Degrees of Freedom within Groups
$df_{wn} = N - k = 20 - 4 = 16$

Once the sum of squares and degrees of freedom have been computed, we can compute the mean square values for between groups and within groups. The mean square values represent the final step in partitioning the variance, and serve as the basis for the computation of the *F* statistic.

Computing the Mean Square between Groups

$$MS_{bn} = SS_{bn}/df_{bn} = 3,238.6/3 = 1,079.53$$

Computing the Mean Square within Groups

$$MS_{wn} = SS_{wn}/df_{wn} = 4,250.4/16 = 265.65$$

Finally, we are at a point where we can compute the value of the *F* statistic. The *F* statistic in analogous to the t-test or *Z* test for significant differences between two means. In the case of the *F* statistic, we are testing a null hypothesis stating that there are no statistically significant differences between any pairs of means associated with the treatment levels of the independent variable. The research hypothesis states that there is a statistically significant difference between at least one pair of means associated with the treatment levels of the independent variable.

The *F* statistic is obtained by dividing the mean square between groups (the variance associated with group membership) by the mean square within groups (the variance associated with random error). The higher the value of the *F* statistic, the greater the proportion of the total variance that is explained by group membership, and the more likely that some pairs of group means will be statistically significantly different. The obtained value of the F statistic is compared to a critical value taken from the F table (critical F values at the $\alpha = .01$, and $\alpha = .05$ may be found in Table 5 in Appendix A).

Computing the *F* Statistic

$$F = MS_{bn} / MS_{wn} = 1,079.53/265.65 = 4.06$$

The ANOVA summary table

The ANOVA summary table is the conventional manner of presenting the results of an ANOVA procedure, given all of the steps involved in leading up to the computation of the F statistic. A traditional ANOVA summary table is organized in the following manner:

ANOVA Summary Table

Source	SS	df	MS	*F*
Between	3,238.6	3	1,079.53	4.06
Within	4,250.4	16	265.65	
Total	7,489	19		

Critical values for the **F distribution** are provided in Table 5 in the Appendix. The F distribution is similar to the t distribution in that it really represents a set of distributions based on the observed degrees of freedom, but there are a few important differences. The F distribution has a mean of 1.00, and has a minimum value of 0.00. In addition, the distribution is skewed to the right (to the higher values of F). Recall that the null hypothesis predicts no significant differences between any of the pairs of means associated with levels of the independent variable. Given the computation of the observed F statistic ($F = MS_{bn}/MS_{wn}$), an observed value of $F = 0.00$ would indicate that there is no variation between groups, and all of the variance is within groups. An observed value of $F = 1.00$ would indicate an equal amount of variance between groups and within groups. The greater the value of the observed F statistic, the greater the amount of variance between groups, and the more important group membership is in predicting the value of the dependent variable. Just as with previous statistical tests, the critical value from the table tells us how large the observed F statistic must be in order for us to reject the null hypothesis, and conclude that there is at least one statistically significant difference among the means associated with each level of the independent variable.

Obtaining the *F* critical values

Critical values for the F statistic at the $\alpha = .05$ level (presented in regular print), and the $\alpha = .01$ level (presented in bold print) for selected degrees of freedom are provided in Table 5 in the Appendix. You may have noticed that we compute degrees of freedom for both the between group computation ($df_{bn} = k\text{-}1$), and the within group computation ($df_{wn} = N\text{–}k$). Both degrees of freedom are used to find the appropriate critical F value in the table. The critical value of the F statistic is often presented as $F_{\alpha(dfbn,\ dfwn)}$, with the alpha level and the degrees of freedom in parentheses as subscripts. Also notice that we list the degrees of freedom between groups first followed by the degrees of freedom within groups.

Degrees of freedom between groups are listed across the top row of Table 5 in Appendix A, and degrees of freedom within groups are listed down the first column. To find the appropriate critical value, locate the degrees of freedom between groups across the top row, and then the degrees of freedom within groups down the first column. The intersection of these two points in the table contains the critical values for F at the $\alpha = .05$ and $\alpha = .01$ levels. In our example, we have $df_{bn} = 3$, and $df_{wn} = 16$. The two F critical values are found at the intersection of those two points, and are $F_{.05(3,16)} = 3.24$, and $F_{.01(3,16)} = 5.29$.

I have arbitrarily chosen the $\alpha = .05$ level for this example, so our observed $F = 4.06$ is greater than the critical value of 3.24 and would be considered statistically significant indicating that there is at least one pair of means among the treatment levels of the independent variable that are statistically significantly different. Further, logic dictates that the two means with the greatest difference between them will be statistically significantly different. In our example, that indicates that the mean income of the "less than high school education" group ($\bar{X} = 20$), and the mean income of the "college graduates" group ($\bar{X} = 55$) are statistically significantly different. It is possible that there are additional significant differences among the other pairs of means; however, we will need to perform a post hoc (Latin for "after the fact") test to determine if any additional comparisons are statistically significantly different.

Post Hoc Tests for Significant Differences

There are several post hoc procedures that can be used to identify additional significant differences among all possible comparisons of sample means. We will examine two such tests: **Tukey's HSD multiple comparisons test** and **Fisher's protected t-test**. Tukey's HSD (HSD stands for "honestly significant difference") test is only appropriate when the sample size "n" of each treatment level is the same (as it is in our example where $n = 5$ in each group). **Fisher's protected t-test** can be used whether the sample size "n" is equal or not across the treatment levels. Fisher's protected t-test is more versatile because it can be used in the absence of equal sample sizes, but it is a more cumbersome procedure that requires us to conduct a modified t-test for each possible pair of means. We will conduct the **Tukey's HSD test** for the present example, and use Fisher's protected t-test for the second example of ANOVA where the sample sizes across the treatment levels are not equal.

Tukey's HSD multiple comparison test

Tukey's HSD multiple comparison test is a type of convoluted t-test. Rather than computing multiple t-values and turning to a table of critical "t" values to see if each observed test statistic is significant, Tukey's HSD provides us with a single value indicating how far apart any two sample means must be in order to be considered statistically significantly different at the chosen level of significance. Conventionally, one will choose the same alpha level for the **post hoc test** as was used to test the F statistic in the ANOVA procedure. In our example, we will use $\alpha = .05$.

Formula for Tukey's HSD

$$\text{HSD} = (q_k)\sqrt{\frac{\text{MS}_{\text{wn}}}{n}}$$

where

q_k = the value of the studentized range statistic
MS_{wn} = the value of the mean square within
n = the sample size of each treatment level.

Computing Tukey's HSD begins by obtaining a value for the studentized range statistic q_k, which can be found in Table 6 in the Appendix. The table is organized much like the table of critical values for the F statistic. The top row contains values for "k," the total number of means in the ANOVA procedure. In our case, $k = 4$. The first column contains the degrees of freedom within groups ($df_{wn} = 16$ for our example). The appropriate value for q_k is found at the intersection of "k," and df_{wn}. The two values in the table represent the q_k values at the $\alpha = .05$ level (presented in regular print), and the $\alpha = .01$ level (presented in bold print). The q_k values for $k = 4$, $df_{wn} = 16$ are $q_{k.05} = 4.05$, and $q_{k.01} = 5.19$. Because we conducted the F test at the $\alpha = .05$ level, we will use the $q_{k.05} = 4.05$ value for the Tukey's HSD procedure.

To complete the HSD computation, we simply multiply the q_k value times the square root of the mean square within value ($MS_{wn} = 265.65$) divided by the treatment level group sample size ($n = 5$) as follows:

$$\begin{aligned} \text{HSD} &= (4.05)\sqrt{\frac{265.65}{5}} \\ &= (4.05) \times \sqrt{53.13} \\ &= 4.05 \times 7.29 = 29.52 \end{aligned}$$

The resulting HSD value of HSD = 29.52 tells us the minimum difference that must be found between any two means among the treatment levels to indicate a statistically significant difference. All that remains is to compute the difference between each possible pair of means, and compare the difference to the size of the HSD value. The differences between the means are obtained by simple subtraction, and one easy way to display the results is to create a matrix of means with the differences presented in the body of the matrix. For example:

Observed Means $\bar{X}$

$\bar{X}$	LTHS = 20	HS = 30.4	SC = 36.6	CG = 55
CG = 55	35*	24.6	18.4	0.0
SC = 36.6	16.6	6.2	0.0	
HS = 30.4	10.4			
LTHS = 20	0.0			

*Represents a statistically significant difference at the $\alpha = .05$ level.

The matrix containing the difference between each possible pair of means may be read in the following manner. The numbers in the body of the matrix represent the absolute difference between the two means that intersect at that point. For example, the number 6.2 near the center of the matrix represents the absolute difference in the mean income of the "high school graduate" group ($\bar{X} = 30.4$) and the "some college" group ($\bar{X} = 36.6$). Any paired difference of 29.52 or greater indicates a statistically significant difference at the $\alpha = .05$ level.

The results of the Tukey's HSD multiple comparison test indicate that the "college graduate" mean income ($\bar{X} = 55$) is statistically significantly greater than the "less than high school" mean ($\bar{X} = 20$). There are no other statistically significant results. (The data analyzed in this example are based on the small hypothetical sample of $n = 20$, but the data are approximately equal to the current U.S. income levels by educational attainment.)

Variance explained by the independent variable: Eta Squared and Omega Squared

ANOVA models and associated post hoc procedures not only allow us to discover any statistically significant differences between all possible pairs of means in a set of data, but we can also determine the amount of variance in the dependent variable that is explained by the independent variable. This last step in the process of computing the amount of variance explained is the same concept we have seen previously in the chapter on variance, and the chapters on correlation, and regression. The question is one of "How much can we improve our prediction of the dependent variable by knowing group membership?" Predicting variance with ANOVA models is very similar to the Point Biserial Correlation Coefficient from Chapter 9 where we saw that in the absence of any other information our best predictor of the level of the dependent variable for individuals is the group mean. Of course we will be wrong in our prediction, and we can measure the amount of our error as variance. The Point Biserial Correlation Coefficient indicates how much we can improve our prediction by using information on group membership, but we are limited to situations where the independent variable (group membership) is dichotomous. The square of the Point Biserial Correlation Coefficient gave us our estimate of the amount of "variance explained."

ANOVA models also allow to estimate the amount of variance explained in the dependent variable by using knowledge of the independent variable as the basis for our prediction. The only difference is that with ANOVA models we are not limited to a dichotomous independent variable. There are two measures of association that are commonly used with ANOVA models: **Eta Squared** (η^2), and **Omega Squared** (ω^2). Each measure of association provides us with an estimate of the amount of variance in the dependent variable that is accounted for by the independent variable. Eta Squared is a descriptive statistic, that is, it provides an estimate of the amount of variance explained in the sample under investigation. Omega Squared is a parametric statistic. Omega Squared uses the sample data to provide an estimate of the amount of variance in the dependent variable in the population that can be explained by the independent variable. Both statistics are relatively simple to compute.

Computational Formula for Eta Squared

Eta Squared is simply the sum of squares between groups divided by the sum of squares total. The resulting proportion indicates the amount of variance explained by the independent variable.

$$\eta^2 = \frac{SS_{bn}}{SS_{tot}}$$

In our example, Eta Squared is as follows:

$$\eta^2 = \frac{3,238.6}{7,489} = .432$$

The resulting value of $\eta^2 = .432$ indicates that 43.2% of the variance in income can be explained by level of educational attainment among our sample of $N = 20$ individuals.

Computational Formula for Omega Squared

Omega Squared provides a similar interpretation to that of Eta Squared, except that Omega Squared estimates the amount of variance explained in the population from which our sample of $N = 20$ was drawn. The formula is similar, and contains the same components of the sum of squares between and within, but the proportion is weighted by the degrees of freedom between groups and the mean square within value as follows:

$$\omega^2 = \frac{SS_{bn} - (df_{bn} \times MS_{wn})}{SS_{tot} + MS_{wn}}$$

In our example, the Omega Squared computation is as follows:

$$\begin{aligned}\omega^2 &= \frac{3,238.6 - (3 \times 265.65)}{7,489 + 265.65} \\ &= \frac{3,238.6 - 796.95}{7,754.65} \\ &= \frac{2,441.65}{7,754.65} \\ &= .315\end{aligned}$$

The resulting value of $\omega^2 = .315$ indicates that 31.5% of the variance in income can be accounted for by educational attainment in the population from which our sample of $N = 20$ individuals was drawn.

An example of ANOVA with unequal sample sizes

A medical researcher is interested in evaluating a new drug expected to reduce the number of psychotic episodes in mental patients. The trial is comprised $N = 20$ patients divided into three different groups. Group 1 ($n = 5$) receive a placebo dosage of zero strength; Group 2 ($n = 8$) receive a low level-dosage of 30 mg per day, and Group 3 ($n = 7$) receive a high-level dosage of 60 mg per day. The number of psychotic episodes experienced during the 6-week trial period is reported below.

Number of Psychotic Episodes

Placebo	Low Dose	High Dose	
10	8	2	
8	8	7	
12	10	3	
12	6	1	
6	5	2	
	3	4	
	3	2	
	8		
$\Sigma X = 48$	$\Sigma X = 51$	$\Sigma X = 21$	$\Sigma X_{tot} = 120$
$n = 5$	$n = 8$	$n = 7$	$N = 20$
$\bar{X} = 9.6$	$\bar{X} = 6.375$	$\bar{X} = 3$	$\bar{X}_{tot} = 6$
$\Sigma X^2 = 488$	$\Sigma X^2 = 371$	$\Sigma X^2 = 87$	$\Sigma X^2_{tot} = 946$

Computing the Sum of Squares Total

$$SS_{tot} = \Sigma X^2_{tot} - \left(\frac{(\Sigma X_{tot})^2}{N}\right)$$
$$= 946 - \left(\frac{(120)^2}{20}\right)$$
$$= 946 - \left(\frac{14,400}{20}\right)$$
$$= 946 - 720 = 226$$

Computing the Sum of Squares between Groups

$$SS_{bn} = \Sigma\left(\frac{(\Sigma X_i)^2}{n_i}\right) - \frac{(\Sigma X_{tot})^2}{N}$$
$$SS_{bn} = \left(\frac{(48)^2}{5} + \frac{(51)^2}{8} + \frac{(21)^2}{7}\right) - \frac{(120)^2}{20}$$
$$= \left(\frac{2,304}{5} + \frac{2,601}{8} + \frac{441}{7}\right) - \frac{14,400}{20}$$
$$= (440.8 + 325.125 + 63) - 720$$
$$= 848.925 - 720 = 128.925$$

Computing the Sum of Squares within Groups

$$SS_{wn} = SS_{tot} - SS_{bn} = 226 - 128.925 = 97.075$$

Computing the Degrees of Freedom

Degrees of Freedom Total
$df_{tot} = N - 1 = 20 - 1 = 19$
Degrees of Freedom between Groups
$df_{bn} = k - 1 = 3 - 1 = 2$
Degrees of Freedom within Groups
$df_{wn} = N - k = 20 - 3 = 17$

Computing the Mean Square between Groups

$$MS_{bn} = SS_{bn} / df_{bn} = 128.925/2 = 64.46$$

Computing the Mean Square within Groups

$$MS_{wn} = SS_{wn} / df_{wn} = 97.075/17 = 5.71$$

Computing the F Statistic

$$F = MS_{bn} / MS_{wn} = 64.46/5.71 = 11.29$$

The ANOVA Summary Table

ANOVA Summary Table

Source	SS	df	MS	F
Between	128.925	2	64.46	11.29
Within	97.075	17	5.71	
Total	226	19		

Our observed *F* statistic is 11.29, and we will need to obtain the appropriate critical value from Table 5 in Appendix A for comparison. We will test the null hypothesis that all group means are equal at the $\alpha = .01$ level with df = (2, 17). The critical value is $F_{.01(2,17)} = 6.11$. The observed value of the *F* statistic is greater than the critical value, so the correct decision is to reject the null hypothesis and conclude that at least two of the sample means are statistically significantly different. Further, logic indicates that the two means with the greatest difference will be statistically significantly different, so we may conclude that there is a statistically significant difference between the placebo group ($\bar{X} = 9.6$) and the high-dose group ($\bar{X} = 3$). A post hoc procedure will reveal any additional differences.

Fisher's protected t-test

In this situation, we will use Fisher's protected t-test for our post hoc procedure, because our three treatment levels are based on unequal sample sizes. Fisher's protected t-test is a variation of the t-test for two independent sample means that we examined in the previous chapter. The numerator is the same consisting of the difference between the two means, but the traditional pooled variance term in the denominator is replaced with the mean square within term. The result is a "t" value interpreted as a two-tail test with N-k degrees of freedom (the df_{wn} value). Critical values may be obtained from Table 2 in the Appendix for the two-tail situation at either the $\alpha = .05$, or $\alpha = .01$ levels of significance. What is important about Fisher's protected t-test is that repeated applications of it across all possible mean pairs avoids the increased risk of a Type I Error that we would encounter with repeated applications of the traditional t-test.

Computational Formula for Fisher's Protected t-Test

$$t = \frac{\bar{X}_1 - \bar{X}_2}{MS_{wn}\left(\frac{1}{n_1} + \frac{1}{n_2}\right)}$$

where

$\bar{X}_1$ = the mean of the first group
$\bar{X}_2$ = the mean of the second group
MS_{wn} = the mean square within value
n_1 = the sample size of the first group
n_2 = the sample size of the second group.

The means associated with our three treatment levels are presented below:

Placebo $\bar{X} = 9.6$
Low-Dose Group $\bar{X} = 6.375$
High-Dose Group $\bar{X} = 3$

We can be assured from the results of the ANOVA procedure that there is a statistically significant difference between the two means with the largest observed difference (the placebo group and the high-dose group). However, we will compute Fisher's protected t-test on that comparison as an illustration, and then on all other possible pairs of means to uncover any additional significant differences.

Comparing the Placebo Group and the High-Dose Group

$$t = \frac{9.6 - 3}{\sqrt{5.71\left(\frac{1}{5}+\frac{1}{7}\right)}}$$

$$= \frac{6.6}{\sqrt{5.71\,(.2+.143)}}$$

$$= \frac{6.6}{\sqrt{5.71\,(.343)}}$$

$$= \frac{6.6}{\sqrt{1.959}}$$

$$= \frac{6.6}{1.40} = 4.71$$

The observed t value is 4.71. Again we will use the same alpha level of $\alpha = .01$ for our post hoc procedure as we did for the ANOVA procedure. The critical value for a two-tailed t-test with df = 17 may be obtained from Table 2 in the Appendix, $t_{.01} = \pm 2.898$. Our observed $t = 4.71$ is larger than the critical value, so the appropriate decision is to reject the null hypothesis and to conclude that the two sample means are statistically significantly different.

Two other mean comparisons are possible: the placebo group with the low-dose group and the low-dose group with the high-dose group. These comparisons are presented below in a slightly abbreviated form.

Comparing the Placebo Group and the Low-Dose Group

$$t = \frac{9.6 - 6.375}{\sqrt{5.71\left(\frac{1}{5}+\frac{1}{8}\right)}}$$

$$= \frac{3.225}{\sqrt{5.71(.2+.143)}}$$

$$= \frac{3.225}{\sqrt{5.71\,(0.325)}}$$

$$= \frac{3.225}{1.362} = 2.368$$

The observed t value of $t = 2.368$ does not equal or exceed the critical value of $t_{.01} = \pm 2.898$, so the correct decision is to fail to reject the null hypothesis. The mean of the placebo group is not statistically significantly different from the mean of the low-dose group.

Comparing the Low-Dose Group and the High-Dose Group

$$t = \frac{6.375 - 3}{\sqrt{5.71\left(\frac{1}{8} + \frac{1}{7}\right)}}$$
$$= \frac{3.375}{\sqrt{5.71(.268)}}$$
$$= \frac{3.375}{1.237}$$
$$= 2.728$$

The observed t value of $t = 2.728$ does not equal or exceed the critical value of $t_{.01} = \pm\, 2.898$, so the correct decision is to fail to reject the null hypothesis. The mean of the low-dose group is not statistically significantly different from the mean of the high-dose group.

Computing Eta Squared and Omega Squared to Estimate Variance Explained

To finish the second example of ANOVA, we will compute the values of Eta Squared, providing us with an estimate of the amount of variance explained by the independent variable in the sample, and Omega Squared, providing us with an estimate of the amount of variance explained by the independent variable in the population. The Eta Squared computation proceeds as follow:

$$\eta^2 = \frac{SS_{bn}}{SS_{tot}}$$

In our example, Eta Squared is as follows:

$$\eta^2 = \frac{128.925}{226} = .57$$

The Eta Squared value of $\eta^2 = .57$ indicates that 57% of the variance in psychotic episodes can be explained by membership in the medication dosage groups.

The computation of the parametric estimate of variance, Omega Squared is as follows:

$$\omega^2 = \frac{SS_{bn} - (df_{bn} \times MS_{wn})}{SS_{tot} + MS_{wn}}$$

In our example, the Omega Squared computation is as follows:

$$\omega^2 = \frac{128.925 - (2 \times 5.71)}{226 + 5.71}$$
$$= \frac{128.925 - 11.42}{231.71}$$
$$= \frac{117.505}{231.71} = .507$$

The Omega Squared value of $\omega^2 = .507$ indicates that 50.7% of the variance in psychotic episodes in the population from which the sample was drawn can be explained by membership in the medication dosage groups.

Variations on a Theme in Analysis of Variance

In this chapter, we have examined one-way ANOVA, but other types of ANOVA models are available for use. One common alternative model is called two-way ANOVA, which takes its name from the fact that two independent variables are examined for their relationship to a single dependent variable. For example, our first one-way ANOVA model examined the relationship between educational attainment and income. We might have also added a second independent variable such as age. The resulting analysis would indicate the effect of educational attainment on income, the effect of age on income, and also any interaction effect of age and educational attainment on income. An interaction effect occurs when the effects of two variables together are greater than the expected additive effects of the two variables alone. For example, low educational attainment might result in a 10% reduction in income, and young age might result in a 10% reduction in income. Individuals with both low educational attainment and young age would then be expected to have a 20% reduction in income; however, if we find that individuals with both low educational attainment and young age on average have a 30% reduction in income, then it suggests the presence of an interaction effect. The logic of two-way ANOVA follows that of the one-way ANOVA model we have examined here, but the computations are needlessly tedious to illustrate by hand. Two-way ANOVA models are easily performed with most basic computer statistical analysis packages.

A second variation of the basic ANOVA model is a multivariate approach called multivariate analysis of variance or MANOVA. MANOVA models allow us to examine two or more dependent variables simultaneously, and most often with more than a single independent variable. Again, the analysis includes both independent effects of the variables taken by themselves and interaction terms. MANOVA models are not as commonly used as some other multivariate techniques such as Path Analysis, Factor Analysis, and Structural Equations models, discussions of which are far beyond the objectives of this text, but those of you with intentions of graduate study might want to remember those terms. You will likely hear them again in your graduate statistics courses.

Computer Applications

(1) Load the GSS data set and conduct an ANOVA of education ("Educ") by region of the country ("Region"). Interpret the results.

How to do it

Load the GSS data set. Click on "Analyze" "Compare Means" and "One Way ANOVA" to open the ANOVA dialog box. Highlight "Educ" and move it to the Dependent List Variables selected box. Highlight "Region" and move it to the Factor box. Click on "Define Range" and supply the range of values associated with the region variable (the range is from 1 to 9). Click on "Post Hoc" and select two post hoc analysis tests (choose the least significant difference test and Tukey's honestly significant difference test). Click on "Options" and select Display labels to make the output easier to interpret. Click on "OK" to compete the procedure. Interpret the results.

Summary of Key Points

ANOVA is an important tool in statistical analysis in the behavioral sciences, and one that is used quite often. The chief purpose of ANOVA is to examine a group of means for statistically significant differences. The primary advantage of ANOVA over multiple t-tests is that ANOVA allows us to conduct a single test to determine if any of the possible paired comparisons of means will be statistically significantly different. Multiple t-tests are not only time consuming, but also run the increased risk of committing Type I Error (rejecting the null hypothesis, when in fact it is true) with each additional t-test that is conducted.

If the results of the ANOVA are significant, we can be sure that at least the smallest and largest means are statistically significantly different. Additional statistically significant differences may be discovered as the result of a post hoc procedure. Both of the post hoc procedures covered in this chapter, Tukey's HSD (for situations when all group sample sizes are equal) and Fisher's protected t-test (for groups of any sample size), allow us to examine all possible pairs of means for a statistically significant difference without increasing the risk of Type I Error.

Analysis of Variance (ANOVA)—A procedure used to examine a group of means to determine if any of the pairs of means are statistically significantly different.
Factor—A term for the independent variable in an analysis of variance procedure.
Treatment—A given level associated with the independent variable in an analysis of variance procedure. The number of treatment levels is equal to the number of group means (k) being compared.
Variance within Groups—The variance in a sample due to random error.
Variance between Groups—The variance in a sample due to group membership.
Sum of Squares—The sum of the squared deviations from the mean used in analysis of variance. Sum of Squares are computed for the total sample, and then partitioned into that due to group membership (sum of squares between groups), and that due to random error (sum of squares within groups).
Mean Square—The value resulting from dividing the sum of squares by its associated degrees of freedom. Mean Square values are computed for between Groups and within Groups.
F Distribution—The distribution for the F statistic which is used in analysis of variance. A statistically significant F value indicates that at least one of the possible pairs of means is statistically significant. The F statistic is equal to the mean square between divided by the mean square within.
Post Hoc Test—One of several procedures used to uncover all of the statistically significantly different pairs of means in an ANOVA procedure.
Tukey's HSD Test—An ANOVA post hoc test appropriate when all groups have the same sample size.
Fisher's Protected t-Test—An ANOVA post hoc test appropriate for groups of differing or equal sample sizes.
Eta Squared—A descriptive statistic indicating the amount of variance in the dependent variable explained by the independent variable for the sample under investigation.
Omega Squared—A parametric statistic indicating the amount of variance in the dependent variable explained by the independent variable for the population from which the sample under investigation was drawn.

Questions and Problems for Review

1. Identify and discuss the advantages of conducting ANOVA instead of multiple t-tests.

2. What is a post hoc procedure, and why do we need to use them in ANOVA?

3. Find a journal article appropriate to your field that reports the results of an ANOVA procedure. Summarize the results including the post hoc procedure.

4. Conduct an ANOVA procedure for the following data. Construct the ANOVA summary table, test the null hypothesis at the $\alpha = .05$ level, and interpret the results.

Data

Number of Traffic Accidents Per Week by Interstate Speed Limit

55 MPH	65 MPH	75 MPH
12	18	12
20	15	10
10	12	14
8	36	8
6	13	12
15	25	6
18	30	5

5. Conduct the appropriate post hoc procedure for the ANOVA procedure in problem 4, and interpret the results. How much variance does the interstate speed limit explain in the number of traffic accidents (conduct a descriptive statistic procedure).

6. Complete the ANOVA summary table below, and test the F statistic for statistical significance at the $\alpha = .05$ level.

ANOVA Summary Table

Source	SS	df	MS	F
Between	2,455.5			
Within		20	65.63	
Total	3,768.1	24		

7. Assume all groups in problem 6 have the same sample size of $n = 5$. What is the minimum difference between any two means required for a statistically significant result at the $\alpha = .05$ level?

8. Construct the ANOVA summary table given the following summary information. Test the F statistic at the $\alpha = .01$ level, and interpret the results.

Group A: $\Sigma X = 82$; $\Sigma X^2 = 1025$; $n = 5$

Group B: $\Sigma X = 94$; $\Sigma X^2 = 1586$; $n = 8$

Group C: $\Sigma X = 65$; $\Sigma X^2 = 645$; $n = 10$

9. Perform the appropriate post hoc test and interpret the results, and estimate the amount of variance explained by the independent variable in the population from which the sample was drawn for the ANOVA procedure from problem 8.

10. A very bright student of statistics understands the Type I Error problem associated with performing multiple t-tests, but is not interested in performing the additional work required in conducting an ANOVA procedure. The student suggests a better way, "Why not begin by conducting a t-test on the two group means with the largest difference? If a significant result is found, continue with the two group means with the next largest difference, and so on, until a nonsignificant t-test results." What is wrong with this suggestion?

CHAPTER 13

Nonparametric Statistics

Key Concepts

Contingency Table
Chi Square Test for Independence
Coefficient of Contingency
Phi Coefficient
Lambda
Gamma
Mann–Whitney U Test
Wilcoxon T Test

Introduction

The past several chapters have dealt with a variety of inferential statistical procedures in which we use statistics based on sample data to make inferences regarding population parameters from which the sample was drawn. These statistical procedures were referred to as **parametric statistics** because they require certain assumptions to be met regarding the distribution of the raw scores in the population. One common assumption was that the raw scores in the population be normally distributed for a variable of interest. Another common assumption was that an interval level or ratio level of measurement be used for the dependent variable allowing us to legitimately compute a mean.

The statistical procedures in this chapter are also inferential statistics, but they are referred to as **nonparametric statistics**. Nonparametric statistics include a variety of inferential procedures that do not require us to meet the strict assumptions associated with parametric statistics. Most commonly, nonparametric statistics are used when the raw scores of a dependent variable do not meet the conditions of normality in the population (the distribution is skewed instead of normally distributed), or the dependent variable has been measured at the nominal level or ordinal level preventing us from legitimately computing a mean. Usually, nonparametric statistics are applied to data where the median is the more appropriate measure of central tendency, or to data presented in contingency tables in a nominal or ordinal level format.

In this chapter, we will examine some of the nonparametric statistical techniques commonly used in behavioral science research. In many cases, the statistical techniques we will examine represent the nonparametric alternative to parametric statistical tests we have seen in earlier chapters. For example, the **Mann–Whitney U** test is a nonparametric alternative to the t-test for independent samples, and the **Wilcoxon T** test is a nonparametric alternative for the t-test for related samples. Other statistical procedures we will cover in this chapter are primarily used to analyze data presented in contingency tables. Included in this group will be the **chi square** (χ^2) test for independence, and some popular measures of association such as the **Coefficient of Contingency** (C), the **Phi Coefficient** (φ), **Guttman's Coefficient** of **Predictability**, **Lambda** (λ), and **Goodman's and Kruskal's Gamma** (γ). We will begin with a discussion of the chi square test for goodness of fit after a brief discussion on the construction of a contingency table.

The Construction and Presentation of Data in a Contingency Table

We briefly introduced the concept of a contingency table and its relationship to the Cartesian Coordinate system in Chapter 9. In this chapter, we will examine the logic of constructing a contingency table and how to interpret the results. Contingency tables are used to illustrate the relationship between two categorical variables, just as the Cartesian coordinate system is used to demonstrate the relationship between two continuous variables. In Chapters 9 and 10, we used the Cartesian Coordinate system to illustrate the relationship between education and income when both variables were measured at the ratio level. In this chapter, we will use the same data on education and income as ordinal level dichotomies in a contingency table as the basis for our first example of the chi square test in the next section.

We saw previously that the Cartesian coordinate system allows us to graphically depict the relationship between two continuous variables. The horizontal axis is used to represent the independent variable "*X*" with values getting larger as we move from left to right, and the vertical axis is used to represent the dependent variable "*Y*" with values getting larger as we move from bottom to top (see Figure 13.1). Data from a sample of $n = 25$ individuals from whom we have collected information on

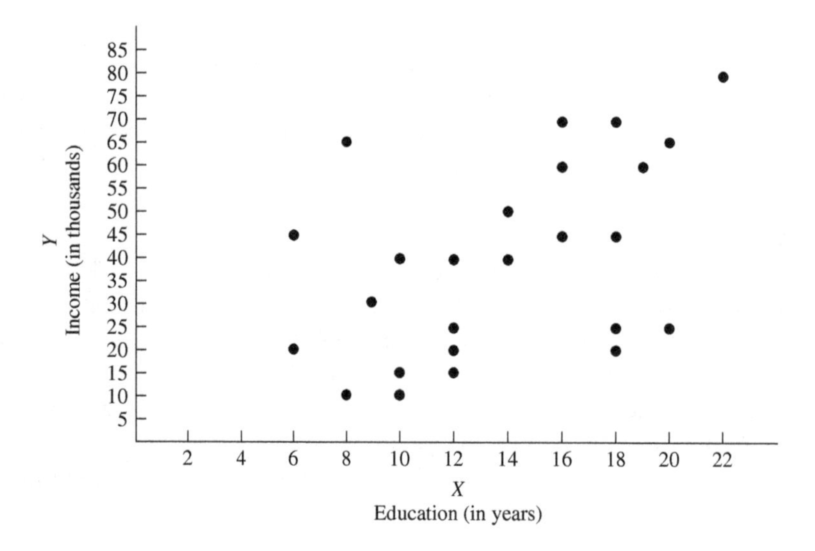

Figure 13.1 Bivariate Plot of Education and Income

level of educational attainment (in years) and level of income (in thousands of dollars) would appear as follows when presented on a Cartesian Coordinate system (see Table 13.1).

As you should recall from Chapter 9 on correlation and chapter 10 on linear regression, the closer the data appear as a straight line the stronger the relationship between the two variables. In this case, we obviously do not have a perfect relationship, but there is a suggestion of a positive relationship between the two variables. In general, as education increases, there is a corresponding increase in income. We can "see" the relationship by the pattern evident in the scatterplot of the data. In the general case, the farther to the right an individual is on the horizontal axis (representing education) the farther toward the top the individual is on the vertical axis (representing income).

A Cartesian coordinate system is a useful tool for presenting data, but it does not fit every need and is not suitable for data below the interval level. Nominal and ordinal level data are often presented in a table format referred to as a contingency table. There are also times when interval or ratio level data are

Table 13.1 Data on Education and Income

Individual	Education (in years)	Income (in thousands)
1	12	20
2	12	40
3	16	45
4	8	10
5	14	50
6	16	70
7	6	45
8	10	15
9	16	60
10	18	25
11	10	10
12	20	25
13	19	60
14	22	80
15	12	25
16	9	30
17	18	70
18	14	40
19	18	45
20	20	65
21	10	40
22	8	65
23	12	15
24	6	20
25	18	10

converted to ordinal level data, and then presented in a contingency table. For example, many of the more powerful parametric procedures presented in earlier chapters require that the dependent variable meet the assumption of being normally distributed. If we are unable to meet that assumption, we might reduce the data to the categorical level and use one of the nonparametric procedures presented in this chapter. To illustrate the technique, we will use the data on education and income from above to demonstrate the proper construction and interpretation of a contingency table.

The first step in presenting the data on education and income for our sample of $n = 25$ individuals is to construct ordinal level categories for the two variables. For this illustration, we will create two categories for each of the two variables. Individuals will be classified as being in either a low education or a high education category, and as being in either a low income or a high income category. A contingency table featuring two variables with two categories each is often referred to as a "2×2" (read as two by two) table.

There are any number of ways to create two categories for a variable. In the case of the education variable, the most logical cutting point (dividing line between low education and high education) is at the "12-year" point in the data. The question becomes, "will 12 years of education be part of the low education category or part of the high education category?" In this case, we will define low education as 12 years or less, and high education as anything over 12 years. The income variable must also be converted to ordinal level data with two categories, and the "35 thousand dollar" level appears to be one possible logical cutting point in these data. We will define low income as anything below 35 thousand dollars and high income as 35 thousand dollars or more. Figure 13.2 presents the original scatterplot of education and income with the cutting points added at the appropriate places.

As you can see in Figure 13.2, the cutting points provide us with four logical areas or categories on the scatterplot: an area of low education and high income; an area of high education and high income; an area of low education and low income; and, an area of high education and low income. With the cutting points in place, you can also see the number of cases from the sample that fall into each of the four possible categories:

Low Education and High Income: $n = 4$
High Education and High Income: $n = 10$
Low Education and Low Income: $n = 8$
High Education and Low Income: $n = 3$

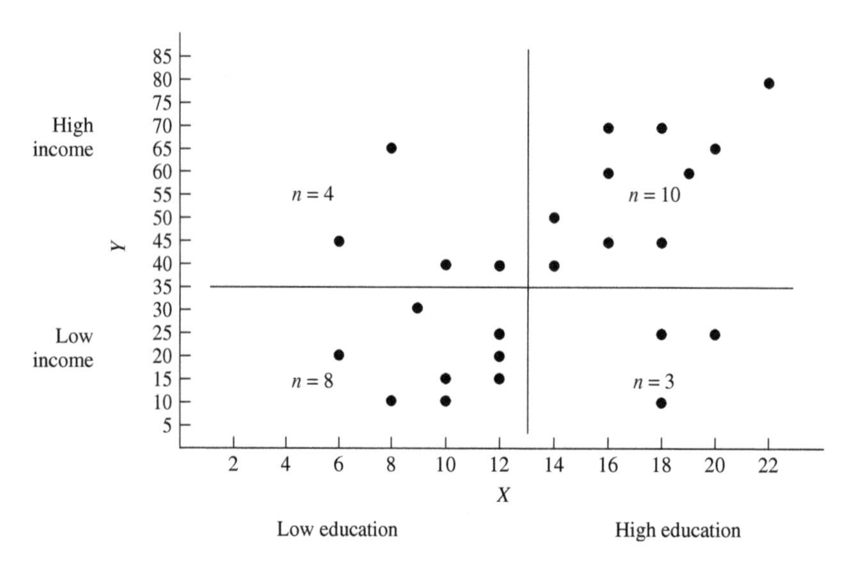

Figure 13.2 Bivariate Plot of Education and Income with Cutting Points Added

A contingency table is constructed following the same logic as the Cartesian coordinate system. Just as the independent variable appears on the horizontal axis with values increasing as you move from left to right, the independent variable in a contingency table is presented on the horizontal with values increasing as you move from left to right. The dependent variable appears on the vertical axis on the Cartesian coordinate system with values increasing as you move from the bottom to the top, and so too in a contingency table the dependent variable is presented on the vertical with values increasing as you move from the bottom to the top. The resulting table should appear as follows (Table 13.2):

Table 13.2 Income by Education

		Education		
		Low	**High**	**Total**
	High	4	10	14
Income	**Low**	8	3	11
	Total	12	13	25

The interpretation of the nature of a relationship between two variables evident in a contingency table follows the same logic as that of a coordinate system. In a positive relationship, individuals who are in the low category for the independent variable should also be found in the low category for the dependent variable. Similarly, individuals who are in the high category for the independent variable should also be found in the high category for the dependent variable. Just as a positive relationship had a certain "look" on a coordinate system with the data points in a linear trend with a positive slope, a similar picture will emerge in the contingency table. Two blank 2×2 contingency tables are presented below (Tables 13.3and 13.4) with the four cells of the table labeled "A," "B," "C," and "D." When two variable have a positive relationship, you will find most of the sample in cells "B" and "C." When two variables have a negative relationship, you will find most of the sample in cells "A" and "D." When two variables have no relationship, you will find most of the sample randomly distributed among the four cells of the table.

If you examine the two blank tables below, you can see that when most of the sample is located in cells "B" and "C," it suggests a linear trend with a positive slope. Imagine drawing a line connecting the two highlighted cells symbolizing most of the cases. The line would have a positive slope when you connect cells "B" and "C," and, the line would have a negative slope when you connect cells "A" and "D."

Table 13.3 A Positive Relationship

		X *Independent Variable*		
		Low	**High**	**Total**
	High	A	**B**	
Y **Dependent Variable**	**Low**	**C**	D	
	Total			Total

Table 13.4 A Negative Relationship

		X Independent Variable		
		Low	**High**	**Total**
	High	**A**	B	
Y Dependent Variable	**Low**	C	**D**	
	Total			Total

Difficult to Interpret Tables

Some contingency tables can be difficult to interpret, especially when the marginal totals (the individual row and column totals) differ by a wide amount. A useful way to make a table easier to interpret is to convert the number of cases in each cell into percentages. The rule to remember when converting numbers to percentages is to "Percent Down, Compare Across."

To convert cell frequencies into percentages: percent down and compare across

"Percent Down" means that each column total should be the basis for the percentage computation. "Compare Across" means that once the percentages have been computed, you should compare across the values in each row and look for the cell with the highest percentage. Once you have identified the cells with the highest percentage values in each row, you will easily be able to see the direction of the relationship in the table. Consider the example below in Table 13.5.

Interpretation of the table is made difficult by the fact that the marginal totals differ by a great deal, especially the column totals (550 and 150). At first glance it may not appear that there is a positive relationship evident in the data. There are a large number of cases in the "low education, low income" category consistent with a positive relationship, but there are only $n = 100$ cases in the "high education, high income" category compared with $n = 200$ cases in the "low education, high income" category. However, the smaller number of cases in the "high education, high income" category can be misleading. There may only be $n = 100$ cases in the category, but that is 100 cases out of a total of the 150 members of the "high education" category in the sample. There are $n = 200$ "low education, high income" cases, but that is 200 out of a total of 550 members of the "low education" category of the sample. Once the number of cases have been converted to percentages, you will see that the 100 "high education, high income" individuals out of 150 "high education" members of the sample is a much higher percentage than the 200 "low education, high income" individuals out of 550 "low education" members of the sample. Table 13.6 presents the same data with the cell frequencies converted to percentages.

Table 13.5 A Table Difficult to Interpret

		Education		
		Low	**High**	**Total**
	High	200	100	300
Income	**Low**	350	50	400
	Total	550	150	700

Table 13.6 Cell Frequencies Converted to Percentages

		Education	
		Low	High
Income	High	36%	**67%**
	Low	**64%**	33%
	Total	100%	100%

With the cell frequencies converted to percentages, it is easy to see that a strong positive relationship between education and income is evident in these data. Comparing across the "high income" row we see 67% in the "high education, high income" category, which is much larger than the 36% in the "low education, high income" category. Comparing across the "low income" row we see 64% in the "low education, low income" category which is much larger than the 33% in the "high education, low income" category. Most of the cases in the sample are in the "B" and "C" cells of the table indicating a strong positive relationship.

Degrees of Freedom in a Contingency Table

We examined the concept of degrees of freedom earlier in Chapter 5 dealing with measures of variation. At that time, degrees of freedom referred to the number of values in a sample that were "free" to take on any value while still meeting the condition that the sample statistic would be an unbiased estimate of the population parameter. In the case of variance and standard deviation, we saw that there were n-1 degrees of freedom. In the context of a contingency table, the concept of degrees of freedom means the number of cells in the table that are "free" to take on any logical value before the remainder of the cells are determined given the existing marginal totals. Let's look at a simple analogy first. Suppose we have the following equation consisting of three terms "A," "B," and "C," which must total 10:

$$A + B + C = 10$$

In a sense, you could say that we have two degrees of freedom in this case. That is, two of the three terms can be assigned any value, but once two of the terms have values assigned to them then value of the remaining term is determined because the total of the three must equal 10. For example, if we allow "A" to equal 5 and "B" to equal 3, then "C" must equal 2. Similarly, if we allow "A" to equal 4 and "C" to equal 6, then "B" must equal 0. In this case, two of the three terms are free to take on any value before the remaining term is determined. A similar situation exists in a contingency table. Consider the blank 2×2 table below with a given set of marginal totals.

Table 13.7 has four cells, how many of them are free to take on any value before the remaining cells will be determined? Keep in mind that any value you try to assign to a cell must be logical given the existing marginal totals. These are cell frequencies so it is not logical for there to be any number in a cell smaller than zero, although zero itself is a logical possibility. Also note that each cell is part of two marginal totals, its row total and its column total. For example, the logical possibilities for cell "A" range between "0 and 80." No negative numbers are permissible, and "80" would be the maximum logical value because the column total for cell "A" is 80. Given the restriction of assigning only logical values, how many degrees of freedom are there in a 2×2 table?

A 2×2 table has only one degree of freedom. We can arbitrarily assign a value to only one of the four cells of the table. Once a single cell has been assigned a value, the remaining three cells will be

Table 13.7 A 2×2 Table with Marginal Totals

		X		
		Low	**High**	**Total**
	High	A	B	130
Y	**Low**	C	D	70
	Total	80	120	200

Table 13.8 A 2×2 Table with One Degree of Freedom

		X		
		Low	**High**	**Total**
	High	30	**100**	130
Y	**Low**	50	20	70
	Total	80	120	200

automatically determined. For example, in Table 13.8, if cell "B" is assigned the value "100," then cell "A" must equal "30" to conform to the row total of "130." Cell "D" must equal "20" to conform to the column total of "120," and if cell "A" must equal "30" then cell "C" must equal "50" to conform to the column total of "80." You can carry out any number of trials using any of the four cells and arbitrarily assigning any logical number, but once any cell has been assigned a logical value, the other three cells are determined. Any 2×2 table will have only one degree of freedom.

How many degrees of freedom are there in larger tables? Suppose we had a 3×3 table consisting of a total of nine cells. Table 13.9 presents a typical 3×3 table based on categories of "low," "medium," and "high." How many degrees of freedom are there in the table?

A 3×3 table will have a total of four degrees of freedom. With the same assumption as before that only logical values may be assigned, we will be able to arbitrarily assign four of the cells with values before the remaining cells will be determined by the respective marginal totals. We can arbitrarily assign one cell in each of the three rows with a value, and then assign any additional cell with a value for our total of four degrees of freedom. After those four cells have been assigned values, the remaining cells will be determined given the respective marginal totals as illustrated in Table 13.9. Cells "B," "D," "E," and "I" have been assigned values, and once we know that cells "D" and "E" each have "60" frequencies, then we know that cell "F" must have "0" frequencies. Once cell "F" is determined, there are at least two cells with assigned values in all three of the columns, and the remaining cell will be determined by the respective column total.

Table 13.9 A 3×3 Table with Four Degrees of Freedom

		X			
		Low	**Medium**	**High**	**Total**
	High	20	**50**	60	130
Y	**Medium**	**60**	**60**	0	120
	Low	70	90	**40**	200
	Total	150	200	100	450

How to Calculate Degrees of Freedom for Any Contingency Table

You may calculate the number of degrees of freedom for any contingency table rather easily. The number of degrees of freedom will be equal to the product of the number of rows minus one times the number of columns minus one.

$$\text{Degrees of Freedom (df)} = (\text{rows} - 1) \times (\text{columns} - 1)$$

The number of rows and columns refers to the number of categories for the independent and dependent variables. The row and column marginal totals do not count for purposes of degrees of freedom. We can use the formula to illustrate the degrees of freedom computation for our 2×2 table and 3×3 table. For the 2×2 table we would calculate degrees of freedom as:

$$\text{df} = (2-1) \times (2-1) = 1 \times 1 = 1$$

For the 3×3 table we would calculate degrees of freedom as:

$$\text{df} = (3-1) \times (3-1) = 2 \times 2 = 4$$

The Chi Square Test for Independence

Now that we have examined the logic of constructing and interpreting a contingency table, we are ready to examine some of the statistical procedures that enable us to determine if any type of relationship is present in the data. The chi square (χ^2) test for independence is one of the most commonly used statistical tests for examining data presented in a tabular format. The chi square test is a little different than the statistical tests we have examined in the previous chapters in that chi square does not allow us to test directional hypotheses regarding the relationship between two variables. For example, in the case of our table of income by education, a chi square test will not tell us if there is a statistically significant positive relationship between the two variables. Instead, the chi square test will tell us if there is some type of relationship in the data.

The chi square test is more accurately referred to as a test of independence. Independence refers to how well the observed pattern of results (what we actually see in the table) conforms to the pattern of results we would expect by chance (the pattern that would exist in the table if there were no relationship at all between the two variables). The computation of the chi square test statistic is based on the difference between the observed results (O) actually obtained, and the expected results (E) that we would see assuming no relationship between the two variables.

Assumptions for the Chi Square Test

Appropriate application of the chi square test assumes the following:

1. A random or probability-based sampling technique.
2. The independent and dependent variables must be capable as being presented as categories.
3. The categories for the variables are mutually exclusive and independent (a given member of the sample can logically be placed in one and only one category, and one person's placement does not effect any other person's placement).
4. The sample size should range from approximately a minimum of 25 to a maximum of 250 (extreme sample sizes in either direction can have distorting effects on the value of the chi square result).
5. The *expected value* of each cell of the table should be greater than or equal to 5.

A Computational Example of Chi Square

The 2 × 2 contingency table created for income by education from our sample of $n = 25$ individuals is presented below (Table 13.10).

The data appear to suggest a positive relationship, but are the results significantly different from the pattern that we would observe by chance? The first step in answering this question is to compute the expected results.

Table 13.10 Income by Education

		Education		
		Low	**High**	**Total**
	High	4	10	14
Income	**Low**	8	3	11
	Total	12	13	25

The Logic of Computing the Expected Results

The contingency table consists of three parts: the cells or categories for the data; the row and column totals called the marginal totals (because they appear on the margin of the table); and, the grand total representing the total sample size. In a completely balanced table, one in which the marginal totals are all equal, computing the expected results is quite easy. The expected results for each cell can be found by dividing the grand total by the number of cells in the table. Consider the balanced table below (Table 13.11).

Table 13.11 A Balanced Table of Equal Marginal Totals

		Independent Variable *X*		
		Low	**High**	**Total**
	High	?	?	50
Dependent Variable *Y*	**Low**	?	?	50
	Total	50	50	Total

What values would we expect to find in each of the four cells of the table if there were no relationship at all between the independent and dependent variable? Assuming no relationship, the $n = 100$ cases of the sample would be evenly distributed across the four cells of the table, so we could compute the expected results by simply dividing the grand total (total sample size) by the number of cells in the table. In this case, the expected results (E) in each cell would be

$$E = \frac{100}{4} = 25$$

Computing expected results when the table is not balanced is a little more difficult. Consider our table of income by education (Table 13.12). What would the expected results be?

Table 13.12 Marginal Totals for Income by Education

		Education		
		Low	**High**	**Total**
	High	?	?	14
Income	**Low**	?	?	11
	Total	12	13	25

Dividing the total sample size (n = 25) by the number of cells in the table (cells = 4) will not work in this case, because the marginal totals are not all equal. **The problem is that the expected cell values must conform to the observed marginal totals.** If we divide 25 by 4, the result is 6.25, and an expected value of E = 6.25 in each of the four cells of the table will not provide the required marginal totals.

The key to computing expected values is to realize that they represent what the cell values would equal under the assumption of no relationship. In that case, the distribution of the sample in the cells should follow the same distribution as the marginal totals. In our table, there are 12 individuals (or .48 of the total) in the "low education" category and 13 individuals (or .52 of the total) in the "high education" category. If there is no relationship between education and income, then we should expect the same proportional distribution of income in each column of the table. For example, the first column (low education) contains a proportional total of .48 of the sample (12/25 = .48) and the second column (high education) contains a proportional total of .52 of the sample (13/25 = .52). The top row of the table (the high income category) has a marginal total of 14 individuals. How many should we expect to be in each column? Assuming no relationship between the two variables, we would expect .48 of the 14 individuals to be in the first column (.48 × 14 = 6.72) and .52 of the 14 individuals to be in the second column (.52 × 14 = 7.28). The low income row of the table has a marginal total of 11 individuals. Again, assuming no relationship between education and income, we would expect .48 of the 11 individuals to be in the first column (.48 × 11 = 5.28) and .52 of the individuals to be in the second column (.52 × 11 = 5.72).

Because of the symmetrical nature of a contingency table, we could derive the same expected values working with the marginal row totals. For example, the top row (high income) contains a proportional total of .56 of the sample (14/25 = .56) and the bottom row (low income) contains a proportional total of .44 of the sample (12/25 = .44). The first column of the table (the low education category) has a marginal total of 12 individuals. How many should we expect to be in each row? Assuming no relationship between the two variables, we would expect .56 of the 12 individuals to be in the top row (.56 × 12 = 6.72) and .44 of the 12 individuals to be in the bottom row (.44 × 12 = 5.28). The second column of the table (the high education category) has a marginal total of 13 individuals. Again, assuming no relationship between education and income, we would expect .56 of the 13 individuals to be in the top row (.56 × 13 = 7.28) and .44 of the individuals to be in the bottom row (.44 × 13 = 5.72).

The Easier Way to Compute Expected Results

It is important to be exposed to the logic behind statistical procedures such as computing expected results in a contingency table, because it helps you to understand the process. Without understanding, statistical analysis becomes just a mechanical process of completing steps 1 through 5 to generate an answer, but if you do not understand the process, you will probably not understand the answer either. However, once you have an understanding of the process, there are usually some shortcuts or easier ways to generate the results. In the case of computing expected results for a contingency table, there is an easier way than to logically analyze the table.

The expected results for any cell can be computed by multiplying the cell's row total by its column total, and then dividing the result by the grand total. We can use this method to generate the expected results for our table of income by education as illustrated below (Table 13.13).

$$\text{Expected Results}\,(E) = \frac{(\text{Row total} \times \text{Column total})}{\text{Grand total}}$$

The Formula for the Chi Square Test for Independence

The chi square (χ^2) test for independence tells us if a given set of observed results differs from what we would expect by chance. The test is a two-tail test of the null hypothesis, which assumes that the observed results are equal to what we would expect by chance, indicating no relationship between the two variables. The alternative research hypothesis states a difference between the observed results and the expected results indicating some type of relationship between the two variables. Both the research hypothesis and the null hypothesis are presented symbolically below.

$$H_1 : \text{Observed results} \neq \text{Expected results}$$

$$H_0 : \text{Observed results} = \text{Expected results}$$

The formula for the chi square test is as follows:

$$\chi^2 = \sum \frac{(O - E)^2}{E}$$

where

O is the observed frequency in the cell and
E is the expected frequency of the cell.

The chi square value is the sum of the observed result minus the expected result quantity squared, divided by the expected result computed cell by cell. Each cell makes a contribution to the final value of the chi square statistic, and because the difference between the observed result and the expected result is squared, none of the cell contributions to the chi square value can be negative. We will use the 2×2 contingency table presenting the cross tabulation of income by education as our first example of the chi square statistic. Both the observed results and the expected results are presented below, along with the computation of the chi square test statistic (Tables 13.14 and 13.15).

Table 13.13 Computation of Expected Results

		Education		
		Low	**High**	**Total**
	High	(14 × 12)/25 = **6.72**	(14 × 13)/25= **7.28**	14
Income	**Low**	(11 × 12)/25 = **5.28**	(11 × 13)/25 = **5.72**	11
	Total	12	13	25

Table 13.14 Income by Education Observed Results

		Education		
		Low	**High**	**Total**
	High	4	10	14
Income	**Low**	8	3	11
	Total	12	13	25

Table 13.15 Income by Education Expected Results

		Education		
		Low	**High**	**Total**
	High	6.72	7.28	14
Income	**Low**	5.28	5.72	11
	Total	12	13	25

Computing the Chi Square Statistic

$$\chi^2 = \sum \frac{(O-E)^2}{E}$$

$$\chi^2 = \frac{(4-6.72)^2}{6.72} + \frac{(10-7.28)^2}{7.28} + \frac{(8-5.28)^2}{5.28} + \frac{(3-5.72)^2}{5.72}$$

$$\chi^2 = \frac{(-2.72)^2}{6.72} + \frac{(2.72)^2}{7.28} + \frac{(2.72)^2}{5.28} + \frac{(-2.72)^2}{5.72}$$

$$\chi^2 = \frac{(7.3984)}{6.72} + \frac{(7.3984)}{7.28} + \frac{(7.3984)}{5.28} + \frac{(7.3984)}{5.72}$$

$$\chi^2 = \frac{(7.3984)}{5.28} + \frac{(7.3984)}{5.72}$$

$$\chi^2 = 1.10 + 1.02 + 1.40 + 1.29$$

$$\chi^2 = 4.81$$

The observed value of the chi square test statistic is $\chi^2 = 4.81$. At this point, we will do what we have done so often in the past, obtain a critical value from a table in the appendix, and reach a decision regarding the null hypothesis.

Critical Values from the Chi Square Distribution

The chi square distribution, like the "t" distribution and the "F" distribution, is actually a set of distributions based on the degrees of freedom associated with the contingency table being evaluated. Table 7 in the appendix contains the critical values associated with the $\alpha = .01$ and $\alpha = .05$ levels for selected degrees of freedom. Degrees of freedom (df) for the contingency table can easily be calculated by the following formula:

$$\text{df} = (r-1) \times (c-1)$$

where
r is the number of rows in the table and
c is the number of columns in the table.

Note: that the rows and columns refer to the number of categories of the dependent and independent variables, and that the "total" row and column do not count in the degrees of freedom computation.

Our 2×2 table of income by education has one degree of freedom:

$$\text{df} = (2-1) \times (2-1) = 1.$$

Arbitrarily choosing the $\alpha = .05$ level of significance to test the null hypothesis, we find a critical value of $\alpha_{.05} = 3.84$ (Figure 13.3).

The observed chi square value of 4.81 is in the critical region, so the correct decision is to REJECT THE NULL HYPOTHESIS. The observed results are statistically significantly different from what we would expect by chance.

Our finding of a statistically significant result does not mean that level of income is positively related to level of education among our sample of $n = 25$ individuals. All that the statistically significant chi square value indicates is that in some way the observed results differ from what would be expected by chance. The fact is that in this particular case, there does appear to be a positive relationship between education and income, as we would expect. However, if we were to conduct a chi square analysis on a second sample of $n = 25$ individuals with the exact opposite relationship between level of income and level of education (a negative relationship where those with high levels of education had low levels of income, and those with low levels of education had high levels of income), we would observe the exact same chi square value of $\chi^2 = 4.81$!

The fact that the same chi square value would be observed for two exactly opposite results is why the chi square test is referred to as a test for independence, and not as a directional hypothesis test such as a t-test or a Z test. A statistically significant chi square test simply tells us that something in the table is different from what would be expected by chance, and then it is up to us to analyze the distribution

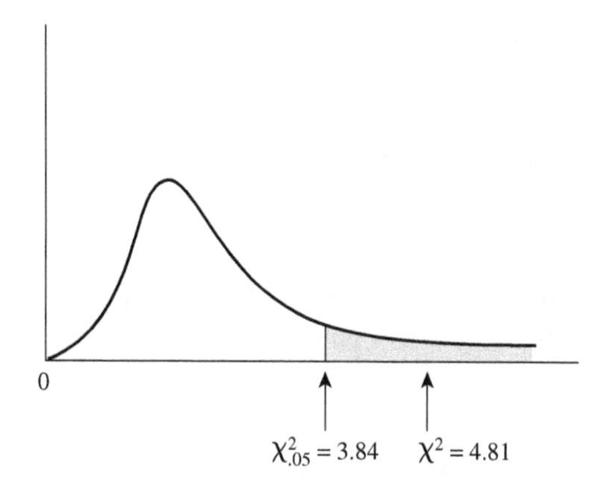

Figure 13.3 Observed and Critical Values of Chi Square

of frequencies in the table and interpret what type of relationship, if any, may be present. Keep in mind that because none of the individual cell component contributions to the overall chi square value can be negative, it is possible for the observed results of a single cell to result in a statistically significant chi square value for the entire table. In which case there may be no overall relationship evident in the data, but simply one extreme cell value.

Some Additional Issues Regarding Chi Square

One of the assumptions of chi square is that the expected value of each cell of the table should be greater than or equal to 5. One drawback of the chi square test is that the value of the observed result is especially susceptible to sample size. Due to the computation of the chi square, small sample sizes can lead to inflated values because we divide the square of the difference between the observed and expected result by the expected result. The smaller the value of the expected result the larger the value of the chi square. Ironically, large sample sizes have a similar result, but for a different reason. The larger the sample size the larger the resulting value of the chi square, because the squared difference between the observed result and the expected result becomes larger.

Our first example of a chi square involving income by education examined a sample of $n = 25$, and resulted in an observed chi square value of 4.81. Had the sample size been 10 times larger at $n = 250$, the exact same proportional distribution of the sample would have resulted in an observed chi square value 10 times larger of 48.1. Had the sample size been 20 times larger with the same proportional distribution, the resulting chi square value would have been 20 times larger at 96.2!

Yates' Correction for Continuity for 2 × 2 Tables

Yate's correction for continuity is an adjustment for the chi square test used for small sample sizes of approximately $n = 25$ up to $n = 75$ for data presented in 2×2 tables. The adjustment results in a slightly smaller observed value for the chi square statistic, and is intended to account for the inflated results that are often associated with small sample sizes. The Yates' correction for continuity is a simple procedure that requires us to subtract .5 from the difference between the observed result and the expected result before squaring the term, and dividing by the expected result. Applying Yates' correction for continuity to our chi square analysis would have resulted in the following:

Chi square computation with yates' correction for continuity Original computation

$$\chi^2 = \frac{(4-6.72)^2}{6.72} + \frac{(10-7.28)^2}{7.28} + \frac{(8-5.28)^2}{5.28} + \frac{(3-5.72)^2}{5.72}$$

$$\chi^2 = \frac{(-2.72)^2}{6.72} + \frac{(2.72)^2}{7.28} + \frac{(2.72)^2}{5.28} + \frac{(-2.72)^2}{5.72}$$

Reducing the Size of the difference between the observed and expected results by .50 yields Yates' correction computation

$$\chi^2 = \frac{(-2.22)^2}{6.72} + \frac{(2.22)^2}{7.28} + \frac{(2.22)^2}{5.28} + \frac{(-2.22)^2}{5.72}$$

$$\chi^2 = \frac{(4.9284)}{6.72} + \frac{(4.9284)}{7.28} + \frac{(4.9284)}{5.28} + \frac{(4.9284)}{5.72}$$

$$\chi^2 = .733 + .677 + .933 + .862$$

$$\chi^2 = 3.205$$

Notice that the corrected chi square value is smaller than the chi square critical value associated with the $\alpha = .05$ level of significance. Applying Yates' correction for continuity to the data reverses our conclusion regarding the null hypothesis, and the correct decision would be to FAIL TO REJECT H0! The observed chi square value adjusted for Yates' correction for continuity is routinely reported along with the unadjusted chi square value in many computer statistical packages. Note that the Yates' correction for continuity is only for 2×2 tables and small sample sizes.

Measures of Association for Contingency Tables

A statistically significant chi square statistic indicates that some pattern of frequency distribution is evident in the contingency table that deviates from what we would expect by chance. Our own analysis of the table can then tell us if that pattern is consistent with a positive or negative relationship between the two variables. There are then several measures of association available for analyzing the contingency table, which can tell us how strong the relationship is between the two variables. This section of the chapter does not begin to provide an exhaustive list of the available measures of association, but does cover several of the more commonly used statistics including the **Coefficient of Contingency** (C); the **Phi Coefficient** (φ); **Guttman's Coefficient of Predictability**, **Lambda** (λ); and, **Goodman's and Kruskal's Gamma** (γ).

We have already examined several measures of association for continuous data in Chapter 9, such as the Pearson r, the Spearman r_s, and the Point Biserial r_{pb}. As you should recall from the earlier discussion, an observed correlation coefficient of 0.00 indicates no relationship between the two variables and an observed correlation coefficient of 1.00 indicates a perfect relationship with the sign indicating the direction of the relationship. The measures of association we examine here are similar to those in Chapter 9, but there are some important differences. First, not all of the measures of association in this chapter can take on negative values providing an indication of the direction of the relationship. Second, not all of the measures of association can achieve an absolute value of 1.00 indicating a perfect relationship, no matter how perfect the actual relationship may appear in the contingency table. We will begin with the two simpler measures of association, the coefficient of contingency (C) and the phi coefficient (φ). Each of these measures of association is easy to apply, and each is based on the observed value of the chi square test statistic.

The Coefficient of Contingency (C)

The coefficient of contingency is a general measure of association used to indicate the strength of a relationship between two variables in a contingency table. It may be applied to tables of any size, which is probably its greatest advantage. Because the coefficient of contingency is based on the observed value of the Chi Square test statistic, it cannot take on negative values although it can achieve a value of zero. A major disadvantage of C is that it can never achieve a value of 1.00 no matter how strong the underlying relationship is in the table. In fact, the value of C is tied to the size of the table upon which the chi square is based. For a 2×2 table, the maximum value that C can achieve is .707. In some cases, a corrected value of C is presented based on dividing the observed value of C by the maximum value it can achieve for a given table size. **The formula for C is presented below:**

$$C = \sqrt{\frac{\chi^2}{N + \chi^2}}$$

where

χ^2 is the observed value of the chi square statistic and
N is the total sample size for the table.

Computing the value of the coefficient of contingency for our first chi square example yields the following:

$$C = \sqrt{\frac{4.81}{25 + 4.81}}$$

$$C = \sqrt{\frac{4.81}{29.81}}$$

$$C = \sqrt{.1614} = .4017$$

There is no separate test of significance for the measure of association. Generally a measure of association based on the chi square is assumed to be statistically significant at the same level of significance as the underlying chi square statistic. Naturally, if the underlying chi square statistic is not statistically significant, then there is no need to compute any measure of association. The interpretation for the observed value of C is based on the rule of thumb guidelines for correlation coefficients in general. The closer the observed value is to 0.00 the weaker the relationship, and the closer the observed value to 1.00 the stronger the relationship. An observed value of $C = .4017$ would indicate a moderate relationship.

The Phi Coefficient

The phi coefficient is similar to the coefficient of contingency in that it is based on the observed value of the chi square statistic, but there are some important differences. First, the phi coefficient may only be applied to a 2×2 table. Second, the phi coefficient can achieve a value of 1.00 indicating the presence of a perfect relationship without any adjustment. Third, the phi coefficient has an interpretation similar to that of the Pearson r correlation coefficient. Just as we can compute an r^2 value as a measure of the amount of variance explained, we can also compute a φ^2 to indicate the amount of variance in the dependent variable explained by the independent variable. The variance explained (or proportional reduction in error—PRE) characteristic of the phi coefficient makes it a preferred measure of association over the coefficient of contingency for a 2×2 table.

The formula for the phi coefficient (φ) is presented below:

$$\varphi = \sqrt{\frac{\chi^2}{N}}$$

where

$\chi^2$is the observed value of the chi square statistic and
N is the total sample size for the table.

Computing the value of the phi coefficient for our first chi square example yields the following:

$$\varphi = \sqrt{\frac{4.81}{25}}$$

$$\varphi = \sqrt{.1924} = .4386$$

Just as was the case with the coefficient of contingency, one assumes statistical significance of the phi coefficient at the same level as the underlying chi square statistic. The advantage of the phi coefficient is the variance explained interpretation indicated by the φ^2 value. Phi squared is not only easy to compute, it is equal to an intermediate value in the computation of the phi coefficient itself. Phi squared is the value under the radical sign once the chi square value has been divided by the sample size. So in our case, the value of φ^2 is .1924, indicating that just over 19% of the variance in the level of income can be explained by the level of education. As a cautionary note, keep in mind that the variance explained interpretation involves using the category of the independent variable to predict the category of the dependent variable. The resulting "variance explained" interpretation is not exactly equivalent to the more powerful r^2 value obtained from the Pearson r when examining interval or ratio level data.

Guttman's Coefficient of Predictability, Lambda

Another commonly used measure of association for data presented in a contingency table is Guttman's coefficient of predictability, or lambda (λ). Like the phi coefficient, lambda has a variance explained interpretation, but lambda is more flexible in that it can be applied to tables of any size. Lambda may also be applied to data reported at any level including nominal level data. **The formula for lambda is as follows:**

$$\lambda = \frac{\Sigma f_r + \Sigma f_c - (F_r + F_c)}{2N - (F_r + F_c)}$$

where

Σf_r is the sum of the largest frequency appearing in each row,
Σf_c is the sum of the largest frequency appearing in each column,
F_r is the largest marginal total observed among the rows,
F_c is the largest marginal total observed among the columns, and
N is the total sample size.

In the table of income by education the appropriate data are as follows:

Largest frequency in each row: 10 and 8
Largest frequency in each column: 8 and 10
Largest marginal row total: 14
Largest marginal column total: 13
Sample size: 25

Computing the value of lambda yields the following:

$$\lambda = \frac{(10+8)+(8+10)-(14+13)}{2(25)-(14+13)}$$

$$\lambda = \frac{18+18-27}{50-27}$$

$$\lambda = \frac{9}{23} = .3913$$

The observed value of lambda, $\lambda = .3913$, may be interpreted directly as the amount of variance in one variable that can be explained by the other variable. Notice that we do not interpret lambda as the amount of variance that may be explained in the dependent variable by the independent variable. Lambda as it is computed above is considered a **symmetrical** statistic, that is, it is a measure of the variance explained by using each variable as a predictor of the other.

It is possible to compute an **asymmetrical** version of lambda in which the predictive ability of a particular variable for the other variable is calculated. The formula for the asymmetrical version of lambda (λ_a) in which the independent variable is used to predict the dependent variable is as follows:

$$\lambda_a = \frac{\Sigma f_i - F_d}{N - F_d}$$

where

Σf_i is the largest observed frequency for each category of the independent variable,
F_d is the largest marginal total for the dependent variable, and
N is the total sample size.

In the table of income by education the appropriate data are as follows:

Largest frequency for each category of the independent variable: 8 and 10
Largest marginal total for the dependent variable: 14
Sample size: 25

Computing the value of asymmetrical lambda with *level of education used to predict level of income* yields the following:

$$\lambda_a = \frac{(8+10)-14}{25-14} = \frac{18-14}{11} = \frac{4}{11} = .3636$$

The observed value of the asymmetrical lambda using level of education to predict level of income is .3636, indicating that 36.36% of the variance in category of income can be explained by category of education. It is possible to compute another asymmetrical value of lambda using level of income to predict level of education. The subscripts of the formula would be adjusted, so that we would sum the largest cell frequency for each level of the dependent variable, and use the largest marginal total for the independent variable.

Computing the value of asymmetrical lambda with *level of income used to predict level of education* yields the following:

$$\lambda_a = \frac{(10+8)-13}{25-13} = \frac{18-13}{12} = \frac{5}{12} = .4167$$

The two values of the asymmetrical lambda indicate that level of income is a better predictor of education (explaining just over 41% of the variance) than level of education is as a predictor of income (explaining just over 36% of the variance). Most computer statistical packages will routinely compute the symmetrical version of lambda, and both asymmetrical versions. In research situations in which two variables are cross tabulated, but there is no logical reason for one to be considered the independent variable and the other to be considered the dependent variable, the symmetrical version of lambda is the preferred statistic. In situations such as our example, where one variable is logically the independent variable and the other is logically the dependent variable, then the asymmetrical version of lambda measuring the predictive ability of the independent variable on the dependent variable is the preferred statistic.

Goodman's and Kruskal's Gamma

The final measure of association that we will examine in this chapter is Goodman's and Kruskal's Gamma. Gamma is a symmetrical measure of association with a PRE or "variance explained" interpretation (subject to the same cautionary note as above). Appropriate application of Gamma requires random or probability sampling, and also a contingency table presenting two ordinal level variables.

In the general case, gamma is reasonably easy to compute. The statistic is based on the number of "concordant" or similar classifications in the table and the number of "discordant" or dissimilar classifications in the table. In a 2 × 2 table, such as our table of level of income by level of education, the frequencies in cells "B" and "C" represent concordant or similar classifications (i.e., high education and high income, or low education and low income), and the frequencies in cells "A" and "D" represent discordant or dissimilar classifications (i.e., high education and low income or low education and high income).

The Formula for Gamma is as follows:

$$\gamma = \frac{NC - ND}{NC + ND}$$

where

NC is the number of the concordant frequencies and
ND is the sum of the discordant frequencies.

Computing the value of gamma for large tables can be tedious, but gamma is rather easy to compute for a 2 × 2 table. The number of concordant frequencies is the product of the two cells along the main, positive sloped diagonal (cell "C" × cell "B"), and the number of discordant frequencies is the product of the two cells along the main, negative sloped diagonal (cell "A" × cell "D"). For our table of level of income by level of education the data are as follows:

Frequencies of concordant cells (cells "C" and "B") = 10 × 8 = 80
Frequencies of discordant cells (cells "A" and "D") = 4 × 3 = 12

The value of gamma can then be computed as follows:

$$\gamma = \frac{(8 \times 10) - (4 \times 3)}{(8 \times 10) + (4 \times 3)} = \frac{80 - 12}{80 + 12} = \frac{68}{92} = .739$$

The observed value of gamma is .739, indicating that almost 74% of the variance between the two variables can be explained. We cannot specify a direction in terms of variance explained because gamma is a symmetrical measure of association. You might notice that the value of gamma is rather high compared with the other measures of association we have computed for the same table. An explanation of why there is so much difference in the estimate of variance explained in the same contingency table is presented below.

Why are the Variance Explained Values Different for Measures of Association Applied to the Same Table?

At this point, we have already examined several measures of association, and five of them (phi, symmetrical lambda, the two asymmetrical lambdas, and gamma) have each provided a different estimate of the amount of variance explained in the same contingency table! A legitimate question to ask is "why are the variance explained estimates different?" The answer is that the variance explained estimates are based on different assumptions and different ways of computing prediction error. We can demonstrate how the variance explained estimates can vary by focusing on the two asymmetrical versions of lambda.

The asymmetrical versions of lambda are based on using the independent variable to predict the dependent variable, and then the dependent variable to predict the independent variable. We have seen how to compute each of these values using the formula, but we can also compute the values intuitively. We will begin by using the independent variable (level of education) to predict the dependent variable (level of income). If you examine the marginal column totals for the table of income by income, you will see that there are a total of 11 cases in the low-income category, and 14 cases in the high income category. Knowing nothing else about the sample, our best strategy would be to guess that all 25 cases belonged in the modal category (the category with the higher number of frequencies). In this case, high income is the modal category, and the number of cases in the other category would represent prediction error, or in this case we would have a total of 11 errors. The table suggests a positive relationship (there are more cases in cells "B" and "C" than in cells "A" and "D" as we have identified earlier). If we had a perfect positive relationship between level of education and level of income, then we would predict that all of the cases would fall into cells "B" and "C." Therefore, the number of cases that falls into cells "A" and "D" represent error when using level of education to predict level of income. In this case, there are a total of seven cases in those two cells. We began with 11 errors by using the modal category as the basis for our prediction, using education as our predictor reduced the error to seven (or in other words, we were able to explain four errors). The four errors we were able to explain represent a reduction of 4/11 = .3636, which is exactly the value of the asymmetrical lambda when education is used to predict income.

We can use the same logic to intuitively derive the value of asymmetric lambda when level of income is used to predict level of education. Again, knowing nothing else about our sample, our best strategy is to guess that all 25 cases belong in the modal category (in this case the modal category would be the high education category). As a result, we would have a total of 12 errors (those individuals in the low education category). We still have a total of seven errors in the table, represented by the number of cases in cells "A" and "D." Using level of income as our predictor variable allows us to reduce the number of errors by five, and expressing the reduction in error as a proportion of the original error (5/12 = .4167) results in exactly the value we computed for the other asymmetrical lambda statistic.

We will not go through a detailed derivation of how each of the measures of association computes the reduction in error. The important point is that the estimates of variance explained can vary widely because the assumptions on which they are based vary. Given the differences in the estimates of variance explained, one important question is "which measure of association should be reported?" There are a number of ways to answer that question. As a general rule, your selection of a particular measure of association should be based on the answers to the following two questions. (1) Do your data meet the assumptions of the measure of association? (2) Is the measure of association consistent with the logic of your research design? For example, if you have nominal level data, then you should not be reporting a value for Goodman's and Kruskal's Gamma that requires ordinal level data. Further, if your design includes a prediction with one variable serving as the independent variable and another variable serving as the dependent variable, then you should use the value of an asymmetrical statistic consistent with your prediction even if the other asymmetrical statistic results in a higher value of variance explained.

A Second Example of Chi Square and Measures of Association

Our second example of computing a chi square statistic and the measures of association will be with a slightly larger table consisting of three rows and three columns, or a 3 × 3 table. We will use actual data from the General Social Survey data set, and examine respondent's willingness to pay taxes to help the environment (the "GRNTAXES" for green taxes variable) as the dependent variable, and educational level (the "EDUC" variable) as the independent variable. Each variable is operationalized with three categories, but it should be noted that neither chi square nor the measures of association that we have covered require a balanced table with an equal number of rows and columns. We could just as easily work with a 2 × 3 table or a 5 × 4 table. (In fact, a table consisting of a single row or column may be analyzed with a version of the chi square statistic called the chi square test for "goodness of fit," but we are not presenting that version of the chi square here).

A 3 × 3 contingency table is presented below, and we will use this table as a second example for computing chi square and the related measures of association. Willingness to pay taxes to help the environment is operationalized into three categories consisting of Willing (a combination of the original "very willing" and "fairly willing" categories); Neutral (the original "neither willing nor unwilling category); and, Unwilling (a combination of the original "not very willing" and "not at all willing" categories). Educational level is operationalized into three categories consisting of LTHS (less than high school); HS (high school); and, GTHS (greater than high school). We will test the null hypothesis that the observed results are equal to the expected results at the $\alpha = .01$ level (Table 13.16). The research and null hypotheses are presented symbolically below:

$$H1: \text{Observed} \neq \text{Expected}$$

$$H0: \text{Observed} = \text{Expected}$$

The table is based on a total sample size of n = 1,332, and if you convert the cell frequencies to percentages using the "percent down, compare across" rule, you will see evidence of a slight positive relationship. The question at this point is, "do the observed results differ from what we would expect by chance?" The chi square test can answer that question, and our first step is to compute the expected frequencies of each cell by multiplying the cell's row total times its column total, and then dividing the product by the total sample size. Expected frequencies are presented below (note that the expected frequencies have been rounded to whole numbers) (Table 13.17).

Table 13.16 Willingness to Pay Environmental Taxes by Educational Attainment Observed Results

		Education			
		LTHS	**HS**	**GTHS**	**Total**
	High	68	120	270	458
Willingness to Pay Environmental Taxes	**Medium**	47	83	156	286
	Low	114	218	256	588
	Total	229	421	682	1332

Table 13.17 Willingness to Pay Environmental Taxes by Educational Attainment Expected Results

		Education			
		LTHS	**HS**	**GTHS**	**Total**
	High	79	145	234	458
Willingness to Pay Environmental Taxes	**Medium**	49	90	146	286
	Low	101	186	301	588
	Total	229	421	682	1332

The Formula for Chi Square:

$$\chi^2 = \Sigma \frac{(O-E)^2}{E}$$

Computing the value of chi square

$$\chi^2 = \frac{(68-79)^2}{79} + \frac{(120-145)^2}{145} + \frac{(270-234)^2}{234} + \frac{(47-49)^2}{49} + \frac{(83-90)^2}{90}$$
$$+ \frac{(156-147)^2}{147} + \frac{(114-101)^2}{101} + \frac{(218-186)^2}{186} + \frac{(256-301)^2}{301}$$

$$\chi^2 = \frac{121}{79} + \frac{625}{145} + \frac{1296}{234} + \frac{4}{49} + \frac{49}{90} + \frac{81}{147} + \frac{169}{101} + \frac{1024}{186} + \frac{2025}{301}$$

$$\chi^2 = 1.53 + 4.31 + 5.53 + .08 + .54 + .55 + 1.67 + 5.51 + 6.73 = 26.45$$

Our observed chi square value is 26.45. A 3 × 3 table has four degrees of freedom ($df = (3-1) \times (3-1) = 2 \times 2 = 4$), so the appropriate critical value is $\alpha_{.01} = 13.28$. Our observed chi square value is well into the critical region, so the correct decision is to REJECT H0; the observed results are significantly different than the expected results. As a cautionary note, our table is based on a sample size of $n = 1{,}332$, so a large chi square value is not unusual.

Computing the Measures of Association, How Strong Is the Relationship?

If you converted the observed cell frequencies to percentages, you saw evidence of a positive relationship in the data. The various measures of association can provide an indication of how strong the relationship is. We will begin in the same manner as before, computing the coefficient of contingency, which is based on the observed chi square value. Note that we will not be computing the value of the phi coefficient, because it is only appropriate for 2 × 2 tables.

Computing *C*

$$C = \sqrt{\frac{26.45}{1332 + 26.45}}$$

$$C = \sqrt{\frac{26.45}{1358.45}}$$

$$C = \sqrt{.0195} = .1396$$

Computing Symmetrical Lambda

Largest frequency in each row: 270, 156, and 256
Largest frequency in each column: 114, 218, and 270
Largest marginal row total: 588
Largest marginal column total: 682
Sample size: 1332

Computing the value of lambda yields the following:

$$\lambda = \frac{(270 + 156 + 256) + (114 + 218 + 270) - (588 + 682)}{2(1332) - (588 + 682)}$$

$$\lambda = \frac{682 + 602 - 1270}{2664 - 1270}$$

$$\lambda = \frac{14}{1394} = .01004$$

Computing Asymmetrical Lambdas

Largest frequency for each category of the independent variable: 114, 218, and 270
Largest marginal total for the dependent variable: 588
Sample size: 1332

Computing the value of asymmetrical lambda with X used to predict Y yields the following:

$$\lambda_a = \frac{(114+218+270)-588}{1332-588} = \frac{602-588}{744} = \frac{14}{744} = .0188$$

Computing the value of asymmetrical lambda with *Y* used to predict *X* yields the following:

$$\lambda_a = \frac{(270+156+256)-682}{1332-682} = \frac{682-682}{650} = \frac{0}{650} = 0.00$$

Computing Gamma

We saw earlier that computing gamma was relatively simple for a 2 × 2 table. The computation is a little more complicated for a larger table such as the current 3 × 3 case. The value of gamma is still based on the ratio of concordant and discordant cells, but the concept is a little more complicated in the larger table. The observed values of our 3 × 3 table are presented below with two additions: each cell has been identified with a lower case letter, and the two main axes have been labeled with an uppercase "C" and "D." The uppercase "C" indicates the lower point of the axis of concordant cells and the uppercase "D" indicates the higher point of the axis of discordant cells (Table 13.18).

Table 13.18 Willingness to Pay Environmental Taxes by Educational Attainment with Cells Labeled

		Education			
		LTHS	**HS**	**GTHS**	**Total**
	High	79	145	234	458
		a	b	c	
	Medium	49	90	146	286
Willingness to Pay Environmental Taxes		d	e	f	
	Low	101	186	301	588
		g	h	i	
	Total	229	421	682	1332

The number of concordant frequencies is the product of each cell in the table times the sum of all cells above and to the right of it. We will begin with cell "g," and proceed through the remainder of the table. In this case, the number of concordant frequencies will be:

$$NC = [(g) \times (e + b + f + c)] + [(d) \times (b + c)] + [(h) \times (f + c)] + [(e) \times (c)]$$

The number of discordant frequencies is the product of each cell in the table times the sum of all cells below and to the right of it. Beginning with cell "a," and proceeding through the remainder of the table, the number of discordant frequencies will be:

$$ND = [(a) \times (e + f + h + i)] + [(d) \times (h + i)] + [(b) \times (f + i)] + [(e) \times (i)]$$

The intermediate computations are as follows:
Concordant cells:

$$NC = [(114) \times (83 + 120 + 156 + 270)] + [(47) \times (120 + 270)] + [(218) \times (156 + 270)] + [(83) \times (270)]$$

$$NC = 71,706 + 18,330 + 92,868 + 22,410 = 205,314$$

Discordant cells:

$$ND = [(68) \times (83 + 156 + 218 + 256)] + [(47) \times (218 + 256)] + [(120) \times (156 + 256)] + [(83) \times (256)]$$

$$ND = 48,484 + 22,278 + 49,440 + 21,248 = 141,450$$

The value of gamma may then be computed as:

$$\gamma = \frac{205,314 - 141,450}{205,314 + 141,450} = \frac{63,864}{346,764} = .1842$$

Again, as in the previous example, the estimates of the amount of variance indicated by the measures of association we have computed for the 3 × 3 table differ, ranging from a low of 0.00 for symmetrical lambda using education as the dependent variable to a high of .1847 for gamma. The keys to selecting the measure of association that is most appropriate are to select the one for which you can meet the assumptions given your data and research design and the one consistent with any directional hypotheses you have made. In our case, we could legitimately report gamma because our two variables may be considered at the ordinal level, or we could report the more general coefficient of contingency. Phi would not be appropriate, and lambda is better suited to nominal level data. Keep in mind that we have not presented an exhaustive listing of all measures of association, so other statistics not examined here would also be appropriate.

Nonparametric Tests of Significance

The final section of this chapter covers two nonparametric statistic tests of significance: the **Mann–Whitney U test** and the **Wilcoxon T test.** Neither of these tests is widely used in actual practice, but both should probably be used more than they are. The Mann–Whitney U test is a test for a significant difference between two independent samples and the Wilcoxon T test is a test for significant difference between two related samples. These two statistical tests are the nonparametric alternative to the corresponding t-tests that we examined in Chapter 11. The t-tests of Chapter 11 are widely used in behavioral science research, but too often they are used without meeting the assumptions associated with appropriate application of the tests (such as a true probability sampling frame, being able to legitimately compute a mean, or meeting the assumption of a normal distribution). Oftentimes, the Mann–Whitney U test or the Wilcoxon T test would be a more appropriate choice because each may be applied to ordinal level data, and neither assumes a normal distribution in the population from which the samples were drawn. We will examine the Mann–Whitney U test first.

The Mann–Whitney U Test for Two Independent Samples

The Mann–Whitney U test is used to determine if two independent samples are statistically significantly different. Unlike the independent sample t-test that evaluates the difference in the value of the mean for two samples to determine statistical significance, the Mann–Whitney U test calls for converting raw scores to ranks, and then evaluates the difference in the sum of the ranks associated with each sample. Two samples with a similar sum of ranks suggests that the two samples come from the same population. The greater the difference between the sum of ranks for two samples the greater the likelihood that the samples come from two different populations.

The null hypothesis associated with the Mann–Whitney U test is that the sum of ranks for population 1 is equal to the sum of ranks for population 2. The research hypothesis predicts a difference in the sum of ranks between the two populations and may be stated as a one-tail or a two-tail test. The procedures associated with the Mann–Whitney U test are similar to those of the other tests of significance we examined in Chapter 11. We will compute a test statistic using a formula, obtain a critical value based on the sample size from the appropriate table in the appendix, and then reach a decision regarding the null hypothesis. There is one major difference in applying the Mann–Whitney U test compared to the statistical tests we have used in the past. To reject the null hypothesis, the OBSERVED VALUE of the test statistic should be equal to or LESS THAN the critical value from the table. For the Mann–Whitney U test (and the same will be true of the Wilcoxon T test), the smaller the observed value of the test statistic the greater the likelihood of a finding of statistical significance!

The Mann–Whitney U test can be performed on data collected at the ordinal level or higher, but the test is based on rank order, so interval level or ratio level data must be converted to a rank order score. The Mann–Whitney U test is usually applied to small sample sizes of approximately $n = 20$ or less in each group. There are alternative nonparametric tests for larger sample sizes, but we will not examine them here. Larger sample sizes usually provide one with the option of choosing the alternative parametric t-test because in general, the larger the sample size the easier it is to violate some of the assumptions associated with parametric tests, and still achieve valid results (we spoke of this property as **robustness** in an earlier chapter).

The Mann–Whitney U test proceeds in the following sequence:

1. Rank scores across both samples by assigning rank 1 to the lowest score, rank 2 the next lowest score, and so on, irrespective of which sample any particular score is located. Resolve tied ranks by the method discussed in Chapter 9 (see the section on the Spearman rank order correlation coefficient).
2. Compute the sum of the ranks (ΣR) and note the sample size for each group.
3. Compute the value of the U statistic for each sample by the following formulae:

$$U_1 = (n_1)(n_2) + \frac{n_1(n_1 + 1)}{2} - \Sigma R_1$$

$$U_2 = (n_1)(n_2) + \frac{n_2(n_2 + 1)}{2} - \Sigma R_2$$

4. Select the observed value of the U statistic as follows:

   For a two-tailed test, U_{obs} is the *smaller* of the two computed U values

   For a one-tail test, U_{obs} is the computed U value consistent with the prediction of the research hypothesis

5. Find the appropriate critical value from the table of critical values in the appendix. Table 8 in the appendix contains critical values for U for both one-tail or two-tail tests at the $\alpha = .05$ and $\alpha = .01$ levels.

6. Compare the observed value of U to the critical value and reach a decision regarding H_0. To reject H_0, the null hypothesis of no difference between the ranks of the two populations, the observed value of U must be equal to or SMALLER than the critical value from the table.

An Application of the Mann–Whitney U Test

Scores on an IQ test for two samples of $n1 = 7$ and $n2 = 9$ have been ranked as indicated below. A rank of 1 indicates the lowest score across both groups, a rank of 2 indicates the second lowest score across both groups, and so on, up to a rank of 16 indicating the highest score across both groups. The research and null hypotheses are stated symbolically below, and we will test the null hypothesis at the $\alpha = .05$ level.

$$H_1 : \Sigma R_1 \neq \Sigma R_2$$

$$H_0 : \Sigma R_1 = \Sigma R_2$$

The ranked scores are as follows:

Group 1	Group 2
1	3
2	5
4	8
6	9
7	11
10	12
15	13
	14
	16
$\Sigma R_1 = 45$	$\Sigma R_2 = 91$
$n_1 = 7$	$n_2 = 9$

Computing the Value of $U1$ and $U2$

$$U_1 = (7)(9) + \frac{7(7+1)}{2} - 45$$

$$U_1 = 63 + \frac{56}{2} - 45$$

$$U_1 = 63 + 28 - 45 = 46$$

$$U_2 = (7)(9) + \frac{9(9+1)}{2} - 91$$

$$U_2 = 63 + \frac{90}{2} - 91$$

$$U_2 = 63 + 45 - 91 = 17$$

The research hypothesis indicates a two-tailed test, so $U_{obs} = 17$, the smaller of the two computed U values. We are evaluating the null hypothesis at the $\alpha = .05$ level for sample sizes of $n_1 = 7$ and $n_2 = 9$. The appropriate critical value from Table 8 in the appendix is $U_{.05} = 12$. The correct decision in this case is to FAIL TO REJECT H_0, the observed value of the test statistic is not equal to or smaller than the critical value. The rank on IQ scores is not statistically significantly different between the two groups.

Our example of the Mann–Whitney U test called for a two-tailed test of the null hypothesis, so our observed value of U was the smaller of the two computed U statistics. A one-tail test may be conducted, and the observed value of U is the computed U statistic consistent with the prediction of the research hypothesis. The computed value of U decreases as the sum of the ranks for a group increases. For example, had the research hypothesis predicted that Group 1 would have the higher ranks we would automatically select U_1 as the observed value of U. However, noticing that U_1 is not the smaller of the two computed U values would allow us to fail to reject H0 immediately without comparison to the critical value. The data are not consistent with a one-tail test predicting that Group 1 has the higher ranks. The data would be consistent with a prediction that Group 2 has the higher ranks (because Group 2 has the smaller of the two computed U values), and had that been the direction of our one-tail hypothesis we would have to find the appropriate critical value from the one-tail section of Table 8 in Appendix A. The correct decision would still have been to fail to reject H0, because the critical value for a one-tail test at the $\alpha = .05$ level is $U_{.05} = 15$.

The Wilcoxon T Test for Two Related Samples

The final statistical procedure we will examine in this chapter (and the book!) is the Wilcoxon T test for two related samples. The Wilcoxon T test is the nonparametric equivalent to the t-test for related samples, and usually involves a pretest and posttest comparison. The Wilcoxon T test involves a comparison of ranked scores, and tests a null hypothesis that the pretest ranks will equal the posttest ranks. The research hypothesis may be either a one-tail test or a two-tail test. Just as the case with the Mann–Whitney U test, the null hypothesis is rejected in the Wilcoxon T test if the observed value of T is equal to or SMALLER than the critical value of the test statistic. The procedure for the Wilcoxon T is outlined below.

1. Compute a difference measure "D" for the pretest and posttest scores (you may subtract in either direction to compute "D," but you must be consistent).
2. Rank the nonzero difference scores ignoring the sign of the difference with rank 1 assigned to the smallest difference, rank 2 to the next smallest difference, and so on, until all nonzero scores have been ranked.

3. Separate the ranked scores into two groups: $R+$ for all ranks assigned to a positive difference; and $R-$ for all ranks assigned to a negative difference.
4. Sum the ranks for each of the two groups: $\Sigma R+$ and $\Sigma R-$.
5. Select the observed value of the T statistic as follows:

   For a two-tailed test, T_{obs} is the smaller of the two computed sum of ranks values

   For a one-tail test, T_{obs} is the sum of ranks value consistent with the prediction of the research Hypothesis.

6. Obtain the appropriate critical value from Table 9 in Appendix A. Critical values are provided for both a one-tail test and a two-tail test at the $\alpha = .05$ and $\alpha = .01$ levels of significance. **Note that when selecting the appropriate critical value, the sample size is the number of nonzero ranks, not the original number of subjects.**
7. Compare the observed value of T to the critical value, and reach a decision regarding H0. To reject H0, the null hypothesis of no difference between the ranks of the two populations, the observed value of T must be equal to or SMALLER than the critical value from the table.

An Application of the Wilcoxon T Test

Pretest and posttest data on sales performance are presented for a sample of $n = 8$ individuals below. The pretest data reflect sales performance prior to participation in a sales motivational program and the posttest data reflect sales performance for a similar period of time following the program. The research hypothesis predicts that sales performance will improve following the sales motivational program, so posttest sales are expected to be higher. The null hypothesis of no difference between the pretest and posttest ranks will be tested at the $\alpha = .05$ level.

Pretest	Posttest	Difference	Rank	$R+$	$R-$
20	24	4	4	4	
25	25	0			
15	21	6	6	6	
18	18	0			
10	15	5	5	5	
12	11	−1	1		1
30	28	−2	2		2
20	23	3	3	3	
				$\Sigma R+ = 18$	$\Sigma R- = 3$

The research hypothesis calls for a one-tail test, so our observed T value is the sum of ranks consistent with the prediction of H_1. Choosing the appropriate sum of ranks may take some thought, because the choice depends not only on the prediction of the research hypothesis but also on the direction one subtracts when computing the difference measure. We expect the posttest scores to be higher than the pretest scores, because the sales motivational program is intended to improve sales performance. Further, in this case, the difference measure is obtained by subtracting the pretest score from the posttest score, so we would expect more of the difference scores to be positive. Therefore, we would expect the sum of ranks of the negative scores to be the smaller value, and in this case it is ($\Sigma R+ = 18$, and the $\Sigma R- = 3$), so our observed value of the test statistic is $T_{obs} = 3$. The data are consistent with the prediction of the research hypothesis, and comparing the observed value of T to the appropriate critical value will tell us if the difference is statistically significant.

Table 9 in Appendix A contains the critical values for the Wilcoxon T statistic for both one-tail and two-tail tests of the null hypothesis. We are testing the null hypothesis at the $\alpha = .05$ level for a one-tail test, and our sample size is now $n = 6$ (the number of scores with a nonzero difference). The appropriate critical value is $T_{.05} = 2$. Our observed test statistic, $T_{obs} = 3$, is not equal to or smaller than the critical value, so the correct decision is to FAIL TO REJECT H_0. The ranks associated with the posttest sales performance are not statistically significantly different from the ranks associated with the pretest sales performance.

Computer Applications

1. Load the GSS data set and conduct a chi square analysis of attitudes on capital punishment by race. Interpret the results.
2. Input the data from problem number 6 at the end of the chapter and conduct a Mann-Whitney U test. Interpret the results.
3. Input the data from problem number 7 and conduct a Wilcoxon T test. Interpret the results.

How to do it

Load the GSS data set. Click on "Analyze," "Descriptive," and "Crosstabs" to open the crosstabs dialog box. Highlight "Cappun" and move it to the Rows (dependent variable) selected box. Highlight "Race" and move it to the Column (independent variable) selected box. Click on "Statistics" to request the chi square and desired measures of association (choose the Contingency Coefficient, Phi and Cramer's V, and Lambda). Click on "Cells" and request column percentages in addition to the default observed counts output. Click on "OK" to compete the procedure. Interpret the results.

Use the "File," "New," "Data" command to clear the GSS data set and obtain a clear data editor screen. Input the ranked data from problem number 6 at the end of the chapter. Note that you will need to enter all of the data for both groups into one column as a single variable. You will then need to create a second variable assigning a "1" to members of Group 1 and a "2" to members of Group 2. Then click on "Statistics," "Nonparametric Tests," and "2 Independent Samples" to open the Two Independent Samples Tests dialog box. Select the variable containing both groups' ranking and move it to the Test Variable List. Select the group identification variable and move it to the Grouping Variable box. Click on "Define Groups" to input the values of "1" and "2" which identify group membership. Make sure that the Mann–Whitney U test is selected under the Test Type option, and then click on "OK" to complete the procedure. Interpret the results.

Input the data from problem 7 using one variable for the pretest scores and one variable for the posttest scores. Click on "Analyze," "Nonparametric Tests," and "two Related Samples" to open the Two Related Samples Tests dialog box. Highlight the two variables containing the pretest and posttest scores and move them to the Test Pairs List by clicking on the appropriate direction arrow. Make sure the Wilcoxon test is selected under the Test Type option, and then click on "OK" to complete the procedure. Interpret the results.

Summary of Key Points

This chapter has covered several of the more frequently used nonparametric statistical procedures. Nonparametric statistical procedures are important, because they allow us to conduct a statistical analysis when our data or research design do not allow us to meet the more rigorous assumptions associated with the parametric statistical procedures we have examined in earlier chapters. The chi square test for independence is a popular technique used to evaluate a contingency table, and allows us to determine if the observed results differ from what we would expect by chance. A statistically significant chi square provides evidence of some type of relationship in the data. Our own analysis of the table is required to determine the type of relationship that might exist, and then we can compute one or more of the measures of association to tell us how strong the relationship between the two variables is.

The final section dealt with two nonparametric tests of significance: the Mann–Whitney U test for two independent samples and the Wilcoxon T test for two related samples. These two statistical tests are not as widely used as their parametric equivalents (the t-test for two independent samples and the t-test for two related samples), but each should probably be applied more often than it is given the occasional tendency for some researchers to apply statistical tests without meeting the assumptions required of the more popular parametric procedures.

Asymmetrical Measure of Association—A statistic indicating the amount of variance one variable can explain in another variable, usually the amount of variance in the dependent variable *Y* that is explained by the independent variable *X*.

Contingency Table—A method of presenting the relationship between two categorical variables by cross tabulating one variable with the other. The logic behind the construction of the contingency table is similar to that of the Cartesian coordinate system for continuous variables.

Chi Square Test—A nonparametric statistical procedure allowing us to determine if the pattern of observed results in a contingency table is statistically significantly different from what would be expected by chance.

Coefficient of Contingency—A measure of association indicating the strength of the relationship between two variables based on the observed value of the chi square statistic. The coefficient of contingency may be applied on contingency tables of any size.

Phi Coefficient—A measure of association indicating the strength of the relationship between two variables based on the observed value of the chi square statistic. The phi coefficient may only be applied on 2×2 tables, and has a proportional reduction in error (PRE), or variance explained interpretation.

Guttman's Coefficient of Predictability, Lambda—A measure of association for two nominal level variables. Lambda has a PRE interpretation, and can be computed as a symmetrical measure of association (mutual predictability between the two variables) or as an asymmetrical measure of association (one variable used to predict the other).

Goodman's and Kruskal's Gamma—A measure of association for two ordinal level variables. Gamma is a symmetrical measure of association, and has a PRE interpretation.

Mann-Whitney U Test—A nonparametric test of significance for two independent samples. The test is based on the difference of the sums of ranks associated with each sample. Unlike most statistical tests, the null hypothesis is rejected when the observed value of *U* is equal to or less than the critical value.

Symmetrical Measure of Association—A statistic indicating the amount of mutual predictability between two variables; i.e. the amount of variance in the dependent variable *Y* that is explained by the independent variable *X*, and the amount of variance in the independent variable *X* that is explained by the dependent variable *Y*.

Wilcoxon T Test—A nonparametric test of significance for two related samples. The test is based on the sum of positive and negative ranks associated with the difference in pretest and posttest scores for a sample. Like the Mann–Whitney U test, the null hypothesis is rejected when the observed value of *T* is equal to or less than the critical value.

Questions and Problems for Review

1. What are nonparametric statistics, and in what research situations should they be used?

2. Below are data on (X) Family Income and (Y) First Year College G.P.A. for a sample of $n = 50$ individuals.

X Family Income (in thousands)	Y First Year G.P.A.	X Family Income (in thousands)	Y First Year G.P.A.
22	3.25	60	0.15
39	2.30	67	2.28
14	3.50	72	3.35
99	3.72	39	1.47
8	3.95	10	1.59
45	2.00	98	3.79
43	2.85	40	2.50
47	1.28	18	2.36
23	2.21	47	3.39
11	2.35	28	2.78
72	1.25	17	3.20
95	3.22	33	3.75
28	4.00	78	3.44
47	2.75	59	2.02
23	3.10	34	3.18
17	2.75	75	2.15
56	3.08	60	1.99
36	1.10	30	3.04
52	2.75	45	3.60
37	3.31	29	3.32
84	2.85	55	2.17
79	2.95	90	3.79
70	3.36	85	1.15
44	3.75	22	3.01
18	3.45	24	3.73

a. Categorize family income into three groups of Low Income (\$0 - \$24,999); Middle Income (\$25,000 - \$59,999); and, High Income (\$60,000 or higher). Dichotomize G.P.A. into two categories of Low (0.00 - 2.99); and, High (3.00 or higher). Create a 2×3 contingency table containing the results.

b. Conduct a chi square analysis to see if the observed results are statistically significantly different from what you would expect by chance. Evaluate the null hypothesis at the $\alpha = .01$ level.

c. Compute the following measures of association: Coefficient of contingency (C) and Gamma (γ).

3. Read a journal article in your major field of interest that reports the results of a chi square analysis of a contingency table (do not be confused by the fact that versions of chi square are also used in connection with more complex statistical procedures such as log linear regression, and structural equations models). If the data permit, verify the computation of the chi square and the measures of association.
4. Given the following data on ranked scores:

Group 1 Ranks	Group 2 Ranks
1	2
3	5
4	9
6	10
7	11
8	13
12	14
	15
	16
	17
	18
	19
	20

   a. Compute the Mann–Whitney U test, and test the null hypothesis at the $\alpha = .01$ level (assume a two-tail test of the hypothesis).
5. Given the following contingency table:

		Education		
		Low	High	Total
	High	35	65	100
Income	Low	105	45	150
	Total	140	110	250

   a. Compute the chi square statistic for the table, and test the null hypothesis at the $\alpha = .01$ level.
   b. Compute the following measures of association: Phi coefficient and Lambda (both symmetrical and asymmetrical).
6. Convert the following raw scores indicating level of anxiety into ranked scores (the higher the raw score, the higher the level of anxiety):

Group 1 Scores	Group 2 Scores
91	64
83	75
75	89
62	71
87	92

Group 1 Scores	Group 2 Scores
98	63
82	74
	55
	56
	71
	50

a. Evaluate the ranked scores using the Mann–Whitney U test (assume a one-tail test with the research hypothesis predicting that Group 1 has the higher ranks). Test the null hypothesis at the $\alpha = .05$ level.

7. The pretest scores below reflect scores on the Scholastic Aptitude Test (SAT) for a group of $n = 10$ students. The students then participate in a study program intended to increase their performance. Posttest scores indicate the students' performance following the study program.

Pretest	Posttest
1240	1100
900	1050
800	950
1000	1200
1200	975
1350	1350
680	650
1150	1325

a. Compute the Wilcoxon T statistic for the data, and test the null hypothesis at the $\alpha = .05$ level. Assume a one-tail test with the posttest scores expected to be higher.

8. Read a journal article in your field that reports the results of a Mann–Whitney U test or a Wilcoxon T test. Why did the researchers choose one of these statistics rather than the parametric alternative?

9. Given the observed results below:

Observed Results

		Independent Variable *X*			
		Low	Med	High	Total
	High	30	20	10	60
Dependent Variable *Y*	Med	30	50	10	90
	Low	10	20	40	70
	Total	70	90	60	220

a. Compute the chi square statistic, and test the null hypothesis of no difference between the observed results and expected results at the $\alpha = .01$ level.

b. Convert the cell frequencies to percentages (percent down and compare across), and interpret the nature of the relationship in the table.

10. Compute the following measures of association for the table in problem 9: coefficient of contingency, lambda (symmetrical), and gamma.

11. How many degrees of freedom are associated with the following contingency tables?

a. 3×2

b. 2×3

c. 3×4

d. 5×5

e. 4×3

12. Complete the cell frequencies for the following contingency table.

		Independent Variable X			
		Low	Med	High	Total
	High	20	30		80
Dependent Variable Y	Med			40	120
	Low		50		100
	Total	75	125	100	300

APPENDIX A

Statistical Tables

Table 1 Proportional Areas under the Standard Normal Curve: The *Z* Tables
Table 2 Critical Values of *t* for Two-Tail and One-Tail Tests: The *t* Tables
Table 3 Critical Values of the Pearson Correlation Coefficient: The *r* - Tables
Table 4 Critical Values of the Spearman Rank-Order Correlation Coefficient: The r_s - Tables
Table 5 Critical Values of the *F* Statistic: The *F* Tables
Table 6 Values of the Studentized Range Statistic, q_k
Table 7 Critical Values of the Chi Square: The χ^2 Tables
Table 8 Critical Values of the Mann-Whitney *U*
Table 9 Critical Values of the Wilcoxon *T*

Table 1 Proportional Areas under the Standard Normal Curve: The *Z* Tables

Column A contains the absolute value of the *Z* score.
Column B contains the proportion of the area of the normal curve between the mean ($\mu = 0$) and the *Z* score.
Column C contains the proportion of the area of the normal curve from the *Z* score to the tail of the curve.

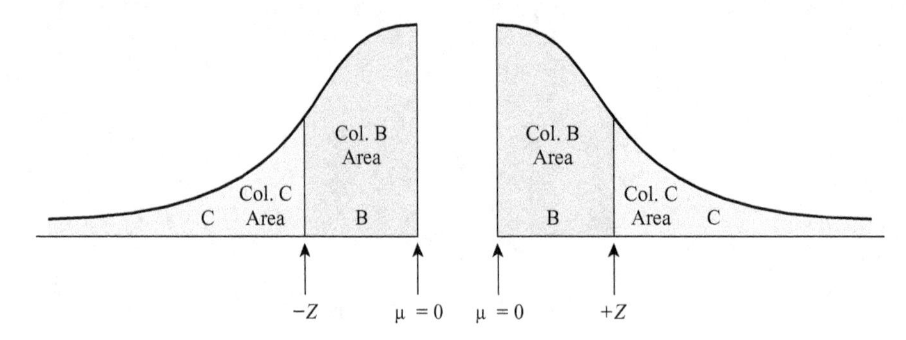

±Z A	B	C	±Z A	B	C	±Z A	B	C
0.00	0.0000	0.5000	0.37	0.1443	0.3557	0.74	0.2704	0.2296
0.01	0.0040	0.4960	0.38	0.1480	0.3520	0.75	0.2734	0.2266
0.02	0.0080	0.4920	0.39	0.1517	0.3483	0.76	0.2764	0.2236
0.03	0.0120	0.4880	0.40	0.1554	0.3446	0.77	0.2794	0.2206
0.04	0.0160	0.4840	0.41	0.1591	0.3409	0.78	0.2823	0.2177
0.05	0.0199	0.4801	0.42	0.1628	0.3372	0.79	0.2852	0.2148
0.06	0.0239	0.4761	0.43	0.1664	0.3336	0.80	0.2881	0.2119
0.07	0.0279	0.4721	0.44	0.1700	0.3300	0.81	0.2910	0.2090
0.08	0.0319	0.4681	0.45	0.1736	0.3264	0.82	0.2939	0.2061
0.09	0.0359	0.4641	0.46	0.1772	0.3228	0.83	0.2967	0.2033
0.10	0.0398	0.4602	0.47	0.1808	0.3192	0.84	0.2995	0.2005
0.11	0.0438	0.4562	0.48	0.1844	0.3156	0.85	0.3023	0.1977
0.12	0.0478	0.4522	0.49	0.1879	0.3121	0.86	0.3051	0.1949
0.13	0.0517	0.4483	0.50	0.1915	0.3085	0.87	0.3078	0.1922
0.14	0.0557	0.4443	0.51	0.1950	0.3050	0.88	0.3106	0.1894
0.15	0.0596	0.4404	0.52	0.1985	0.3015	0.89	0.3133	0.1867
0.16	0.0636	0.4364	0.53	0.2019	0.2981	0.90	0.3159	0.1841
0.17	0.0675	0.4325	0.54	0.2054	0.2946	0.91	0.3186	0.1814
0.18	0.0714	0.4286	0.55	0.2088	0.2912	0.92	0.3212	0.1788
0.19	0.0753	0.4247	0.56	0.2123	0.2877	0.93	0.3238	0.1762
0.20	0.0793	0.4207	0.57	0.2157	0.2843	0.94	0.3264	0.1736
0.21	0.0832	0.4168	0.58	0.2190	0.2810	0.95	0.3289	0.1711
0.22	0.0871	0.4129	0.59	0.2224	0.2776	0.96	0.3315	0.1685
0.23	0.0910	0.4090	0.60	0.2257	0.2743	0.97	0.3340	0.1660
0.24	0.0948	0.4052	0.61	0.2291	0.2709	0.98	0.3365	0.1635
0.25	0.0987	0.4013	0.62	0.2324	0.2676	0.99	0.3389	0.1611
0.26	0.1026	0.3974	0.63	0.2357	0.2643	1.00	0.3413	0.1587
0.27	0.1064	0.3936	0.64	0.2389	0.2611	1.01	0.3438	0.1562
0.28	0.1103	0.3897	0.65	0.2422	0.2578	1.02	0.3461	0.1539
0.29	0.1141	0.3859	0.66	0.2454	0.2546	1.03	0.3485	0.1515
0.30	0.1179	0.3821	0.67	0.2486	0.2514	1.04	0.3508	0.1492
0.31	0.1217	0.3783	0.68	0.2517	0.2483	1.05	0.3531	0.1469
0.32	0.1255	0.3745	0.69	0.2549	0.2451	1.06	0.3554	0.1446
0.33	0.1293	0.3707	0.70	0.2580	0.2420	1.07	0.3577	0.1423
0.34	0.1331	0.3669	0.71	0.2611	0.2389	1.08	0.3599	0.1401
0.35	0.1368	0.3632	0.72	0.2642	0.2358	1.09	0.3621	0.1379
0.36	0.1406	0.3594	0.73	0.2673	0.2327	1.10	0.3643	0.1357

±Z A	B	C	±Z A	B	C	±Z A	B	C
1.11	0.3665	0.1335	1.65	0.4505	0.0495	2.20	0.4861	0.0139
1.12	0.3686	0.1314	1.66	0.4515	0.0485	2.21	0.4864	0.0136
1.13	0.3708	0.1292	1.67	0.4525	0.0475	2.22	0.4868	0.0132
1.14	0.3729	0.1271	1.68	0.4535	0.0465	2.23	0.4871	0.0129
1.15	0.3749	0.1251	1.69	0.4545	0.0455	2.24	0.4875	0.0125
1.16	0.3770	0.1230	1.70	0.4554	0.0446	2.25	0.4878	0.0122
1.17	0.3790	0.1210	1.71	0.4564	0.0436	2.26	0.4881	0.0119
1.18	0.3810	0.1190	1.72	0.4573	0.0427	2.27	0.4884	0.0116
1.19	0.3830	0.1170	1.73	0.4582	0.0418	2.28	0.4887	0.0113
1.20	0.3849	0.1151	1.74	0.4591	0.0409	2.29	0.4890	0.0110
1.21	0.3869	0.1131	1.75	0.4599	0.0401	2.30	0.4893	0.0107
1.22	0.3888	0.1112	1.76	0.4608	0.0392	2.31	0.4896	0.0104
1.23	0.3907	0.1093	1.77	0.4616	0.0384	2.32	0.4898	0.0102
1.24	0.3925	0.1075	1.78	0.4625	0.0375	2.33	0.4901	0.0099
1.25	0.3944	0.1056	1.79	0.4633	0.0367	2.34	0.4904	0.0096
1.26	0.3962	0.1038	1.80	0.4641	0.0359	2.35	0.4906	0.0094
1.27	0.3980	0.1020	1.81	0.4649	0.0351	2.36	0.4909	0.0091
1.28	0.3997	0.1003	1.82	0.4656	0.0344	2.37	0.4911	0.0089
1.29	0.4015	0.0985	1.83	0.4664	0.0336	2.38	0.4913	0.0087
1.30	0.4032	0.0968	1.84	0.4671	0.0329	2.39	0.4916	0.0084
1.31	0.4049	0.0951	1.85	0.4678	0.0322	2.40	0.4918	0.0082
1.32	0.4066	0.0934	1.86	0.4686	0.0314	2.41	0.4920	0.0080
1.33	0.4082	0.0918	1.87	0.4693	0.0307	2.42	0.4922	0.0078
1.34	0.4099	0.0901	1.88	0.4699	0.0301	2.43	0.4925	0.0075
1.35	0.4115	0.0885	1.89	0.4706	0.0294	2.44	0.4927	0.0073
1.36	0.4131	0.0869	1.90	0.4713	0.0287	2.45	0.4929	0.0071
1.37	0.4147	0.0853	1.91	0.4719	0.0281	2.46	0.4931	0.0069
1.38	0.4162	0.0838	1.92	0.4726	0.0274	2.47	0.4932	0.0068
1.39	0.4177	0.0823	1.93	0.4732	0.0268	2.48	0.4934	0.0066
1.40	0.4192	0.0808	1.94	0.4738	0.0262	2.49	0.4936	0.0064
1.41	0.4207	0.0793	1.95	0.4744	0.0256	2.50	0.4938	0.0062
1.42	0.4222	0.0778	1.96	0.4750	0.0250	2.51	0.4940	0.0060
1.43	0.4236	0.0764	1.97	0.4756	0.0244	2.52	0.4941	0.0059
1.44	0.4251	0.0749	1.98	0.4761	0.0239	2.53	0.4943	0.0057
1.45	0.4265	0.0735	1.99	0.4767	0.0233	2.54	0.4945	0.0055
1.46	0.4279	0.0721	2.00	0.4772	0.0228	2.55	0.4946	0.0054
1.47	0.4292	0.0708	2.01	0.4778	0.0222	2.56	0.4948	0.0052
1.48	0.4306	0.0694	2.02	0.4783	0.0217	2.57	0.4949	0.0051
1.49	0.4319	0.0681	2.03	0.4788	0.0212	2.575	0.4950	0.0050
1.50	0.4332	0.0668	2.04	0.4793	0.0207	2.58	0.4951	0.0049
1.51	0.4345	0.0655	2.05	0.4798	0.0202	2.59	0.4952	0.0048
1.52	0.4357	0.0643	2.06	0.4803	0.0197	2.60	0.4953	0.0047
1.53	0.4370	0.0630	2.07	0.4808	0.0192	2.61	0.4955	0.0045
1.54	0.4382	0.0618	2.08	0.4812	0.0188	2.62	0.4956	0.0044
1.55	0.4394	0.0606	2.09	0.4817	0.0183	2.63	0.4957	0.0043
1.56	0.4406	0.0594	2.10	0.4821	0.0179	2.64	0.4959	0.0041
1.57	0.4418	0.0582	2.11	0.4826	0.0174	2.65	0.4960	0.0040
1.58	0.4429	0.0571	2.12	0.4830	0.0170	2.66	0.4961	0.0039
1.59	0.4441	0.0559	2.13	0.4834	0.0166	2.67	0.4962	0.0038
1.60	0.4452	0.0548	2.14	0.4838	0.0162	2.68	0.4963	0.0037
1.61	0.4463	0.0537	2.15	0.4842	0.0158	2.69	0.4964	0.0036
1.62	0.4474	0.0526	2.16	0.4846	0.0154	2.70	0.4965	0.0035
1.63	0.4484	0.0516	2.17	0.4850	0.0150	2.71	0.4966	0.0034
1.64	0.4495	0.0505	2.18	0.4854	0.0146	2.72	0.4967	0.0033
1.645	0.4500	0.0500	2.19	0.4857	0.0143	2.73	0.4968	0.0032

±Z A	B	C	±Z A	B	C	±Z A	B	C
2.74	0.4969	0.0031	2.95	0.4984	0.0016	3.16	0.4992	0.0008
2.75	0.4970	0.0030	2.96	0.4985	0.0015	3.17	0.4992	0.0008
2.76	0.4971	0.0029	2.97	0.4985	0.0015	3.18	0.4993	0.0007
2.77	0.4972	0.0028	2.98	0.4986	0.0014	3.19	0.4993	0.0007
2.78	0.4973	0.0027	2.99	0.4986	0.0014	3.20	0.4993	0.0007
2.79	0.4974	0.0026	3.00	0.4987	0.0013	3.21	0.4993	0.0007
2.80	0.4974	0.0026	3.01	0.4987	0.0013	3.22	0.4994	0.0006
2.81	0.4975	0.0025	3.02	0.4987	0.0013	3.23	0.4994	0.0006
2.82	0.4976	0.0024	3.03	0.4988	0.0012	3.24	0.4994	0.0006
2.83	0.4977	0.0023	3.04	0.4988	0.0012	3.25	0.4994	0.0006
2.84	0.4977	0.0023	3.05	0.4989	0.0011	3.26	0.4994	0.0006
2.85	0.4978	0.0022	3.06	0.4989	0.0011	3.27	0.4994	0.0006
2.86	0.4979	0.0021	3.07	0.4989	0.0011	3.28	0.4994	0.0006
2.87	0.4980	0.0020	3.08	0.4990	0.0010	3.29	0.4994	0.0006
2.88	0.4980	0.0020	3.09	0.4990	0.0010	3.30	0.4995	0.0005
2.89	0.4981	0.0019	3.10	0.4990	0.0010	3.40	0.4997	0.0003
2.90	0.4981	0.0019	3.11	0.4991	0.0009	3.50	0.4998	0.0008
2.91	0.4982	0.0018	3.12	0.4991	0.0009	3.60	0.4998	0.0002
2.92	0.4983	0.0017	3.13	0.4991	0.0009	3.70	0.4999	0.0001
2.93	0.4983	0.0017	3.14	0.4992	0.0008			
2.94	0.4984	0.0016	3.15	0.4992	0.0008			

Critical Z Values

Alpha α	*One-tail* + or –	*Two-tail* + and –
0.10	1.28	1.645
0.05	1.645	1.96
0.025	1.96	2.24
0.01	2.33	2.575
0.001	3.08	3.30

Source: Table prepared by the author.

Table 2 Critical Values of t for Two-Tail and One-Tail Tests: The t Tables

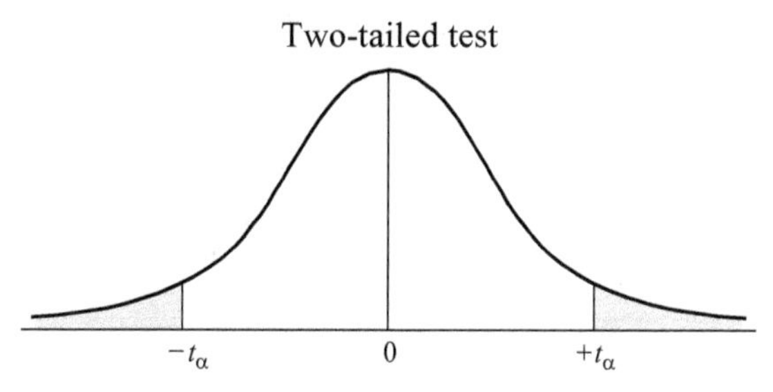

Level of Significance

df	$\alpha = 0.05$	$\alpha = 0.01$
1	12.706	63.656
2	4.303	9.925
3	3.182	5.841
4	2.776	4.604
5	2.571	4.032
6	2.447	3.707
7	2.365	3.499
8	2.306	3.355
9	2.262	3.250
10	2.228	3.169
11	2.201	3.106
12	2.179	3.055
13	2.160	3.012
14	2.145	2.977
15	2.131	2.947
16	2.120	2.921
17	2.110	2.898
18	2.101	2.878
19	2.093	2.861
20	2.086	2.845
21	2.080	2.831
22	2.074	2.819
23	2.069	2.807
24	2.064	2.797
25	2.060	2.787
26	2.056	2.779
27	2.052	2.771
28	2.048	2.763
29	2.045	2.756
30	2.042	2.750
35	2.030	2.724
40	2.021	2.704
45	2.014	2.690
50	2.009	2.678
75	1.992	2.643
100	1.984	2.626
1000	1.962	2.581
∞	1.960	2.576

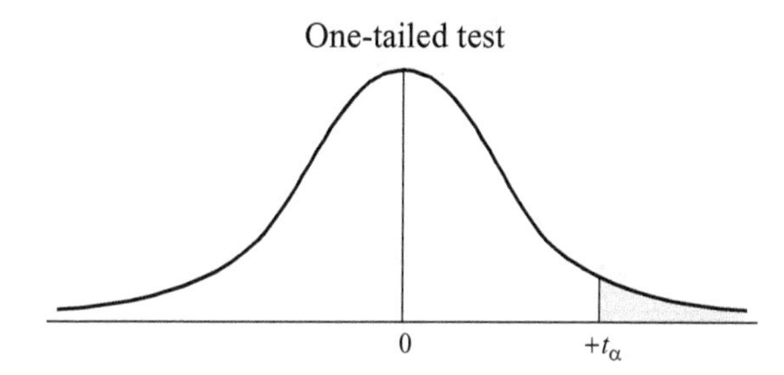

Level of Significance

df	$\alpha = 0.05$	$\alpha = 0.01$
1	6.314	31.821
2	2.920	6.965
3	2.353	4.541
4	2.132	3.747
5	2.015	3.365
6	1.943	3.143
7	1.895	2.998
8	1.860	2.896
9	1.833	2.821
10	1.812	2.764
11	1.796	2.718
12	1.782	2.681
13	1.771	2.650
14	1.761	2.624
15	1.753	2.602
16	1.746	2.583
17	1.740	2.567
18	1.734	2.552
19	1.729	2.539
20	1.725	2.528
21	1.721	2.518
22	1.717	2.508
23	1.714	2.500
24	1.711	2.492
25	1.708	2.485
26	1.706	2.479
27	1.703	2.473
28	1.701	2.467
29	1.699	2.462
30	1.697	2.457
35	1.690	2.438
40	1.684	2.423
45	1.679	2.412
50	1.676	2.403
75	1.665	2.377
100	1.660	2.364
1000	1.646	2.330
∞	1.645	2.326

Source: Table prepared by the author using Microsoft Excel's t inverse function. (Excel is a registered trademark of the Microsoft Corporation.)

Table 3 Critical Values of the Pearson Correlation Coefficient: The r Tables

Two-tailed test

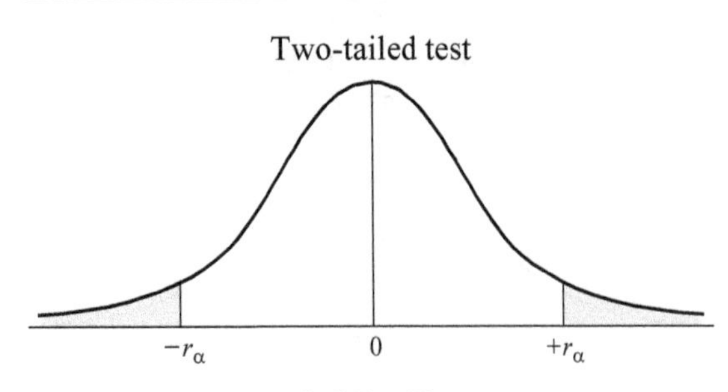

One-tailed test

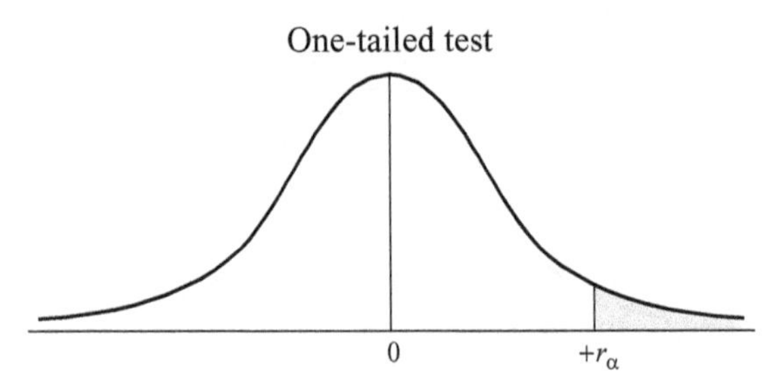

	Two-tailed test: Level of Significance		One-tailed test: Level of Significance	
df ($n-2$)	$\alpha = .05$	$\alpha = .01$	$\alpha = .05$	$\alpha = .01$
1	.997	.999	.988	.999
2	.950	.990	.900	.980
3	.878	.959	.805	.934
4	.811	.917	.729	.882
5	.754	.874	.669	.833
6	.707	.834	.622	.789
7	.666	.798	.582	.750
8	.632	.765	.549	.716
9	.602	.735	.521	.685
10	.576	.708	.497	.658
11	.553	.684	.476	.634
12	.532	.661	.458	.612
13	.514	.641	.441	.592
14	.497	.623	.426	.574
15	.482	.606	.412	.558
16	.468	.590	.400	.542
17	.456	.575	.389	.528
18	.444	.561	.378	.516
19	.433	.549	.369	.503
20	.423	.537	.360	.492
21	.413	.526	.352	.482
22	.404	.515	.344	.472
23	.396	.505	.337	.462
24	.388	.496	.330	.453
25	.381	.487	.323	.445
26	.374	.479	.317	.437
27	.367	.471	.311	.430
28	.361	.463	.306	.423
29	.355	.456	.301	.416
30	.349	.449	.296	.409
35	.325	.418	.275	.381
40	.304	.393	.257	.358
45	.288	.372	.243	.338
50	.273	.354	.231	.322
60	.250	.325	.211	.295
70	.232	.302	.195	.274
80	.217	.283	.183	.256
90	.205	.267	.173	.242
100	.195	.254	.164	.230

Source: Table prepared by the author using Microsoft Excel's t inverse function and conversion of t values to r values. (Excel is a registered trademark of the Microsoft Corporation.)

Table 4 Critical Values of the Spearman Rank-Order Correlation Coefficient: The r_s Tables

Two-tailed test

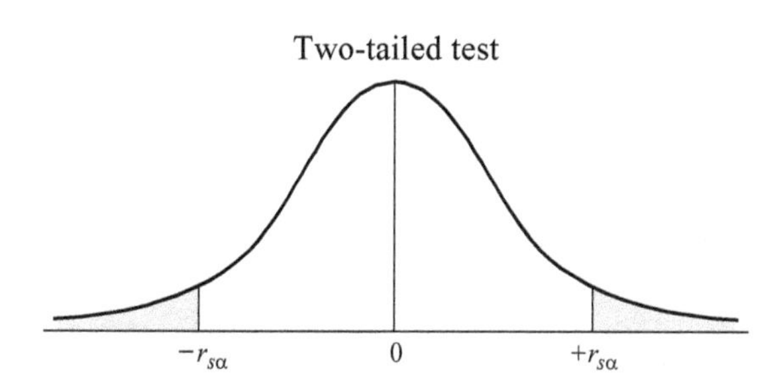

One-tailed test

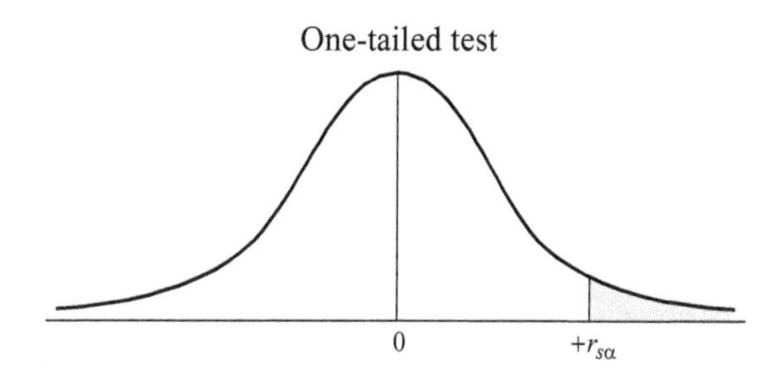

Two-tailed test — *Level of Significance*

N	$\alpha = .05$	$\alpha = .01$
5	1.00	—
6	.886	1.00
7	.786	.929
8	.738	.881
9	.683	.833
10	.648	.794
12	.591	.777
14	.544	.715
16	.506	.665
18	.475	.625
20	.450	.591
22	.428	.562
24	.409	.537
26	.392	.515
28	.377	.496
30	.364	.478

One-tailed test — *Level of Significance*

N	$\alpha = .05$	$\alpha = .01$
5	.900	1.00
6	.829	.943
7	.714	.893
8	.643	.833
9	.600	.783
10	.564	.746
12	.506	.712
14	.456	.645
16	.425	.601
18	.399	.564
20	.377	.534
22	.359	.508
24	.343	.485
26	.329	.465
28	.317	.448
30	.306	.432

Source: From E. G. Olds (1949), "The 5 Percent Significance Levels of Sums of Squares of Rank Differences and a Correction," *Annals of Mathematical Statistics*, 20, 117–118, and E. G. Olds (1938), "Distribution of Sums of Squares of Rank Differences for Small Numbers of Individuals," *Annals of Mathematical Statistics*, 9, 133–148. Reprinted with permission from the Institute of Mathematical Statistics.

Table 5 Critical Values of the *F* Statistic: The *F* Tables

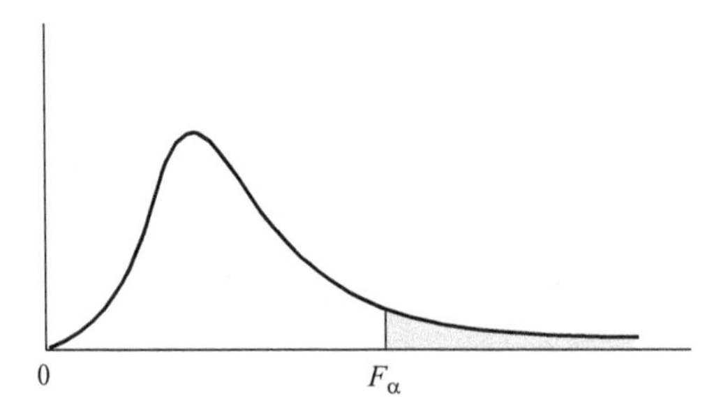

Critical values for $\alpha = .05$ in regular text
Critical values for $\alpha = .01$ in **bold text**

Degrees of Freedom within Groups ($df_{wn} = N - k$)	α	Degrees of Freedom Between Groups ($df_{bn} = k - 1$) 1	2	3	4	5	6	7
1	.05	161	200	216	225	230	234	237
	.01	**4052**	**4999**	**5404**	**5624**	**5764**	**5859**	**5928**
2	.05	18.51	19.00	19.16	19.25	19.30	19.33	19.35
	.01	**98.50**	**99.00**	**99.16**	**99.25**	**99.30**	**99.33**	**99.36**
3	.05	10.13	9.55	9.28	9.12	9.01	8.94	8.89
	.01	**34.12**	**30.82**	**29.46**	**28.71**	**28.24**	**27.91**	**27.67**
4	.05	7.71	6.94	6.59	6.39	6.26	6.16	6.09
	.01	**21.20**	**18.00**	**16.69**	**15.98**	**15.52**	**15.21**	**14.98**
5	.05	6.61	5.79	5.41	5.19	5.05	4.95	4.88
	.01	**16.26**	**13.27**	**12.06**	**11.39**	**10.97**	**10.67**	**10.46**
6	.05	5.99	5.14	4.76	4.53	4.39	4.28	4.21
	.01	**13.75**	**10.92**	**9.78**	**9.15**	**8.75**	**8.47**	**8.26**
7	.05	5.59	4.74	4.35	4.12	3.97	3.87	3.79
	.01	**12.25**	**9.55**	**8.45**	**7.85**	**7.46**	**7.19**	**6.99**
8	.05	5.32	4.46	4.07	3.84	3.69	3.58	3.50
	.01	**11.26**	**8.65**	**7.59**	**7.01**	**6.63**	**6.37**	**6.18**
9	.05	5.12	4.26	3.86	3.63	3.48	3.37	3.29
	.01	**10.56**	**8.02**	**6.99**	**6.42**	**6.06**	**5.80**	**5.61**
10	.05	4.96	4.10	3.71	3.48	3.33	3.22	3.14
	.01	**10.04**	**7.56**	**6.55**	**5.99**	**5.64**	**5.39**	**5.20**
11	.05	4.84	3.98	3.59	3.36	3.20	3.09	3.01
	.01	**9.65**	**7.21**	**6.22**	**5.67**	**5.32**	**5.07**	**4.89**
12	.05	4.75	3.89	3.49	3.26	3.11	3.00	2.91
	.01	**9.33**	**6.93**	**5.95**	**5.41**	**5.06**	**4.82**	**4.64**

	a	*1*	*2*	*3*	*4*	*5*	*6*	*7*
13	.05	4.67	3.81	3.41	3.18	3.03	2.92	2.83
	.01	**9.07**	**6.70**	**5.74**	**5.21**	**4.86**	**4.62**	**4.44**
14	.05	4.60	3.74	3.34	3.11	2.96	2.85	2.76
	.01	**8.86**	**6.51**	**5.56**	**5.04**	**4.69**	**4.46**	**4.28**
15	.05	4.54	3.68	3.29	3.06	2.90	2.79	2.71
	.01	**8.68**	**6.36**	**5.42**	**4.89**	**4.56**	**4.32**	**4.14**
16	.05	4.49	3.63	3.24	3.01	2.85	2.74	2.66
	.01	**8.53**	**6.23**	**5.29**	**4.77**	**4.44**	**4.20**	**4.03**
17	.05	4.45	3.59	3.20	2.96	2.81	2.70	2.61
	.01	**8.40**	**6.11**	**5.19**	**4.67**	**4.34**	**4.10**	**3.93**
18	.05	4.41	3.55	3.16	2.93	2.77	2.66	2.58
	.01	**8.29**	**6.01**	**5.09**	**4.58**	**4.25**	**4.01**	**3.84**
19	.05	4.38	3.52	3.13	2.90	2.74	2.63	2.54
	.01	**8.18**	**5.93**	**5.01**	**4.50**	**4.17**	**3.94**	**3.77**
20	.05	4.35	3.49	3.10	2.87	2.71	2.60	2.51
	.01	**8.10**	**5.85**	**4.94**	**4.43**	**4.10**	**3.87**	**3.70**
21	.05	4.32	3.47	3.07	2.84	2.68	2.57	2.49
	.01	**8.02**	**5.78**	**4.87**	**4.37**	**4.04**	**3.81**	**3.64**
22	.05	4.30	3.44	3.05	2.82	2.66	2.55	2.46
	.01	**7.95**	**5.72**	**4.82**	**4.31**	**3.99**	**3.76**	**3.59**
23	.05	4.28	3.42	3.03	2.80	2.64	2.53	2.44
	.01	**7.88**	**5.66**	**4.76**	**4.26**	**3.94**	**3.71**	**3.54**
24	.05	4.26	3.40	3.01	2.78	2.62	2.51	2.42
	.01	**7.82**	**5.61**	**4.72**	**4.22**	**3.90**	**3.67**	**3.50**
25	.05	4.24	3.39	2.99	2.76	2.60	2.49	2.40
	.01	**7.77**	**5.57**	**4.68**	**4.18**	**3.85**	**3.63**	**3.46**
26	.05	4.23	3.37	2.98	2.74	2.59	2.47	2.39
	.01	**7.72**	**5.53**	**4.64**	**4.14**	**3.82**	**3.59**	**3.42**
27	.05	4.21	3.35	2.96	2.73	2.57	2.46	2.37
	.01	**7.68**	**5.49**	**4.60**	**4.11**	**3.78**	**3.56**	**3.39**
28	.05	4.20	3.34	2.95	2.71	2.56	2.45	2.36
	.01	**7.64**	**5.45**	**4.57**	**4.07**	**3.75**	**3.53**	**3.36**
29	.05	4.18	3.33	2.93	2.70	2.55	2.43	2.35
	.01	**7.60**	**5.42**	**4.54**	**4.04**	**3.73**	**3.50**	**3.33**
30	.05	4.17	3.32	2.92	2.69	2.53	2.42	2.33
	.01	**7.56**	**5.39**	**4.51**	**4.02**	**3.70**	**3.47**	**3.30**

	α	*1*	*2*	*3*	*4*	*5*	*6*	*7*
35	.05	4.12	3.27	2.87	2.64	2.49	2.37	2.29
	.01	**7.42**	**5.27**	**4.40**	**3.91**	**3.59**	**3.37**	**3.20**
40	.05	4.08	3.23	2.84	2.61	2.45	2.34	2.25
	.01	**7.31**	**5.18**	**4.31**	**3.83**	**3.51**	**3.29**	**3.12**
45	.05	4.06	3.20	2.81	2.58	2.42	2.31	2.22
	.01	**7.23**	**5.11**	**4.25**	**3.77**	**3.45**	**3.23**	**3.07**
50	.05	4.03	3.18	2.79	2.56	2.40	2.29	2.20
	.01	**7.17**	**5.06**	**4.20**	**3.72**	**3.41**	**3.19**	**3.02**
75	.05	3.97	3.12	2.73	2.49	2.34	2.22	2.13
	.01	**6.99**	**4.90**	**4.05**	**3.58**	**3.27**	**3.05**	**2.89**
100	.05	3.94	3.09	2.70	2.46	2.31	2.19	2.10
	.01	**6.90**	**4.82**	**3.98**	**3.51**	**3.21**	**2.99**	**2.82**
250	.05	3.88	3.03	2.64	2.41	2.25	2.13	2.05
	.01	**6.74**	**4.69**	**3.86**	**3.40**	**3.09**	**2.87**	**2.71**
500	.05	3.86	3.01	2.62	2.39	2.23	2.12	2.03
	.01	**6.69**	**4.65**	**3.82**	**3.36**	**3.05**	**2.84**	**2.68**
1000	.05	3.85	3.00	2.61	2.38	2.22	2.11	2.02
	.01	**6.66**	**4.63**	**3.80**	**3.34**	**3.04**	**2.82**	**2.66**

Critical values for $\alpha = .05$ in regular text
Critical values for $\alpha = .01$ in **bold text**

Degrees of Freedom within Groups ($df_{wn} = N - k$)	*α*	**Degrees of Freedom Between Groups ($df_{bn} = k - 1$)**						
		8	***9***	***10***	***11***	***12***	***13***	***14***
1	.05	239	241	242	243	244	245	245
	.01	**5981**	**6022**	**6056**	**6083**	**6107**	**6126**	**6143**
2	.05	19.37	19.38	19.40	19.40	19.41	19.42	19.42
	.01	**99.38**	**99.39**	**99.40**	**99.41**	**99.42**	**99.42**	**99.43**
3	.05	8.85	8.81	8.79	8.76	8.74	8.73	8.71
	.01	**27.49**	**27.34**	**27.23**	**27.13**	**27.05**	**26.98**	**26.92**
4	.05	6.04	6.00	5.96	5.94	5.91	5.89	5.87
	.01	**14.80**	**14.66**	**14.55**	**14.45**	**14.37**	**14.31**	**14.25**
5	.05	4.82	4.77	4.74	4.70	4.68	4.66	4.64
	.01	**10.29**	**10.16**	**10.05**	**9.96**	**9.89**	**9.82**	**9.77**
6	.05	4.15	4.10	4.06	4.03	4.00	3.98	3.96
	.01	**8.10**	**7.98**	**7.87**	**7.79**	**7.72**	**7.66**	**7.60**
7	.05	3.73	3.68	3.64	3.60	3.57	3.55	3.53
	.01	**6.84**	**6.72**	**6.62**	**6.54**	**6.47**	**6.41**	**6.36**
8	.05	3.44	3.39	3.35	3.31	3.28	3.26	3.24
	.01	**6.03**	**5.91**	**5.81**	**5.73**	**5.67**	**5.61**	**5.56**
9	.05	3.23	3.18	3.14	3.10	3.07	3.05	3.03
	.01	**5.47**	**5.35**	**5.26**	**5.18**	**5.11**	**5.05**	**5.01**
10	.05	3.07	3.02	2.98	2.94	2.91	2.89	2.86
	.01	**5.06**	**4.94**	**4.85**	**4.77**	**4.71**	**4.65**	**4.60**
11	.05	2.95	2.90	2.85	2.82	2.79	2.76	2.74
	.01	**4.74**	**4.63**	**4.54**	**4.46**	**4.40**	**4.34**	**4.29**
12	.05	2.85	2.80	2.75	2.72	2.69	2.66	2.64
	.01	**4.50**	**4.39**	**4.30**	**4.22**	**4.16**	**4.10**	**4.05**
13	.05	2.77	2.71	2.67	2.63	2.60	2.58	2.55
	.01	**4.30**	**4.19**	**4.10**	**4.02**	**3.96**	**3.91**	**3.86**
14	.05	2.70	2.65	2.60	2.57	2.53	2.51	2.48
	.01	**4.14**	**4.03**	**3.94**	**3.86**	**3.80**	**3.75**	**3.70**
15	.05	2.64	2.59	2.54	2.51	2.48	2.45	2.42
	.01	**4.00**	**3.89**	**3.80**	**3.73**	**3.67**	**3.61**	**3.56**
16	.05	2.59	2.54	2.49	2.46	2.42	2.40	2.37
	.01	**3.89**	**3.78**	**3.69**	**3.62**	**3.55**	**3.50**	**3.45**

	α	*8*	*9*	*10*	*11*	*12*	*13*	*14*
17	.05	2.55	2.49	2.45	2.41	2.38	2.35	2.33
	.01	**3.79**	**3.68**	**3.59**	**3.52**	**3.46**	**3.40**	**3.35**
18	.05	2.51	2.46	2.41	2.37	2.34	2.31	2.29
	.01	**3.71**	**3.60**	**3.51**	**3.43**	**3.37**	**3.32**	**3.27**
19	.05	2.48	2.42	2.38	2.34	2.31	2.28	2.26
	.01	**3.63**	**3.52**	**3.43**	**3.36**	**3.30**	**3.24**	**3.19**
20	.05	2.45	2.39	2.35	2.31	2.28	2.25	2.22
	.01	**3.56**	**3.46**	**3.37**	**3.29**	**3.23**	**3.18**	**3.13**
21	.05	2.42	2.37	2.32	2.28	2.25	2.22	2.20
	.01	**3.51**	**3.40**	**3.31**	**3.24**	**3.17**	**3.12**	**3.07**
22	.05	2.40	2.34	2.30	2.26	2.23	2.20	2.17
	.01	**3.45**	**3.35**	**3.26**	**3.18**	**3.12**	**3.07**	**3.02**
23	.05	2.37	2.32	2.27	2.24	2.20	2.18	2.15
	.01	**3.41**	**3.30**	**3.21**	**3.14**	**3.07**	**3.02**	**2.97**
24	.05	2.36	2.30	2.25	2.22	2.18	2.15	2.13
	.01	**3.36**	**3.26**	**3.17**	**3.09**	**3.03**	**2.98**	**2.93**
25	.05	2.34	2.28	2.24	2.20	2.16	2.14	2.11
	.01	**3.32**	**3.22**	**3.13**	**3.06**	**2.99**	**2.94**	**2.89**
26	.05	2.32	2.27	2.22	2.18	2.15	2.12	2.09
	.01	**3.29**	**3.18**	**3.09**	**3.02**	**2.96**	**2.90**	**2.86**
27	.05	2.31	2.25	2.20	2.17	2.13	2.10	2.08
	.01	**3.26**	**3.15**	**3.06**	**2.99**	**2.93**	**2.87**	**2.82**
28	.05	2.29	2.24	2.19	2.15	2.12	2.09	2.06
	.01	**3.23**	**3.12**	**3.03**	**2.96**	**2.90**	**2.84**	**2.79**
29	.05	2.28	2.22	2.18	2.14	2.10	2.08	2.05
	.01	**3.20**	**3.09**	**3.00**	**2.93**	**2.87**	**2.81**	**2.77**
30	.05	2.27	2.21	2.16	2.13	2.09	2.06	2.04
	.01	**3.17**	**3.07**	**2.98**	**2.91**	**2.84**	**2.79**	**2.74**
35	.05	2.22	2.16	2.11	2.07	2.04	2.01	1.99
	.01	**3.07**	**2.96**	**2.88**	**2.80**	**2.74**	**2.69**	**2.64**
40	.05	2.18	2.12	2.08	2.04	2.00	1.97	1.95
	.01	**2.99**	**2.89**	**2.80**	**2.73**	**2.66**	**2.61**	**2.56**
45	.05	2.15	2.10	2.05	2.01	1.97	1.94	1.92
	.01	**2.94**	**2.83**	**2.74**	**2.67**	**2.61**	**2.55**	**2.51**
50	.05	2.13	2.07	2.03	1.99	1.95	1.92	1.89
	.01	**2.89**	**2.78**	**2.70**	**2.63**	**2.56**	**2.51**	**2.46**

	α	*8*	*9*	*10*	*11*	*12*	*13*	*14*
75	.05	2.06	2.01	1.96	1.92	1.88	1.85	1.83
	.01	**2.76**	**2.65**	**2.57**	**2.49**	**2.43**	**2.38**	**2.33**
100	.05	2.03	1.97	1.93	1.89	1.85	1.82	1.79
	.01	**2.69**	**2.59**	**2.50**	**2.43**	**2.37**	**2.31**	**2.27**
250	.05	1.98	1.92	1.87	1.83	1.79	1.76	1.73
	.01	**2.58**	**2.48**	**2.39**	**2.32**	**2.26**	**2.20**	**2.15**
500	.05	1.96	1.90	1.85	1.81	1.77	1.74	1.71
	.01	**2.55**	**2.44**	**2.36**	**2.28**	**2.22**	**2.17**	**2.12**
1000	.05	1.95	1.89	1.84	1.80	1.76	1.73	1.70
	.01	**2.53**	**2.43**	**2.34**	**2.27**	**2.20**	**2.15**	**2.10**

Critical values for $\alpha = .05$ in regular text
Critical values for $\alpha = .01$ in **bold text**

Degrees of Freedom within groups ($df_{wn} = N - k$)	α	**Degrees of Freedom Between Groups ($df_{bn} = k - 1$)**					
		15	***16***	***17***	***18***	***19***	***20***
1	.05	246	246	247	247	248	248
	.01	**6157**	**6170**	**6181**	**6191**	**6201**	**6209**
2	.05	19.43	19.43	19.44	19.44	19.44	19.45
	.01	**99.43**	**99.44**	**99.44**	**99.44**	**99.45**	**99.45**
3	.05	8.70	8.69	8.68	8.67	8.67	8.66
	.01	**26.87**	**26.83**	**26.79**	**26.75**	**26.72**	**26.69**
4	.05	5.86	5.84	5.83	5.82	5.81	5.80
	.01	**14.20**	**14.15**	**14.11**	**14.08**	**14.05**	**14.02**
5	.05	4.62	4.60	4.59	4.58	4.57	4.56
	.01	**9.72**	**9.68**	**9.64**	**9.61**	**9.58**	**9.55**
6	.05	3.94	3.92	3.91	3.90	3.88	3.87
	.01	**7.56**	**7.52**	**7.48**	**7.45**	**7.42**	**7.40**
7	.05	3.51	3.49	3.48	3.47	3.46	3.44
	.01	**6.31**	**6.28**	**6.24**	**6.21**	**6.18**	**6.16**
8	.05	3.22	3.20	3.19	3.17	3.16	3.15
	.01	**5.52**	**5.48**	**5.44**	**5.41**	**5.38**	**5.36**
9	.05	3.01	2.99	2.97	2.96	2.95	2.94
	.01	**4.96**	**4.92**	**4.89**	**4.86**	**4.83**	**4.81**
10	.05	2.85	2.83	2.81	2.80	2.79	2.77
	.01	**4.56**	**4.52**	**4.49**	**4.46**	**4.43**	**4.41**
11	.05	2.72	2.70	2.69	2.67	2.66	2.65
	.01	**4.25**	**4.21**	**4.18**	**4.15**	**4.12**	**4.10**
12	.05	2.62	2.60	2.58	2.57	2.56	2.54
	.01	**4.01**	**3.97**	**3.94**	**3.91**	**3.88**	**3.86**
13	.05	2.53	2.51	2.50	2.48	2.47	2.46
	.01	**3.82**	**3.78**	**3.75**	**3.72**	**3.69**	**3.66**
14	.05	2.46	2.44	2.43	2.41	2.40	2.39
	.01	**3.66**	**3.62**	**3.59**	**3.56**	**3.53**	**3.51**
15	.05	2.40	2.38	2.37	2.35	2.34	2.33
	.01	**3.52**	**3.49**	**3.45**	**3.42**	**3.40**	**3.37**

	α	*15*	*16*	*17*	*18*	*19*	*20*
16	.05	2.35	2.33	2.32	2.30	2.29	2.28
	.01	**3.41**	**3.37**	**3.34**	**3.31**	**3.28**	**3.26**
17	.05	2.31	2.29	2.27	2.26	2.24	2.23
	.01	**3.31**	**3.27**	**3.24**	**3.21**	**3.19**	**3.16**
18	.05	2.27	2.25	2.23	2.22	2.20	2.19
	.01	**3.23**	**3.19**	**3.16**	**3.13**	**3.10**	**3.08**
19	.05	2.23	2.21	2.20	2.18	2.17	2.16
	.01	**3.15**	**3.12**	**3.08**	**3.05**	**3.03**	**3.00**
20	.05	2.20	2.18	2.17	2.15	2.14	2.12
	.01	**3.09**	**3.05**	**3.02**	**2.99**	**2.96**	**2.94**
21	.05	2.18	2.16	2.14	2.12	2.11	2.10
	.01	**3.03**	**2.99**	**2.96**	**2.93**	**2.90**	**2.88**
22	.05	2.15	2.13	2.11	2.10	2.08	2.07
	.01	**2.98**	**2.94**	**2.91**	**2.88**	**2.85**	**2.83**
23	.05	2.13	2.11	2.09	2.08	2.06	2.05
	.01	**2.93**	**2.89**	**2.86**	**2.83**	**2.80**	**2.78**
24	.05	2.11	2.09	2.07	2.05	2.04	2.03
	.01	**2.89**	**2.85**	**2.82**	**2.79**	**2.76**	**2.74**
25	.05	2.09	2.07	2.05	2.04	2.02	2.01
	.01	**2.85**	**2.81**	**2.78**	**2.75**	**2.72**	**2.70**
26	.05	2.07	2.05	2.03	2.02	2.00	1.99
	.01	**2.81**	**2.78**	**2.75**	**2.72**	**2.69**	**2.66**
27	.05	2.06	2.04	2.02	2.00	1.99	1.97
	.01	**2.78**	**2.75**	**2.71**	**2.68**	**2.66**	**2.63**
28	.05	2.04	2.02	2.00	1.99	1.97	1.96
	.01	**2.75**	**2.72**	**2.68**	**2.65**	**2.63**	**2.60**
29	.05	2.03	2.01	1.99	1.97	1.96	1.94
	.01	**2.73**	**2.69**	**2.66**	**2.63**	**2.60**	**2.57**
30	.05	2.01	1.99	1.98	1.96	1.95	1.93
	.01	**2.70**	**2.66**	**2.63**	**2.60**	**2.57**	**2.55**
35	.05	1.96	1.94	1.92	1.91	1.89	1.88
	.01	**2.60**	**2.56**	**2.53**	**2.50**	**2.47**	**2.44**
40	.05	1.92	1.90	1.89	1.87	1.85	1.84
	.01	**2.52**	**2.48**	**2.45**	**2.42**	**2.39**	**2.37**
45	.05	1.89	1.87	1.86	1.84	1.82	1.81
	.01	**2.46**	**2.43**	**2.39**	**2.36**	**2.34**	**2.31**

	α	***15***	***16***	***17***	***18***	***19***	***20***
50	.05	1.87	1.85	1.83	1.81	1.80	1.78
	.01	**2.42**	**2.38**	**2.35**	**2.32**	**2.29**	**2.27**
75	.05	1.80	1.78	1.76	1.74	1.73	1.71
	.01	**2.29**	**2.25**	**2.22**	**2.18**	**2.16**	**2.13**
100	.05	1.77	1.75	1.73	1.71	1.69	1.68
	.01	**2.22**	**2.19**	**2.15**	**2.12**	**2.09**	**2.07**
	.01	**2.11**	**2.07**	**2.04**	**2.01**	**1.98**	**1.95**
500	. 05	1.69	1.66	1.64	1.62	1.61	1.59
	.01	**2.07**	**2.04**	**2.00**	**1.97**	**1.94**	**1.92**
1000	.05	1.68	1.65	1.63	1.61	1.60	1.58
	.01	**2.06**	**2.02**	**1.98**	**1.95**	**1.92**	**1.90**

Source: Table prepared by the author using Microsoft Excel's *F* inverse function. (Excel is a registered trademark of the Microsoft Corporation.)

Table 6 Values of the Studentized Range Statistic, q_k

Note: For a one-way ANOVA the value of *k* is the number of means in the factor.

Critical values for $\alpha = .05$ in regular text
Critical values for $\alpha = .01$ in **bold text**

Degrees of Freedom within Groups (df_{wn} = N – k)	*α*			*k* = **Number of Means Being Compared**			
		2	***3***	***4***	***5***	***6***	***7***
1	.05	18.00	27.00	32.80	37.10	40.40	43.10
	.01	**90.00**	**135.0**	**164.0**	**186.0**	**202.0**	**216.0**
2	.05	6.09	8.30	9.80	10.90	11.70	12.40
	.01	**14.00**	**19.00**	**22.30**	**24.70**	**26.60**	**28.20**
3	.05	4.50	5.91	6.82	7.50	8.04	8.48
	.01	**8.26**	**10.60**	**12.20**	**13.30**	**14.20**	**15.00**
4	.05	3.93	5.04	5.76	6.29	6.71	7.05
	.01	**6.51**	**8.12**	**9.17**	**9.96**	**10.60**	**11.10**
5	.05	3.64	4.60	5.22	5.67	6.03	6.33
	.01	**5.70**	**6.97**	**7.80**	**8.42**	**8.91**	**9.32**
6	.05	3.46	4.34	4.90	5.31	5.63	5.89
	.01	**5.24**	**6.33**	**7.03**	**7.56**	**7.97**	**8.32**
7	.05	3.34	4.16	4.69	5.06	5.36	5.61
	.01	**4.95**	**5.92**	**6.54**	**7.01**	**7.37**	**7.68**
8	.05	3.26	4.04	4.53	4.89	5.17	5.40
	.01	**4.74**	**5.63**	**6.20**	**6.63**	**6.96**	**7.24**
9	.05	3.20	3.95	4.42	4.76	5.02	5.24
	.01	**4.60**	**5.43**	**5.96**	**6.35**	**6.66**	**6.91**
10	.05	3.15	3.88	4.33	4.65	4.91	5.12
	.01	**4.48**	**5.27**	**5.77**	**6.14**	**6.43**	**6.67**
11	.05	3.11	3.82	4.26	4.57	4.82	5.03
	.01	**4.39**	**5.14**	**5.62**	**5.97**	**6.25**	**6.48**
12	.05	3.08	3.77	4.20	4.51	4.75	4.95
	.01	**4.32**	**5.04**	**5.50**	**5.84**	**6.10**	**6.32**
13	.05	3.06	3.73	4.15	4.45	4.69	4.88
	.01	**4.26**	**4.96**	**5.40**	**5.73**	**5.98**	**6.19**
14	.05	3.03	3.70	4.11	4.41	4.64	4.83
	.01	**4.21**	**4.89**	**5.32**	**5.63**	**5.88**	**6.08**
16	.05	3.00	3.65	4.05	4.33	4.56	4.74
	.01	**4.13**	**4.78**	**5.19**	**5.49**	**5.72**	**5.92**

Degrees of Freedom within Groups ($df_{wn} = N - k$)	***α***	***k* = Number of Means Being Compared**					
		2	***3***	***4***	***5***	***6***	***7***
18	.05	2.97	3.61	4.00	4.28	4.49	4.67
	.01	**4.07**	**4.70**	**5.09**	**5.38**	**5.60**	**5.79**
20	.05	2.95	3.58	3.96	4.23	4.45	4.62
	.01	**4.02**	**4.64**	**5.02**	**5.29**	**5.51**	**5.69**
24	.05	2.92	3.53	3.90	4.17	4.37	4.54
	.01	**3.96**	**4.54**	**4.91**	**5.17**	**5.37**	**5.54**
30	.05	2.89	3.49	3.84	4.10	4.30	4.46
	.01	**3.89**	**4.45**	**4.80**	**5.05**	**5.24**	**5.40**
40	.05	2.86	3.44	3.79	4.04	4.23	4.39
	.01	**3.82**	**4.37**	**4.70**	**4.93**	**5.11**	**5.27**
60	.05	2.83	3.40	3.74	3.98	4.16	4.31
	.01	**3.76**	**4.28**	**4.60**	**4.82**	**4.99**	**5.13**
120	.05	2.80	3.36	3.69	3.92	4.10	4.24
	.01	**3.70**	**4.20**	**4.50**	**4.71**	**4.87**	**5.01**
	.05	2.77	3.31	3.63	3.86	4.03	4.17
	.01	**3.64**	**4.12**	**4.40**	**4.60**	**4.76**	**4.88**

Note: For a one-way ANOVA the value of *k* is the number of means in the factor.

Critical values for $\alpha = .05$ in regular text
Critical values for $\alpha = .01$ in **bold text**

Degrees of Freedom within Groups ($df_{wn} = N - k$)	α	**k = Number of Means Being Compared**				
		8	***9***	***10***	***11***	***12***
1	.05	45.40	47.40	49.10	50.60	52.00
	.01	**227.0**	**237.0**	**246.0**	**253.0**	**260.0**
2	.05	13.00	13.50	14.00	14.40	14.70
	.01	**29.50**	**30.70**	**31.70**	**32.60**	**33.40**
3	.05	8.85	9.18	9.46	9.72	9.95
	.01	**15.60**	**16.20**	**16.70**	**17.10**	**17.50**
4	.05	7.35	7.60	7.83	8.03	8.21
	.01	**11.50**	**11.90**	**12.30**	**12.60**	**12.80**
5	.05	6.58	6.80	6.99	7.17	7.32
	.01	**9.67**	**9.97**	**10.20**	**10.50**	**10.70**
6	.05	6.12	6.32	6.49	6.65	6.79
	.01	**8.61**	**8.87**	**9.10**	**9.30**	**9.49**
7	.05	5.82	6.00	6.16	6.30	6.43
	.01	**7.94**	**8.17**	**8.37**	**8.55**	**8.71**
8	.05	5.60	5.77	5.92	6.05	6.18
	.01	**7.47**	**7.68**	**7.87**	**8.03**	**8.18**
9	.05	5.43	5.60	5.74	5.87	5.98
	.01	**7.13**	**7.32**	**7.49**	**7.65**	**7.78**
10	.05	5.30	5.46	5.60	5.72	5.83
	.01	**6.87**	**7.05**	**7.21**	**7.36**	**7.48**
11	.05	5.20	5.35	5.49	5.61	5.71
	.01	**6.67**	**6.84**	**6.99**	**7.13**	**7.26**
12	.05	5.12	5.27	5.40	5.51	5.62
	.01	**6.51**	**6.67**	**6.81**	**6.94**	**7.06**
13	.05	5.05	5.19	5.32	5.43	5.53
	.01	**6.37**	**6.53**	**6.67**	**6.79**	**6.90**
14	.05	4.99	5.13	5.25	5.36	5.46
	.01	**6.26**	**6.41**	**6.54**	**6.66**	**6.77**
16	.05	4.90	5.03	5.15	5.26	5.35
	.01	**6.08**	**6.22**	**6.35**	**6.46**	**6.56**

Degrees of Freedom within Groups ($df_{wn} = N - k$)	α	*k* = Number of Means Being Compared				
		8	***9***	***10***	***11***	***12***
18	.05	4.82	4.96	5.07	5.17	5.27
	.01	**5.94**	**6.08**	**6.20**	**6.31**	**6.41**
20	.05	4.77	4.90	5.01	5.11	5.20
	.01	**5.84**	**5.97**	**6.09**	**6.19**	**6.29**
24	.05	4.68	4.81	4.92	5.01	5.10
	.01	**5.69**	**5.81**	**5.92**	**6.02**	**6.11**
30	.05	4.60	4.72	4.83	4.92	5.00
	.01	**5.54**	**5.56**	**5.76**	**5.85**	**5.93**
40	.05	4.52	4.63	4.74	4.82	4.91
	.01	**5.39**	**5.50**	**5.60**	**5.69**	**5.77**
60	.05	4.44	4.55	4.65	4.73	4.81
	.01	**5.25**	**5.36**	**5.45**	**5.53**	**5.60**
120	.05	4.36	4.48	4.56	4.64	4.72
	.01	**5.12**	**5.21**	**5.30**	**5.38**	**5.44**
	.05	4.29	4.39	4.47	4.55	4.62
	.01	**4.99**	**5.08**	**5.16**	**5.23**	**5.29**

Source: Abridged from H. L. Harter, D. S. Clemm, and E. H. Guthrie, The probability integrals of the range and the studentized range, WADC Technical Report 58-484, Vol. 2, 1959, Wright Air Development Center, Table II. 2, pp. 243–281. Reprinted from B.J. Winer, *Statistical Principles in Experimental Design*, 2E (1965), The McGraw-Hill Companies. Reproduced with permission of the McGraw-Hill Companies.

Table 7 Critical Values of the Chi Square: The χ^2 Tables

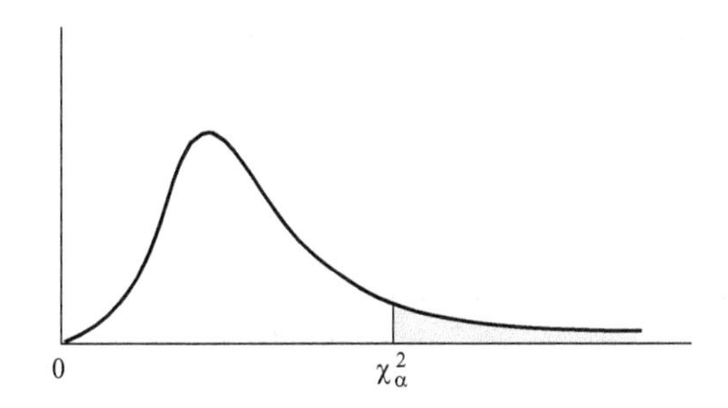

Level of Significance

df	$\alpha = 0.05$	$\alpha = 0.01$
1	3.84	6.64
2	5.99	9.21
3	7.81	11.34
4	9.49	13.28
5	11.07	15.09
6	12.59	16.81
7	14.07	18.48
8	15.51	20.09
9	16.92	21.67
10	18.31	23.21
11	19.68	24.73
12	21.03	26.22
13	22.36	27.69
14	23.68	29.14
15	25.00	30.58
16	26.30	32.00
17	27.59	33.41
18	28.87	34.81
19	30.14	36.19
20	31.41	37.57
25	37.65	44.31
30	43.77	50.89
35	49.80	57.34
40	55.76	63.69
50	67.50	76.15
75	96.22	106.39
100	124.34	135.81

Source: Table prepared by the author using Microsoft Excel's chi inverse function. (Excel is a registered trademark of the Microsoft Corporation.)

Table 8 Critical Values of the Mann-Whitney *U*

Note: The value of U_{obs} is considered statistically significant if it is *less then* or equal to the critical value in the table.

Critical values for $\alpha = .05$ in regular text
Critical values for $\alpha = .01$ in **bold text**

A dash (—) indicates that no decision on H_0 is possible.

Two-Tailed Test

n_2 (number of scores in group 2)	α	n_1 (number of scores in group 1)										
		1	*2*	*3*	*4*	*5*	*6*	*7*	*8*	*9*	*10*	*11*
1	.05	—	—	—	—	—	—	—	—	—	—	—
	.01	**—**	**—**	**—**	**—**	**—**	**—**	**—**	**—**	**—**	**—**	**—**
2	.05	—	—	—	—	—	—	—	0	0	0	0
	.01	**—**	**—**	**—**	**—**	**—**	**—**	**—**	**—**	**—**	**—**	**—**
3	.05	—	—	—	—	0	1	1	2	2	3	3
	.01	**—**	**—**	**—**	**—**	**—**	**—**	**—**	**—**	**0**	**0**	**0**
4	.05	—	—	—	0	1	2	3	4	4	5	6
	.01	**—**	**—**	**—**	**—**	**—**	**0**	**0**	**1**	**1**	**2**	**2**
5	.05	—	—	0	1	2	3	5	6	7	8	9
	.01	**—**	**—**	**—**	**—**	**0**	**1**	**1**	**2**	**3**	**4**	**5**
6	.05	—	—	1	2	3	5	6	8	10	11	13
	.01	**—**	**—**	**—**	**0**	**1**	**2**	**3**	**4**	**5**	**6**	**7**
7	.05	—	—	1	3	5	6	8	10	12	14	16
	.01	**—**	**—**	**—**	**0**	**1**	**3**	**4**	**6**	**7**	**9**	**10**
8	.05	—	0	2	4	6	8	10	13	15	17	19
	.01	**—**	**—**	**—**	**1**	**2**	**4**	**6**	**7**	**9**	**11**	**13**
9	.05	—	0	2	4	7	10	12	15	17	20	23
	.01	**—**	**—**	**0**	**1**	**3**	**5**	**7**	**9**	**11**	**13**	**16**
10	.05	—	0	3	5	8	11	14	17	20	23	26
	.01	**—**	**—**	**0**	**2**	**4**	**6**	**9**	**11**	**13**	**16**	**18**
11	.05	—	0	3	6	9	13	16	19	23	26	30
	.01	**—**	**—**	**0**	**2**	**5**	**7**	**10**	**13**	**16**	**18**	**21**
12	.05	—	1	4	7	11	14	18	22	26	29	33
	.01	**—**	**—**	**1**	**3**	**6**	**9**	**12**	**15**	**18**	**21**	**24**
13	.05	—	1	4	8	12	16	20	24	28	33	37
	.01	**—**	**—**	**1**	**3**	**7**	**10**	**13**	**17**	**20**	**24**	**27**
14	.05	—	1	5	9	13	17	22	26	31	36	40
	.01	**—**	**—**	**1**	**4**	**7**	**11**	**15**	**18**	**22**	**26**	**30**

	α	*1*	*2*	*3*	*4*	*5*	*6*	*7*	*8*	*9*	*10*	*11*
15	.05	—	1	5	10	14	19	24	29	34	39	44
	.01	**—**	**—**	**2**	**5**	**8**	**12**	**16**	**20**	**24**	**29**	**33**
16	.05	—	1	6	11	15	21	26	31	37	42	47
	.01	**—**	**—**	**2**	**5**	**9**	**13**	**18**	**22**	**27**	**31**	**36**
17	.05	—	2	6	11	17	22	28	34	39	45	51
	.01	**—**	**—**	**2**	**6**	**10**	**15**	**19**	**24**	**29**	**34**	**39**
18	.05	—	2	7	12	18	24	30	36	42	48	55
	.01	**—**	**—**	**2**	**6**	**11**	**16**	**21**	**26**	**31**	**37**	**42**
19	.05	—	2	7	13	19	25	32	38	45	52	58
	.01	**—**	**0**	**3**	**7**	**12**	**17**	**22**	**28**	**33**	**39**	**45**
20	.05	—	2	8	13	20	27	34	41	48	55	62
	.01	**—**	**0**	**3**	**8**	**13**	**18**	**24**	**30**	**36**	**42**	**48**

Table 8 Critical Values of the Mann-Whitney U

Note: The value of U_{obs} is considered statistically significant if it is *less then* or equal to the critical value in the table.

Critical values for α = .05 in regular text
Critical values for α = .01 in **bold text**

A dash (—) indicates that no decision on H_0 is possible.

Two-Tailed Test

n_2 (number of scores in group 2)	α	n_1 (number of scores in group 1)								
		12	***13***	***14***	***15***	***16***	***17***	***18***	***19***	***20***
1	.05	—	—	—	—	—	—	—	—	—
	.01	—	—	—	—	—	—	—	—	—
2	.05	1	1	1	1	1	2	2	2	2
	.01	—	—	—	—	—	—	—	**0**	**0**
3	.05	4	4	5	5	6	6	7	7	8
	.01	**1**	**1**	**1**	**2**	**2**	**2**	**2**	**3**	**3**
4	.05	7	8	9	10	11	11	12	13	13
	.01	**3**	**3**	**4**	**5**	**5**	**6**	**6**	**7**	**8**
5	.05	11	12	13	14	15	17	18	19	20
	.01	**6**	**7**	**7**	**8**	**9**	**10**	**11**	**12**	**13**
6	.05	14	16	17	19	21	22	24	25	27
	.01	**9**	**10**	**11**	**12**	**13**	**15**	**16**	**17**	**18**
7	.05	18	20	22	24	26	28	30	32	34
	.01	**12**	**13**	**15**	**16**	**18**	**19**	**21**	**22**	**24**
8	.05	22	24	26	29	31	34	36	38	41
	.01	**15**	**17**	**18**	**20**	**22**	**24**	**26**	**28**	**30**
9	.05	26	28	31	34	37	39	42	45	48
	.01	**18**	**20**	**22**	**24**	**27**	**29**	**31**	**33**	**36**
10	.05	29	33	36	39	42	45	48	52	55
	.01	**21**	**24**	**26**	**29**	**31**	**34**	**37**	**39**	**42**
11	.05	33	37	40	44	47	51	55	58	62
	.01	**24**	**27**	**30**	**33**	**36**	**39**	**42**	**45**	**48**
12	.05	37	41	45	49	53	57	61	65	69
	.01	**27**	**31**	**34**	**37**	**41**	**44**	**47**	**51**	**54**
13	.05	41	45	50	54	59	63	67	72	76
	.01	**31**	**34**	**38**	**42**	**45**	**49**	**53**	**56**	**60**
14	.05	45	50	55	59	64	67	74	78	83
	.01	**34**	**38**	**42**	**46**	**50**	**54**	**58**	**63**	**67**

	α	12	13	14	15	16	17	18	19	20
15	.05	49	54	59	64	70	75	80	85	90
	.01	**37**	**42**	**46**	**51**	**55**	**60**	**64**	**69**	**73**
16	.05	53	59	64	70	75	81	86	92	98
	.01	**41**	**45**	**50**	**55**	**60**	**65**	**70**	**74**	**79**
17	.05	57	63	67	75	81	87	93	99	105
	.01	**44**	**49**	**54**	**60**	**65**	**70**	**75**	**81**	**86**
18	.05	61	67	74	80	86	93	99	106	112
	.01	**47**	**53**	**58**	**64**	**70**	**75**	**81**	**87**	**92**
19	.05	65	72	78	85	92	99	106	113	119
	.01	**51**	**56**	**63**	**69**	**74**	**81**	**87**	**93**	**99**
20	.05	69	76	83	90	98	105	112	119	127
	.01	**54**	**60**	**67**	**73**	**79**	**86**	**92**	**99**	**105**

Table 8 Critical Values of the Mann-Whitney U

Note: The value of U_{obs} is considered statistically significant if it is *less then* or equal to the critical value in the table.

Critical values for $\alpha = .05$ in regular text
Critical values for $\alpha = .01$ in **bold text**

A dash (—) indicates that no decision on H_0 is possible.

One-Tailed Test

n_2 (number of scores in group 2)	α	n_1 (number of scores in group 1)										
		1	*2*	*3*	*4*	*5*	*6*	*7*	*8*	*9*	*10*	*11*
1	.05	—	—	—	—	—	—	—	—	—	—	—
	.01	**—**	**—**	**—**	**—**	**—**	**—**	**—**	**—**	**—**	**—**	**—**
2	.05	—	—	—	—	0	0	0	1	1	1	1
	.01	**—**	**—**	**—**	**—**	**—**	**—**	**—**	**—**	**—**	**—**	**—**
3	.05	—	—	0	0	1	2	2	3	3	4	5
	.01	**—**	**—**	**—**	**—**	**—**	**—**	**0**	**0**	**1**	**1**	**1**
4	.05	—	—	0	1	2	3	4	5	6	7	8
	.01	**—**	**—**	**—**	**—**	**0**	**1**	**1**	**2**	**3**	**3**	**4**
5	.05	—	0	1	2	4	5	6	8	9	11	12
	.01	**—**	**—**	**—**	**0**	**1**	**2**	**3**	**4**	**5**	**6**	**7**
6	.05	—	0	2	3	5	7	8	10	12	14	16
	.01	**—**	**—**	**—**	**1**	**2**	**3**	**4**	**6**	**7**	**8**	**9**
7	.05	—	0	2	4	6	8	11	13	15	17	19
	.01	**—**	**—**	**0**	**1**	**3**	**4**	**6**	**7**	**9**	**11**	**12**
8	.05	—	1	3	5	8	10	13	15	18	20	23
	.01	**—**	**—**	**0**	**2**	**4**	**6**	**7**	**9**	**11**	**13**	**15**
9	.05	—	1	3	6	9	12	15	18	21	24	27
	.01	**—**	**—**	**1**	**3**	**5**	**7**	**9**	**11**	**14**	**16**	**18**
10	.05	—	1	4	7	11	14	17	20	24	27	31
	.01	**—**	**—**	**1**	**3**	**6**	**8**	**11**	**13**	**16**	**19**	**22**
11	.05	—	1	5	8	12	16	19	23	27	31	34
	.01	**—**	**—**	**1**	**4**	**7**	**9**	**12**	**15**	**18**	**22**	**25**
12	.05	—	2	5	9	13	17	21	26	30	34	38
	.01	**—**	**—**	**2**	**5**	**8**	**11**	**14**	**17**	**21**	**24**	**28**
13	.05	—	2	6	10	15	19	24	28	33	37	42
	.01	**—**	**0**	**2**	**5**	**9**	**12**	**16**	**20**	**23**	**27**	**31**
14	.05	—	2	7	11	16	21	26	31	36	41	46
	.01	**—**	**0**	**2**	**6**	**10**	**13**	**17**	**22**	**26**	**30**	**34**

	α	*1*	*2*	*3*	*4*	*5*	*6*	*7*	*8*	*9*	*10*	*11*
15	.05	—	3	7	12	18	23	28	33	39	44	50
	.01	**—**	**0**	**3**	**7**	**11**	**15**	**19**	**24**	**28**	**33**	**37**
16	.05	—	3	8	14	19	25	30	36	42	48	54
	.01	**—**	**0**	**3**	**7**	**12**	**16**	**21**	**26**	**31**	**36**	**41**
17	.05	—	3	9	15	20	26	33	39	45	51	57
	.01	**—**	**0**	**4**	**8**	**13**	**18**	**23**	**28**	**33**	**38**	**44**
18	.05	—	4	9	16	22	28	35	41	48	55	61
	.01	**—**	**0**	**4**	**9**	**14**	**19**	**24**	**30**	**36**	**41**	**47**
19	.05	0	4	10	17	23	30	37	44	51	58	65
	.01	**—**	**1**	**4**	**9**	**15**	**20**	**26**	**32**	**38**	**44**	**50**
20	.05	0	4	11	18	25	32	39	47	54	62	69
	.01	**—**	**1**	**5**	**10**	**16**	**22**	**28**	**34**	**40**	**47**	**53**

Table 8 Critical Values of the Mann-Whitney U

Note: The value of U_{obs} is considered statistically significant if it is *less than* or equal to the critical value in the table.

Critical values for $\alpha = .05$ in regular text
Critical values for $\alpha = .01$ in **bold text**

A dash (—) indicates that no decision on H_0 is possible.

One-Tailed Test

n_2 (number of scores in group 2)		**n_1 (number of scores in group 1)**								
	α	***12***	***13***	***14***	***15***	***16***	***17***	***18***	***19***	***20***
1	.05	—	—	—	—	—	—	—	0	0
	.01	—	—	—	—	—	—	—	—	—
2	.05	2	2	2	3	3	3	4	4	4
	.01	—	**0**	**0**	**0**	**0**	**0**	**0**	**1**	**1**
3	.05	5	6	7	7	8	9	9	10	11
	.01	**2**	**2**	**2**	**3**	**3**	**4**	**4**	**4**	**5**
4	.05	9	10	11	12	14	15	16	17	18
	.01	**5**	**5**	**6**	**7**	**7**	**8**	**9**	**9**	**10**
5	.05	13	15	16	18	19	20	22	23	25
	.01	**8**	**9**	**10**	**11**	**12**	**13**	**14**	**15**	**16**
6	.05	17	19	21	23	25	26	28	30	32
	.01	**11**	**12**	**13**	**15**	**16**	**18**	**19**	**20**	**22**
7	.05	21	24	26	28	30	33	35	37	39
	.01	**14**	**16**	**17**	**19**	**21**	**23**	**24**	**26**	**28**
8	.05	26	28	31	33	36	39	41	44	47
	.01	**17**	**20**	**22**	**24**	**26**	**28**	**30**	**32**	**34**
9	.05	30	33	36	39	42	45	48	51	54
	.01	**21**	**23**	**26**	**28**	**31**	**33**	**36**	**38**	**40**
10	.05	34	37	41	44	48	51	55	58	62
	.01	**24**	**27**	**30**	**33**	**36**	**38**	**41**	**44**	**47**
11	.05	38	42	46	50	54	57	61	65	69
	.01	**28**	**31**	**34**	**37**	**41**	**44**	**47**	**50**	**53**
12	.05	42	47	51	55	60	64	68	72	77
	.01	**31**	**35**	**38**	**42**	**46**	**49**	**53**	**56**	**60**
13	.05	47	51	56	61	65	70	75	80	84
	.01	**35**	**39**	**43**	**47**	**51**	**55**	**59**	**63**	**67**
14	.05	51	56	61	66	71	77	82	87	92
	.01	**38**	**43**	**47**	**51**	**56**	**60**	**65**	**69**	**73**

	a	*12*	*13*	*14*	*15*	*16*	*17*	*18*	*19*	*20*
15	.05	55	61	66	72	77	83	88	94	100
	.01	**42**	**47**	**51**	**56**	**61**	**66**	**70**	**75**	**80**
16	.05	60	65	71	77	83	89	95	101	107
	.01	**46**	**51**	**56**	**61**	**66**	**71**	**76**	**82**	**87**
17	.05	64	70	77	83	89	96	102	109	115
	.01	**49**	**55**	**60**	**66**	**71**	**77**	**82**	**88**	**93**
18	.05	68	75	82	88	95	102	109	116	123
	.01	**53**	**59**	**65**	**70**	**76**	**82**	**88**	**94**	**100**
19	.05	72	80	87	94	101	109	116	123	130
	.01	**56**	**63**	**69**	**75**	**82**	**88**	**94**	**101**	**107**
20	.05	77	84	92	100	107	115	123	130	138
	.01	**60**	**67**	**73**	**80**	**87**	**93**	**100**	**107**	**114**

Source: From D. Auble. "Extended Tables for the Mann-Whitney Statistic." *Bulletin of the Institute of Educational Research* vol. 1, no. 2 (1953). Copyright Trustees of Indiana University, reprinted with permission.

Table 9 Critical Values of the Wilcoxon T

Note: The value of T_{obs} is considered statistically significant if it is *less than* or equal to the critical value in the table.
N is the number of nonzero differences used to compute T_{obs} (not the original sample size "*n*").
A dash (—) indicates that no decision on H_0 is possible.

	Two-Tailed Test			***One-Tailed Test***	
N	$\alpha = .05$	$\alpha = .01$	*N*	$\alpha = .05$	$\alpha = .01$
5	—	—	5	0	—
6	0	—	6	2	—
7	2	—	7	3	0
8	3	0	8	5	1
9	5	1	9	8	3
10	8	3	10	10	5
11	10	5	11	13	7
12	13	7	12	17	9
13	17	9	13	21	12
14	21	12	14	25	15
15	25	15	15	30	19
16	29	19	16	35	23
17	34	23	17	41	27
18	40	27	18	47	32
19	46	32	19	53	37
20	52	37	20	60	43
21	58	42	21	67	49
22	65	48	22	75	55
23	73	54	23	83	62
24	81	61	24	91	69
25	89	68	25	100	76
26	98	75	26	110	84
27	107	83	27	119	92
28	116	91	28	130	101
29	126	100	29	140	110
30	137	109	30	151	120
31	147	118	31	163	130
32	159	128	32	175	140
33	170	138	33	187	151
34	182	148	34	200	162
35	195	159	35	213	173
36	208	171	36	227	185
37	221	182	37	241	198
38	235	194	38	256	211
39	249	207	39	271	224
40	264	220	40	286	238
41	279	233	41	302	252
42	294	247	42	319	266
43	310	261	43	336	281
44	327	276	44	353	296
45	343	291	45	371	312
46	361	307	46	389	328
47	378	322	47	407	345
48	396	339	48	426	362
49	415	355	49	446	379
50	434	373	50	466	397

Source: From F. Wilcoxon and R. A. Wilcox, *Some Rapid Approximate Statistical Procedures* (New York: Lederle Laboratories, 1964).

APPENDIX B

Answers to Selected Problems

Chapter 1

7a. variable

7b. attribute

7c. attribute

7d. variable

7e. attribute

7f. attribute

7g. attribute

7h. variable

7i. concept

8a. 94

8b. 132

8c. 1,432

8d. 8,836

8e. 3,080

8f. 17,424

8g. 12,408

8h. 11.75

8i. 16.5

8j. 2,015

Chapter 2

2. Stratified random
3. 210
4. Adam-Beth; Adam-Kyle; Adam-Sara; Adam-Sherry; Beth-Kyle; Beth-Sara; Beth-Sherry; Kyle-Sara; Kyle-Sherry; Sara-Sherry

8a. accidental

8b. quota

8c. spatial

8d. systematic

Chapter 3

3.

X	f	Cf	LL	MP	UL
140 – 149	3	30	139.5	144.5	149.5
130 – 139	3	27	129.5	134.5	139.5
120 – 129	2	24	119.5	124.5	129.5
110 – 119	1	22	109.5	114.5	119.5
100 – 109	4	21	99.5	104.5	109.5
90 – 99	7	17	89.5	94.5	99.5
80 – 89	5	10	79.5	84.5	89.5
70 – 79	5	5	69.5	74.5	79.5
	$\Sigma f = 30$				

5. Age 65 = 66.9th percentile; age 42 = 38.5th percentile; age 18 = 16th percentile.
7. The data are suggestive of a normal distribution with a slight skew to the left.
9. X = 30 is 31st percentile; X = 40 is 51st percentile;

   X = 50 is 71st percentile; X = 60 is 91st percentile.

Chapter 4

2. mode = 45; mean = 54; median = 38.5
3. median — due to one extreme value much larger than the rest of the distribution
5. mode = interval 25–29, or 27; mean = 24.13; median = 24.34

7. The author was a sociology major

10. mean = 46.5; mode = interval 40–49 or 44.5; median = 46.5

Chapter 5

1a. 52

1b. 10.5

4. range = 50; interquartile range = 12.88; semi-interquartile range = 6.44

6. class (1) variance = 146.67; standard deviation = 12.11

   class (2) variance = 373.33; standard deviation = 19.32

9a. variance = 11,935.25

9b. mean score male = 535; mean score female = 588.75

10. 5.8% of variance explained

Chapter 6

3a. .4115

3b. .0465

3c. .0113

3d. .9750

5a. 1.00

5b. −1.00

5c. 0.374

5d. 5.00

7a. .7056

7b. .0852

7c. .1129

7d. .0099

10. 242

14a. .0668

14b. .9772

14c. .0062

14d. .1525

16a. 90% C. I. = ($46.71 \leq \mu \leq 53.29$);

   95% C. I. = ($46.08 \leq \mu \leq 53.92$).

16b. 90% C. I. = ($48.03 \leq \mu \leq 51.97$);

   95% C. I. = ($47.65 \leq \mu \leq 52.35$).

Chapter 7

1a. .10

1b. .40

1c. .04

1d. .60

2. .01536

3a. .60

3b. .70

3c. .80

3d. .50

3e. .714

3f. .833

5. s.e. = .0219; margin of error = ± 4.29%

7a. yes

7b. .1056

11a. ± 2.25%

11b. ± 2.70%

11c. > 2.82%

11d. The margin for error increases.

Chapter 8

6a. two-tail

6b. one-tail

6c. one-tail

6d. two-tail

7. $Z_{.05} = \pm 1.96$; reject H_0

9a. $H_1: \mu_{text} > \mu$; $H_0: \mu_{text} = \mu$

9b. $Z_\alpha = 2.33$; $Z_{obs} = 2.50$

9c. Reject H_0; students using this test have a statistically significantly higher mean score than the population at the $\alpha = .01$ level of significance.

10. Just look at the two population means. No, no statistical test is necessary to evaluate to population means for a difference.

Chapter 9

1. $r_s = -.891$; $r_s.05 = -.564$; Reject H_0
3. $r_{pb} = .1314$; $r_{pb}.05 = .549$; Fail to reject H_0; $r^2 = .017$ (1.7% of variance explained).
6. $r = .894$; $r_{.01} = .834$; Reject H_0; $r^2 = .799$ (79.9% of the variance explained).

8a. $r_{xy.z} = -.649$

8b. $r_{xz.y} = .393$

Chapter 10

2a. $b = -1.40$; $a = 15.24$; $Y' = -1.40\ X + 15.24$

2c. $X = 20$; $Y' = -12.76$

$X = -14$; $Y' = 34.84$

$X = 3$; $Y' = 11.04$

2d. $r^2 = .834$ (83.4% variance explained)

4. $b = .624$; $a = 14.4$; $Y' = .624\ X + 14.4$

7a. $b = 3.52$; $a = -17.80$; $Y' = 3.52\ X - 17.80$

7b. $r^2 = .78$ (78% variance explained)

7c. Removing the two data points as outliers had very little effect on the regression analysis.

10a. No, a curvilinear relationship is evident

10b. $b = .0035$; $a = 9.46$; $Y' = .0035\ X + 9.46$

10c. $r^2 = .00004$ (.004% of the variance explained)

Chapter 11

1a. Z test for two independent samples, two-tail test

1b. t-test for a sample mean compared with a population mean; one-tail test

1c. Z test for a sample mean compared with a population mean; two-tail test

1d. t-test for two related samples; two-tail test

1e. Z test for two independent samples; one-tail test

3. $t = -2.00$; $t_{.01} = \pm 2.947$; Fail to reject H_0
4. $t = -2.102$ (or 2.102 depending on how you compute the difference); $t_{.05} = 1.895$; Reject H_0
6. $Z = 2.169$; $Z.05 = \pm 1.96$; Reject H_0; The results of the Z test differ very little from the results of the t-test.
9. $Z = 0.822$; $Z_{.05} = 1.645$; Fail to reject H_0
10. None, just look at the two population means to determine if they are different.

Chapter 12

4.

ANOVA Summary Table

Source	SS	df	MS	F
Between	514.67	2	257.335	6.255
Within	740.57	18	41.14	
Total	1,255.24	20		

$F_{.05} = 3.55$; Reject H_0

5. Tukey's HSD = 8.752; Means for 65 mph and 75 mph are statistically significantly different. (Fisher's protected t-test could also be used, but is more cumbersome.) Eta Squared = .41 (41% of the variance explained).

7. $q_k = 4.23$; Tukey's HSD = 15.325.

8.

ANOVA Summary Table

Source	SS	df	MS	F
Between	346.54	2	173.27	9.02
Within	384.20	20	19.21	
Total	730.74	22		

$F_{.01} = 5.85$; Reject H_0

9. Fisher's protected t-test is appropriate in this case. The critical value of $t_{.01} = \pm 2.845$.

   Group comparisons:

   Group A with group B: t = 1.86; Fail to reject H_0;
   Group A with group C: t = 4.12; Reject H_0;
   Group B with group C: t = 2.53; Fail to reject H_0;

10. The size of the difference between the means is not the only factor that contributes to a statistically significant difference. The size of the respective variances and samples are also important.

Chapter 13

2a. low income, low GPA n = 5; low income, high GPA n = 8; middle income, low GPA n = 12; middle income, high GPA n = 10; high income low GPA n = 8; high income, high GPA n = 7

2b. Chi square observed = 0.940; chi square critical = 9.21; Fail to reject H_0

2c. Coefficient of Contingency = .136; Gamma = -.174

4a. $U_1 = 78$; $U_2 = 13$; Uobs = 13; $U_{.01} = 13$; Reject H_0

7. $T_{obs} = 10$; $T_{.05} = 3$; Fail to Reject H_0

10. Coefficient of Contingency = .448; Lambda = .231; Gamma = -.577

11a. 2

11b. 2

11c. 6

11d. 16

11e. 6

APPENDIX C

The National Opinion Research Center General Social Survey

The 2012 General Social Survey (GSS) is part of a regular, ongoing personal interview survey of U.S. households conducted by the National Opinion Research Center (NORC). The mission of the GSS is to make timely, high-quality, scientifically relevant data available to the social science research community. The 2012 GSS is the 29th independent cross-sectional surveys of the adult household population of the United States since the GSS began in 1972. These surveys have been widely distributed and extensively analyzed by social scientists around the world.

The entire 2012 GSS, along with data from prior years may be downloaded from the NORC website (or the SDA Archive housed at the University of California, Berkley). A portion of the GSS for the years 2000 to 2012 in the form of an SPSS data file is provided as a resource for this text. Variables included in the data file are identified below.

Citation

Smith, Tom W, Peter Marsden, Michael Hout, and Jibum Kim. *General Social Surveys, 1972–2012* [machine-readable data file] /Principal Investigator, Tom W. Smith; Co-Principal Investigator, Peter V. Marsden; Co-Principal Investigator, Michael Hout; Sponsored by National Science Foundation. –NORC ed.– Chicago: National Opinion Research Center [producer]; Storrs, CT: The Roper Center for Public Opinion Research, University of Connecticut [distributor], 2013.

1 data file (57,061 logical records) + 1 codebook (3,432p.). (National Data Program for the Social Sciences, No. 21).

Selected Variable List from 2012 General Social Survey

Age—Age of respondent

Value Label Coded as actual age. Range: 18–89; 99 = Missing

Agewed—Age When First Married

Value Label Coded as actual age. Range: 13–90; NAP = 0; DK = 98; NA = 99 all coded as Missing

Attend—How often *R* attends religious services

Value Label	Value	Frequency	Percent	Value Percent	Cum Percent
NEVER	0	3980	21.0	21.2	100.0
LT ONCE A YEAR	1	1309	6.9	7.0	78.8
ONCE A YEAR	2	2518	13.3	13.4	71.8
SEVERAL TIMES A YEAR	3	2234	11.8	11.9	58.4
ONCE A MONTH	4	1304	6.9	6.9	46.6
2–3X A MONTH	5	1627	8.6	8.7	39.6
NEARLY EVERY WEEK	6	978	5.2	5.2	31.0
EVERY WEEK	7	3413	18.0	18.2	25.8
MORE THAN ONCE WEEK	8	1425	7.5	7.6	7.6
DK,NA	9	157	.8	0.0	
Total		18788	99.2	100.0	
Total		18945	100.0	100.0	

Cappun—Favor or oppose death penalty for murder

Value Label	Value	Frequency	Percent	Valid Percent	Cum Percent
FAVOR	1	9146	48.3	67.3	100.0
OPPOSE	2	4446	23.5	32.7	32.7
DK	8	841	4.4	Missing	
NA	9	129	.7	Missing	
Total		18945	100.0		

Childs—Number of children

Value Label Coded as actual number. Range: 0–8+; 99 = Missing

Class—Subjective Class Identification

Value Label	Value	Frequency	Percent	Valid Percent	Cum Percent
LOWER CLASS	1	1241	6.6	7.2	100.0
WORKING CLASS	2	7782	41.1	45.0	92.8
MIDDLE CLASS	3	7711	40.7	44.5	47.9
UPPER CLASS	4	577	3.0	3.3	3.3
DK	8	76	.4	Missing	
NA	9	40	.2	Missing	
Total		18945	100.0	100.0	

Conbus—Confidence in major companies

Value Label	Value	Frequency	Percent	Valid Percent	Cum Percent
A GREAT DEAL	1	1764	9.3	18.6	100.0
ONLY SOME	2	5943	31.4	62.7	81.4
HARDLY ANY	3	1767	9.3	18.7	18.7
IAP	0	9192	48.5	Missing	
DK	8	260	1.4	Missing	
NA	9	19	.1	Missing	
Total		18945	100.0	100.0	

Conclerg—Confidence in organized religion

Value Label	Value	Frequency	Percent	Valid Percent	Cum Percent
A GREAT DEAL	1	2199	11.6	23.3	100.0
ONLY SOME	2	5029	26.5	53.3	76.7
HARDLY ANY	3	2203	11.6	23.4	23.4
IAP	0	9192	48.5	Missing	
DK	8	280	1.5	Missing	
NA	9	42	.2	Missing	
Total		18945	100.0	100.0	

Coneduc—Confidence in education

Value Label	Value	Frequency	Percent	Valid Percent	Cum Percent
A GREAT DEAL	1	2626	13.9	27.2	100.0
ONLY SOME	2	5544	29.3	57.3	72.8
HARDLY ANY	3	1500	7.9	15.5	15.5
IAP	0	9192	48.5	Missing	
DK	8	71	.4	Missing	
NA	9	12	.1	Missing	
Total		18945	100.0	100.0	

Confed—Confidence in executive branch of fed govt

Value Label	Value	Frequency	Percent	Valid Percent	Cum Percent
A GREAT DEAL	1	1509	8.0	15.9	100.0
ONLY SOME	2	4546	24.0	47.9	84.1
HARDLY ANY	3	3429	18.1	36.2	36.2
IAP	0	9192	48.5	Missing	
DK	8	243	1.3	Missing	
NA	9	26	.1	Missing	
Total		18945	100.0	100.0	

Confinan—Confidence in banks and financial institutions

Value Label	Value	Frequency	Percent	Valid Percent	Cum Percent
A GREAT DEAL	1	2153	11.4	22.4	100.0
ONLY SOME	2	5267	27.8	54.8	77.6
HARDLY ANY	3	2186	11.5	22.8	22.8
IAP	0	9192	48.5	Missing	
DK	8	135	.7	Missing	
NA	9	12	.1	Missing	
Total		18945	100.0	100.0	

Conjudge—Confidence in united states supreme court

Value Label	Value	Frequency	Percent	Valid Percent	Cum Percent
A GREAT DEAL	1	3003	15.9	32.0	100.0
ONLY SOME	2	4936	26.1	52.6	68.0
HARDLY ANY	3	1441	7.6	15.4	36.2
IAP	0	9192	48.5	Missing	
DK	8	351	1.9	Missing	
NA	9	22	.1	Missing	
Total		18945	100.0	100.0	

Coninc—Family income in constatnt dollars

Value Label Coded as actual number. Range: 383–178712; 0; 999998; 999999 = Missing

Conrinc—Respondent income in constatnt dollars

Value Label Coded as actual number. Range: 383–434612; 0; 999998; 999999 = Missing

Divorce—Ever been divorced or separated

Value Label	Value	Frequency	Percent	Valid Percent	Cum Percent
YES	1	2583	13.6	24.5	100.0
NO	2	7946	41.9	75.5	75.5
NAP	0	8375	44.2	Missing	
DK	8	1	.0	Missing	
NA	9	40	.2	Missing	
Total		18945	100.0	100.0	

Educ—Highest year of school completed

Value Label Coded as actual year. Range: 0–20; DK = 98; NA = 99 both coded as Missing

Grass—Should marijuana be made legal

Value Label	Value	Frequency	Percent	Valid Percent	Cum Percent
LEGAL	1	3552	18.7	39.5	100.0
NOT LEGAL	2	5449	28.8	60.5	60.5
NAP	0	9192	48.5	Missing	
DK	8	739	3.9	Missing	
NA	9	13	.1	Missing	
Total		18945	100.0	100.0	

Grntaxes—Pay higher taxes to help envir?

Value Label	Value	Frequency	Percent	Valid Percent	Cum Percent
VERY WILLING	1	154	.8	6.1	100.0
FAIRLY WILLING	2	674	3.6	26.6	93.9
NOT WILL OR UNWILL	3	598	3.2	23.6	67.3
NOT VERY WILLING	4	508	2.7	20.0	43.7
NOT AT ALL WILLING	5	601	3.2	23.7	23.7
NAP	0	16239	85.7	Missing	
DK	8	106	.6	Missing	
NA	9	65	.3	Missing	
Total		18945	100.0	100.0	

Gunlaw—Favor or oppose gun permits

Value Label	Value	Frequency	Percent	Valid Percent	Cum Percent
FAVOR	1	7434	39.2	78.5	100.0
OPPOSE	2	2034	10.7	21.5	21.5
NAP	0	9318	49.2	Missing	
DK	8	142	.7	Missing	
NA	9	17	.1	Missing	
Total		18945	100.0	100.0	

Happy—General happiness

Value Label	Value	Frequency	Percent	Valid Percent	Cum Percent
VERY HAPPY	1	4365	23.0	30.1	100.0
PRETTY HAPPY	2	8179	43.2	56.5	69.9
NOT TOO HAPPY	3	1943	10.3	13.4	13.4
DK	8	20	.1	Missing	
NA	9	55	.3	Missing	
IAP	0	4383	23.1	Missing	
Total		18945	100.0	100.0	

Letdie1—Allow incurable patients to die

Value Label	Value	Frequency	Percent	Valid Percent	Cum Percent
YES	1	5916	31.2	67.9	100.0
NO	2	2794	14.7	32.1	32.4
NAP	0	9882	52.2	Missing	
DK	8	311	1.6	Missing	
NA	9	42	.2	Missing	
Total		18945	100.0	100.0	

Marital—Marital status

Value Label	Value	Frequency	Percent	Valid Percent	Cum Percent
MARRIED	1	8959	47.3	47.3	100.0
WIDOWED	2	1598	8.4	8.4	52.7
DIVORCED	3	2972	15.7	15.7	44.2
SEPARATED	4	662	3.5	3.5	25.8
NEVER MARRIED	5	4741	25.0	25.0	25.0
NA	9	13	.1	Missing	
Total		18945	100.0	100.0	

Polviews—Think of self as liberal or conservative

Value Label	Value	Frequency	Percent	Valid Percent	Cum Percent
EXTREMELY LIBERAL	1	565	3.0	3.7	100.0
LIBERAL	2	1838	9.7	11.9	96.3
SLIGHTLY LIBERAL	3	1775	9.4	11.5	84.4
MODERATE	4	5955	31.4	38.7	72.9
SLIGHTLY CONSERVATIVE	5	2232	11.8	14.5	34.2
CONSERVATIVE	6	2463	13.0	16.0	19.7
EXTREMELY CONSERVATIVE	7	569	3.0	3.7	3.7
DK	8	590	3.1	Missing	
NA	9	93	.5	Missing	
IAP	0	2865	15.1	Missing	
Total		18945	100.0	100.0	

Postlife—Belief in life after death

Value Label	Value	Frequency	Percent	Valid Percent	Cum Percent
YES	1	10141	53.5	81.6	100.0
NO	2	2291	12.1	18.4	18.4
NAP	0	4870	25.7	Missing	
DK	8	1583	8.4	Missing	
NA	9	60	.3	Missing	
Total		18945	100.0	100.0	

Region—Region of interview

Value Label	Value	Frequency	Percent	Valid Percent	Cum Percent
NEW ENGLAND	1	799	4.2	4.2	100.0
MIDDLE ATLANTIC	2	2533	13.4	13.4	95.8
E. NOR. CENTRAL	3	3237	17.1	17.1	82.4
W. NOR. CENTRAL	4	1369	6.7	6.7	65.3
SOUTH ATLANTIC	5	3922	20.7	20.7	58.6
E. SOU. CENTRAL	6	1165	6.1	6.1	37.9
W. SOU. CENTRAL	7	1984	10.5	10.5	31.8
MOUNTAIN	8	1360	7.2	7.2	21.3
PACIFIC	9	2676	14.1	14.1	14.1
Total		18945	100.0	100.0	

Relig—Rs religious preference

Value Label	Value	Frequency	Percent	Valid Percent	Cum Percent
PROTESTANT	1	9721	51.3	51.6	100.0
CATHOLIC	2	4518	23.8	24.0	48.4
JEWISH	3	348	1.8	1.8	24.5
NONE	4	3001	15.8	15.9	22.6
OTHER	5	197	1.0	1.0	6.7
BUDDHISM	6	121	.6	.6	5.7
HINDUISM	7	58	.3	.3	5.0
OTHER EASTERN	8	29	.2	.2	4.7
MOSLEM/ISLAM	9	95	.5	.5	4.6
OROTHODOX-CHRISTIAN	10	86	.5	.5	4.1
CHRISTIAN	11	555	2.9	2.9	3.6
NATIVE AMERICAN	12	21	.1	.1	.7
INTER-NONDENOM.	13	105	.6	.6	.6
DK	98	12	.1		
NA	99	78	.4		
Total		18945	100.0	100.0	

Satjob—Job or housework

Value Label	Value	Frequency	Percent	Valid Percent	Cum Percent
VERY SATISFIED	1	5546	29.3	49.0	100.0
MOD. SATISFIED	2	4308	22.7	38.1	51.0
A LITTLE DISSATISFIED	3	1029	5.4	9.1	12.9
VERY DISSATISFIED	4	435	2.3	3.8	3.8
DK	8	123	.6	Missing	
NA	9	250	1.3	Missing	
IAP	0	7254	38.3	Missing	
Total		18945	100.0	100.0	

SEX—Respondents sex

Value Label	Value	Frequency	Percent	Valid Percent	Cum Percent
MALE	1	8847	44.6	44.6	100.0
FEMALE	2	10498	55.4	55.4	55.4
Total		18945	100.0	100.0	

Sibs—Number of brothers and sisters

Value Label Coded as actual number. Range: 0–55; DK = 98; NA = 99; IAP = −1 Coded as Missing

Size—Size of place in 1000s

Value Label Coded as actual number. Range: 0–8008; DK = 98; NA = 99; IAP = −1 Coded as Missing

Spanking—Favor spanking to discipline child

Value Label	Value	Frequency	Percent	Valid Percent	Cum Percent
STRONGLY AGREE	1	2594	13.7	27.0	100.0
AGREE	2	4346	22.9	45.2	73.0
DISAGREE	3	1969	10.4	20.5	27.8
STRONGLY DISAGREE	4	699	3.7	7.3	7.3
IAP	0	9201	48.6	Missing	
DK	8	116	.6	Missing	
NA	9	20	.1	Missing	
Total		18945	100.0	100.0	

Trust—Can people be trusted

Value Label	Value	Frequency	Percent	Valid Percent	Cum Percent
CAN TRUST	1	3908	20.8	33.5	100.0
CANNOT TRUST	2	7131	37.6	61.2	66.5
DEPENDS	3	626	3.3	5.4	5.4
IAP	0	9192	48.5	Missing	
DK	8	135	.7	Missing	
NA	9	12	.1	Missing	
Total		18945	100.0	100.0	

Tvhours—Hours per day watching tv

Value Label Coded as actual number. Range: 0–24; IAP = −1; DK = 98; NA = 99 All coded as Missing

Year—Gss year for respondnet

Value	Frequency	Percent	Valid Percent	Cum Percent
2012	1974	10.4	10.4	100.0
2010	2044	10.8	10.8	89.6
2008	2023	10.7	10.7	78.8
2006	4510	23.8	23.8	68.1
2004	2812	14.8	14.8	44.3
2002	2765	14.6	14.6	29.5
2000	2817	14.9	14.9	14.9
Total	18945	100.0	100.0	

APPENDIX D

How to Use SPSS

The Statistical Package for the Social Sciences (SPSS) is one of the most widely used computer software packages available on the market today that is dedicated to statistical analysis. SPSS initially appeared as a mainframe software package in the 1960s and may still be found in that format, but it is more likely to be encountered in a personal computer version operating in a Windows environment. Appendix D presents a brief overview of using SPSS and some of its more frequently used options. Appendix D is not intended to be a substitute for the user's manual or the online tutorial, both of which I recommend to you if you really want to improve your proficiency with SPSS, but should enable you to complete "Using the Computer" exercises that appear at the end of most of the chapters in the text.

Basics

The SPSS program is opened like any other Windows application software. Select the start button, and then the programs button to display a list of available programs, and then select SPSS, or simply double-click the SPSS icon if it is available already on your desktop. An SPSS data editor screen will open in the background, and an SPSS window will open in the foreground providing you an opportunity to open an existing SPSS file.

The Data Editor Window

The Data Editor window allows you input data directly into SPSS, or to edit data that you have previously entered. The Data Editor window appears in a spreadsheet type format with each column representing a single variable, and each row representing a single case or member of the sample. The point where a row and column intersect is referred to as a "cell." **Entering numeric data (numbers) into SPSS** is as easy as positioning the cursor on a desired cell, and entering the desired information by using the keypad or the keyboard. The numbers you key will appear in the cell and will be entered into the cell once you press the "enter" key, or one of the directional arrow keys moving the cursor to a new cell location. **Entering alphanumeric data** (combinations of letters such as names or majors and so on) may also be entered in the same fashion. Generally a variable or column will not accept both data formats. SPSS assumes you intend to use the data format that you first enter but you may change a particular variable's format by using the "defining the variable" feature.

To **define a variable** you must first select the variable by moving the cursor to the desired variable and press the left mouse button, or by tapping the touch screen. Then move to the menu bar at the top of the SPSS application window and select the word "Data" and select "Define Variable Properties" from the available options. Move to the "Type" option and select the appropriate choice from the options available. The two primary options for a variable type are "numeric" for data consisting of or represented by numerals, or "string" used for alphanumeric data. For example, if the age of a respondent is represented by actual age in years (18, 25, 38, 57, and so on) then the variable would be numeric. If age is represented in categories of: young (ages 18–34); middle (ages 35–64); and, old (ages 65 and over) then the variable would be a string variable. A third strategy might be to use numerals to represent the categories such as "young" would be represented by the number 1, "middle" would be represented by the number 2, and "old" would be represented by the number 3. As far as SPSS is concerned the variable is a numeric one as it contains numbers, but the data represented would actually be ordinal alphanumeric data. Generally speaking it is always better to gather data at the highest level of measurement possible (in the case of age the ratio level – the respondent's actual age in years). If you prefer to use age categories later you can always recode the variable. If you begin by collecting age as categories and later discover that "middle" consisting of ages 35–64 was actually too broad a range you have no way to convert it to something else.

The Output Window

The Output Window is where any output or results of your statistical analysis will appear. The Output Window is typically in the background behind the Data Editor Window when you first begin using SPSS or will appear after you conduct your first SPSS operation. After a statistical analysis is completed, the results are displayed in the Output Window, and it will appear in the foreground with the Data Editor Window in the background. You may use the "elevator buttons" to scroll up or down the page if your output occupies more than a single screen, and you can also expand the size of the Output Window by clicking on the expansion icon in the upper right-hand corner of the window. At the end of your SPSS session you will have the option to save the output in a separate file for later use.

The SPSS Tool Bar

The SPSS Tool Bar is accessible at the top of both the Data Editor Window and the Output Window, and consists of both a text-based and icon-based Tool bar that can be used to access various SPSS procedures and options. In the next few paragraphs we will be exploring the functions available on the Tool bar. We will first examine the major functions of the Tool bar. There are some slight variations between the options available from the Data Editor Window and the Output Window, for example, the Output Window has a "format" option that does not appear on the Data Editor Window, but the essential functions are very similar on both windows.

The text-based Tool Bar consists of the words: "File; Edit; View; Data, Transform; Analyze; Graphs; Utilities; Window; and, Help." Moving the cursor and clicking or touching any of the words will open a submenu containing a number of additional choices. Many of the Menu bar major options, such as "File, Edit, Window, and Help" are familiar to you if you have worked with other Windows applications, and you will find that they operate in SPSS much as they do in other Windows applications. Keeping in mind that this Appendix is not intended to serve as a substitute for a user's manual or the online tutorial, but we will explore the major functions of each of the options available on the Menu bar.

File

The major functions of "File" are to create a new data file, open an existing data file or output file, save a data file or output file, or to enter the printer menu or printer setup menu. Additional functions exist (such as applying a data dictionary or a template), but they are beyond our present interest of exploring the basic workings of SPSS. To work with any of the "File" options select "File" and a submenu will drop down providing you with the options described above. The three most important options for our purposes are: "Open" to open an existing file; "Save" to save the current file; and "Print" or "Printer Setup" to print some output we have generated from a statistical analysis. **To Open a data file in SPSS** select "File" then "Open." An additional submenu appears giving you several options. Usually you will want to select the first option "Data" and an "Open Data File" dialog box will appear. SPSS data files are saved with the extension "sav," and if you are working on a personal computer SPSS assumes that your data files are stored on the hard drive in a subdirectory. You may have to select the appropriate subdirectory or perhaps identify a portable storage location where you SPSS files are stored. Any data file with the extension "sav" on that drive will appear in a list of available files. If you see the file you want simply select it and its name will appear in the "file name" area. Then select "Open" and the file will then appear in the Data Editor Window. SPSS data files are not the only type of file you may open in SPSS. The other common option is to open a spreadsheet file that may exist as a Lotus 1-2-3 file, a Microsoft Excel file, or a text (txt) file to name a few other types.

Spreadsheet files are easy ways to create a data set, and are very easy to open in SPSS. **To open a spreadsheet file** you will need to select the "File type" field and a variety of alternate file types will appear. Select the desired file format and any spreadsheets in that location should appear. Note: in most spreadsheet applications the first row of the spreadsheet is used to provide names for the column variables. If your spreadsheet is stored in this manner be sure to select the option "Read variable names" just below the "File type" field. This option will import the first row of the spreadsheet as variable names. Once you have indicated all of the desired information (file name, file type, file location, and appropriate options) you are ready to select the "Open" command and your file should appear in the Data Editor Window.

Saving a data file in SPSS is very similar to the process of opening a file. To **save a data file in SPSS** select "File" and then "Save Data," or "Save As" from the drop-down menu. "Save Data" will allow you to save the open data file with the same name it had when you opened the file. The "Save As" option will allow you to save the file under a different name. The "Save As" option is useful if you want to keep the original file as a backup, and use a copy under a different name as the working file. If you select the "Save Data" option the file will save under its current name, and location (the hard drive or portable storage site, for example). If you select the "Save As" option a dialog box opens allowing you to enter the new file name, the file type, and the location where you would like to store the file. Once you have entered all of the appropriate parameters you will need to select the "Save" option to complete the process.

The final major option under the "File" command on the Menu bar allows us to print output generated by a statistical procedure. **To print a copy of output from SPSS** select the word "File" and when the submenu opens select "Print" and a print dialog box will appear. You will have the option of printing all of the output that has been generated, or a selected portion of the output. Output may be selected in the usual method with Windows applications by clicking and dragging the cursor over the desired output. There are also "Page Setup," and "Page Attributes" options available from the "File" command if you wish to customize or format your printed output.

One final command available in the submenu from the "File" option is the "Exit" command. Clicking on the "exit" command is one of several ways that you can end your SPSS work session. The other frequently used method to end your SPSS work session is to select the standard exit icon in the extreme right-hand corner of the SPSS Application window (the red X). Either method will cause SPSS to prompt you with questions about saving your output and data file prior to shut down of the SPSS program.

Edit

The second major option on the text-based menu bar is the "Edit" command. Selecting the "Edit" command will reveal a drop-down menu with a number of options, and many of them will be familiar to you. The "Edit" submenu will allow you to "Cut," "Paste," or "Copy" portions of the file displayed in the Data Editor Window or the Output Window. You can also "Select" text or a portion of the data file from the "Edit" submenu, "Search for text," or "Clear (delete)" a portion of the data file or output file. Additional options allow you to "Search and replace" text, and to add page breaks. There is also an "Options" choice that allows you to alter some of the default settings in SPSS, but you will probably not find it necessary to use this option.

Data

We have already explored one option under the "Data" command in the Menu bar when we saw how to use the "Define Variable Properties" option to specify alphanumeric data for a variable. Additional options in the dialog box for the "**Define Variable Properties**" option will allow you to **assign a custom label to a variable,** which will appear in your output in place of the default variable names such as "Var001, Var002," and so on as originally assigned by SPSS. It is also possible to replace the default variable name such as "Var001" by simply selecting the "View" option and switching from "Data View" to "Variable View." Once in variable view you may simply enter the desired variable name in place of the default. For example, Var001 might represent the "age" of the individual. "Variable View" will allow you to assign the name "age" to that variable, and that name will then appear as the column heading in the Data Editor window, and on any statistical output. You may also toggle back and forth between "Data View" and "Variable View" by selecting the desired choice appearing at the bottom left side of the Data Editor Screen.

The "Define Variable Properties" option also allows you to assign "**Missing Values**" to a variable, to **assign value "Labels,"** and to **change the "Column Format"** for a variable. SPSS will interpret a blank cell assigned to a particular case or individual as a "missing value" (that is, there is no data for that particular person). SPSS will then represent the missing data using the system default of a period. Often times researchers will enter values for a cell that will also represent missing values. The values "9," "99," or "999" (and so on depending on how large the range is associated with the particular variable) are commonly used to represent missing values. The "Missing Values" option will allow you to tell SPSS that these or other values you have used should be considered as missing data, and should therefore not be used in any statistical calculations. The "Labels" option will allow you to assign a label to each numeric value associated with a particular variable. For example, Var002 might represent "sex," and you might use "1" to represent males, and "2" to represent females. The "Labels" option will allow you to assign those labels to the values, and the labels will then be used in any statistical output. You will find that the "Variable Name" and value "Labels" options are very convenient and helpful when interpreting output from statistical procedures. Their use will keep you from having to remember what "Var002" represents, and what the variable values represented. For example, without the value labels you might have a difficult time remembering if "1" represented a male or a female. Finally, the "Column Format" option allows you to alter the width and placement of text in each column. The SPSS default is to allow 8 spaces with right justification for each column, and you might want to reduce a column to only 2 spaces or have the data centered or left justified for some variables.

The "**Define Dates**" option allows you to select a date format different from the SPSS default. Selecting the "**Merge Files**" option opens a submenu that allows you to expand the size of the data file by adding new variables or new cases. The "**Add Cases**" option allows you to add new cases from another SPSS data set into the existing one. For example if you have data collected at different points in time or in different locations the "Add Cases" option will allow you to import those data into the active data file. The "**Add Variables**" option will allow you to add additional variables from a larger file into the active SPSS data file. For example if you have downloaded a subset of the data from a larger data set for a research project and later realize that you did not download a key variable that you needed you can download the single variable you overlooked into a separate SPSS data file and then incorporate that file into the active one. The "Add Variables" option can save you the effort of having to create an entirely new data set that is very useful if you have done a great deal of work to recode a number of variables.

Three final functions we will examine under the "Data" command allow you **to split a data file, select certain cases**, or **weight certain cases**. To split a data file select "Data" and then "Split File," and a dialog box will appear that will allow you to specify the criteria for splinting the file. For example, you might want to take a data file and split it into two files, one for males and one for females. SPSS will also allow you to select certain cases from a file for analysis. To select cases from a data file select "Data" and then on "Select Cases," and a dialog box will appear that will allow you to specify the criteria for selecting certain cases within the file. For example, you might want to analyze only those members of the sample who meet a certain income level. Those individuals may be selected with the "select cases" option. The select cases option will also allow you to select a random sample of cases from the larger data file. You have the option of selecting a specified percentage of the entire file, or selecting a specified number of cases from the entire file. The nonselected cases may either be filtered out (left in the data file but not used in subsequent analysis) or deleted from the data file. You should always be careful any time you select the "delete" option, because you will not be able to recover deleted cases. Finally, SPSS will allow you to weigh selected cases for subsequent analysis. To weight cases in the data file select "Data" and then on "Weight Cases," and a dialog box will appear allowing you to specify the criteria for the weight you will assign to selected cases. Researchers often weigh cases (give some cases a higher level of representation in the analysis) of small minorities. Additional options exist under the "Data" command that allow you to sort cases, merge file, transpose cases, and aggregate variables, but these options are not likely to be of use for our purposes here and will not be discussed.

Transform

The transform command in the text-based menu bar has several options that will be useful for our purposes including: **compute, count, recode, automatic recode, and replace missing values**, along with a few other options that are less likely to be of use to beginners working with SPSS. The compute option allows us to create a new variable based on some type of computation. **To compute a new variable** select "Transform" and then "Compute Variable," and a dialog box appears. The dialog box allows you to specify the name of the new variable, and to enter the numeric expression that serves as the basis for the new variable. Researchers often compute new variables. For example, socioeconomic status (SES) is often a combination of one's educational level and income. Another frequent "computed" variable is an attitudinal scale based on one's answers to a series of questions. For example, we might create a "risk factor" variable based on one's responses to questions regarding cigarette smoking, use of seat belts, unprotected sex with multiple partners, intravenous drug use, and so on.

The count option allows you to count the number of times a particular response occurs in a single variable or in a set of variables. To count cases with a given response select "Transform" and then "Count," and a dialog box appears that will allow you to specify the response to be counted in the variable or variables of interest. An option will allow you to use all cases in the data file, or only those that meet selected criteria.

The recode option allows you to recode or change the values of a selected variable. There are two options of recoding including "**Recode into the selected variables**", or to "**Recode into new variables**." Choose the appropriate option, but note that recoding a variable "into same variables" means that the recoding you specify will replace the existing coding for the variable. In other words, the original coding will be lost once you replace it with the new coding. Recoding a variable "into different variables" means that a new variable will be created with the recoded values, and the original values will be saved under the name of the selected variable. Recoding into different variables is highly recommended because once you recode into the same variable and save your work the original coding cannot be recovered.

To recode a variable into the same variable select the appropriate option and a dialog box will appear. Select the variable to be recoded from the list of available variables, and then select the directional arrow to move it to the selected variable list. Then select the "old and new values" button, and a second dialog box will open allowing you to specify the old value and the way you want to recode it. You may specify exact values to be recoded or a range of values. Be sure to select the "Add" button after you specify each recode or SPSS will not act on the command. For example, we might want to recode age into a categorical variable with individuals from age 18 to 34 recoded as "1" to represent "young," individuals age 35 to 54 recoded as "2" to represent "middle age," and individuals age 55 or greater recoded as "3" to represent "old." Once the recodes have been specified select the "continue" button. You may then select the "OK" button to complete the recode for all cases in the data file, or select the "if" button to complete the recode on cases meeting criteria that you specify. To recode a variable into a different variable select the appropriate choice and follow the same procedure as above. The only difference is that SPSS will allow you to specify the name for the new variable that will be created with the recoded values.

The Transform command also offers an "**Automatic Recode**" option. Automatic recode is used to recode numeric values for a variable into sequential numeric values. Most often the automatic recode option is used to convert a ratio or interval level variable into ordinal level ranks. For example, the lowest observed value of education will be assigned a "1," the next lowest observed value will be assigned a "2," and so on. To automatically recode a variable select "Transform," and then on "Automatic Recode," and a dialog box will open. Specify the variable to be recoded by clicking on it, supply the new name for the recoded variable, and select the option to tell SPSS if you want to begin the recode with the lowest observed value or the highest observed value. Select the "OK" button to complete the procedure.

The final option we will examine under the Transform command allows you to replace missing values. Researchers often want to avoid having missing values in the data set, and will often assign a value to missing data. Common strategies are to assign the overall average (mean) to missing values, or to assign the median. To replace missing values select "Transform" and then "**Replace Missing Values**," and a dialog box will open. Highlight the desired variable by clicking on it and moving it into the selected box, specify the new name for the variable, and then select the method for replacing the missing values. Select the "OK" button to complete the procedure.

Analyze

The Analyze command from the text-based menu bar is one of the more frequently used commands in SPSS. To begin a statistical procedure select "Analyze" and a submenu of options for a number of statistical procedures will open. SPSS is often sold as a base module with additional modules available at additional cost, so the statistical tests available will vary depending on how each system is configured. In this Appendix, I will discuss the basic statistical tests. The more commonly used options included in the Analyze command are: "Reports," "Descriptive Statistics," "Compare Means," "Correlate" and, "Regression."

The "**Reports" procedure** contains several options but only three will be discussed here: list cases, report summaries in rows, and report summaries in columns. Each of these options will provide summary statistics for selected variables from the data file. The more useful option is probably "**List Cases**" that allows you to see the values associated with each member of the sample for selected variables. One common use of the "List Cases" option is to display the values for selected variables for a few cases in the data file to verify that the data file was loaded properly, or to display the entire data file to verify the accuracy of the data. For example, you might want to print the contents of the entire data file and compare it to another source to make sure that it is free of errors. To list cases from the data file select "Analyze" and then "Reports" and then select "List Cases." A dialog box will open allowing you to highlight some or all of the variables, and then move them to the selected list by using the directional arrow. There are several parameters that you may specify including the number of cases to list, and the method of displaying the data. For large data sets you will probably not want to list all cases, although that is an option. An alternative is to list only a few cases (you may specify as many as you want), or to list every ith case where "i" is an interval that you specify. The display options include the entire set of data for each case, or for files with multiple lines you have the option of only displaying the first line. If your purpose is to verify that a file loaded properly, you will probably only want to display a few cases from the file. However, if you plan to check the accuracy of a data file you will need to display the entire file. Once you have selected the variables and specified the parameters, select the "OK" button to complete the procedure. The "**Report Summaries in Rows**," and "**Report Summaries in Columns**" are less frequently used statistical procedures, and are not discussed in detail here. Each offers you the option of reporting some summary data for selected variables for each category of another variable. The output is similar to the "Explore" option discussed below.

To obtain a frequency distribution select "Analyze" then "Descriptive Statistics" and several options are available including: frequency distribution, descriptive, explore, and crosstabs. Select "**Frequencies**" and a dialog box will open and you may select the variables you wish by highlighting them and then using the directional arrow to move them to the selected list. Several options are available with the "Frequencies" option including "Statistics," "Charts," and, "Format." A variety of statistics are available including: quartiles, percentiles, measures of central tendency, measures of variation, and the option to select cutting points for the values of the distribution categories or the number categories to display. The "**Charts**" option will allow you to produce common graphical displays such as bar charts or histograms for the variables, and the "Format" option will allow you specify the presentation of the data (ascending order or descending order). Once you have chosen the variables and specified all of the desired options select the "OK" button to complete the procedure.

The "**Descriptives**" option will produce a variety of descriptive statistics (logically enough). To obtain descriptive statistics select "Analyze" and then move the cursor to "Descriptive Statistics" and then select "Descriptives," and a dialog box will open. Select the variables you want by highlighting them and moving them to the selected list. The "Options" button will allow you to add additional results to the default output that includes: the mean, standard deviation, minimum value, and maximum value. Once you have chosen the variables and specified all of the desired options select the "OK" button to complete the procedure.

The "**Explore**" option allows you to obtain descriptive statistics and stem-and-leaf plots or histograms for a dependent variable by levels of an independent variable. To explore a variable select "Analyze" and then "Descriptive Statistics" and then select "Explore," and a dialog box will open. Highlight the dependent variable you want to explore and move it to the selected list by using the appropriate direction arrow. Highlight the "Factor" variable and move it to the selected list using the appropriate direction arrow. You will receive summary statistics for the dependent variable for each level of the factor variable. For example, if you recode education to consist of three categories of "low," "medium," and "high," and choose it as the factor variable, and then select "income" as the dependent variable, then you will receive descriptive statistics on income for each of the three education groups.

The "**Crosstabs**" option allows you to obtain a cross tabulation (or create a contingency table) for two or more variables. In most cases, you will be creating a two variable table with one dependent variable and one independent variable. To create a contingency table select "Descriptive Statistics" and then select "Crosstabs." A dialog box will open allowing you to specify the row and column variables. The independent variable should be assigned as the "Row" variable, and the dependent variable should be assigned as the "Column" variable. To select the independent variable, highlight it and then move it to the selected box for the "Row" variable by using the appropriate direction arrow. To select the dependent variable, highlight it and then move it to the selected box for the Column variable by using the appropriate direction arrow. You also have the option of creating more elaborate tables by specifying a control variable by using the "Layer" feature. For example, you might have level of education as the row or independent variable and level of income as the column or dependent variable for your basic cross tabulation. The result would be a cross tabulation of income by education for all members of the sample. If you add the variable sex as a layer variable the result would be a separate cross tabulation of income by education for males and a separate cross tabulation of income by education for females.

Several options are available in Crosstabs including: Statistics, Cells, and Format. The "Statistics" option allows you to select the type of statistics you want reported for the contingency table. You will usually want to specify the Chi-square statistic, and one or more of the available measures of association. The "Cells" option allows you to specify the summary information you would like in addition to the default observed cell counts (column percents is usually a very good choice to add). The final option is the "Format" option, and it is one that you will probably need to make use of when creating a contingency table. "Format" allows you to specify the labeling of the table, but the real advantage of the "Format" option is that it allows you to present the row variable in descending order in place of the SPSS default of ascending order. In a properly constructed contingency table, the dependent variable (the row variable) if measured at the ordinal level should have the smallest value at the bottom of the table, and the highest value at the top of the table (descending order). For some reason the SPSS default is to present the row variable in ascending order. Once you have specified the variables and options select the "OK" button to complete the procedure.

The next major statistical procedure available under "Analyze" is "**Compare Means**." The "Compare Means" has several options available including: "Means," "One Sample T Test," "Independent Samples T Test," "Paired Samples T Test," and, "One Way ANOVA." The "Means" option will display the mean value for a selected dependent variable or variables by categories of a selected independent variable. To conduct the "Means" procedure select the "Analyze" option then select "Compare Means" option, and then select "**Means**." A dialog box will open allowing you to highlight one or more dependent variables. Move the variables to the selected list of dependent variables by using the directional arrow. Next highlight one or more independent variables (these should usually be categorical variables) and move them to the selected independent variable list by clicking on the appropriate directional arrow. The default output will display the mean, standard deviation, and number of cases for each dependent variable by each category of the independent variable. The "Options" button will allow you to request additional summary statistics, alter the labels for the variables, and request an Analysis of Variance table or Test of Linearity. Analysis of Variance is also available as a separate option under the "Compare Means" procedure, and is discussed below. As is the case with many procedures in Windows-based applications, there is often more than one way to obtain the same result. Select the "OK" button to complete the analysis.

To conduct a "**One Sample T Test,**" select "Analyze" and then "Compare Means," and then select "One Sample T Test." A dialog box will open allowing you to specify one or more test variables. Highlight the desired variable or variables and move it or them to the selected list by using the directional arrow. For example, in the simplest case of a single variable such as "income" SPSS will conduct a t test to determine if the sample mean is statistically significantly different from zero, or from some other value that you specify in the "Test Value" parameter box. The "Options" button will allow you to change the size of the reported confidence interval, alter the labels that are reported, and choose one of two ways to handle missing values. Select the "OK" button to complete the analysis.

To conduct an "**Independent Samples T Test**" select "Analyze" move the cursor to "Compare Means," and then select "Independent Samples T Test." A dialog box will open allowing you to specify one or more test variables. Highlight the desired variable or variables and move them to the selected box by using the appropriate directional arrow. Then select the "Grouping Variable" by highlighting it and using the directional arrow. The grouping variable in most cases will be a categorical variable that will usually be thought of as an independent variable. For example, we might want to compute a t test to compare the mean income (the test variable) by sex (the grouping variable). The result would be a t test comparing the mean income of males with that of females. The "Options" feature offers the same choices as above. Select the "OK" button to complete the analysis.

To compute a "**Paired Samples T Test**" select "Analyze" move the cursor to "Compare Means," and then select "Paired Samples T Test." A dialog box will open allowing you to specify two variables for comparison. The resulting t test will indicate the presence of a statistically significant difference between the two sample means. The "Paired Samples T Test" is most often used to compare the pretest and posttest results from an experimental or quasi-experimental design. The "Options" feature offers the same choices as above. Select the "OK" button to complete the analysis.

The final option under the "Compare Means" procedure is the "**One Way ANOVA**," which will generate a one-way analysis of variance. One-way analysis of variance allows you the option of conducting a single statistical test to determine in any of several group means are statistically significantly different rather than conducting multiple t tests. To conduct a "One Way ANOVA" select "Analyze" move the cursor to "Compare Means," and then select "One Way ANOVA." A dialog box will open allowing you to specify one or more dependent variables. Highlight the desired dependent variable and move it to the selected list by using directional arrow key. You will then need to select the "Factor" or independent variable whose values will determine the computation of the group means for the dependent variable. Highlight the desired "Factor" variable and move it to the selected list by using the directional arrow. Once the factor has been identified you will need to identify the range of values associated with it for SPSS. For example, "income" might be the dependent variable, and "religion" might be the Factor or independent variable. The result would be an analysis to determine if any of the religious groups have statistically significant mean incomes. Several options are available for the ANOVA procedure including: contrasts, post hoc, and options. The "Contrasts" option allows you to specify a more advanced ANOVA model than is described in the text, so it is not discussed here. The "Post Hoc" option allows you to specify the type of post hoc test to determine the identity of all of the statistically significant comparisons that may exist in the data. For example, the text discusses Tukey's HSD test, and Fisher's exact t test. Tukey's HSD test is an option in SPSS along with several other post hoc procedures. The "Options" button will allow you to specify the summary statistics (including a homogeneity of variance test), a choice between two ways of handling missing values, and an option to display the value labels associated with the variables. Select the "OK" button to complete the analysis.

The next major option under the "Analyze" procedure discussed here is "**Correlate**," which allows you to conduct bivariate or partial correlation coefficients (not all procedures in the base SPSS package are discussed in this brief overview of SPSS). To conduct a bivariate correlation analysis select "Analyze" and move the cursor down to "Correlate" then select "**Bivariate**." A dialog box will open allowing you to specify the desired variables for the correlation analysis. At least two variables must be highlighted and moved to the selected list by clicking on the directional arrow, but more than two variables may be included in the analysis. Several parameters need to be set beginning with the type of correlation coefficient to be computed. The options include: **Pearson r** (the default procedure used for variables measured at the interval or ratio level), Kendall's tau-b (used for two ordinal level variables—Kendall's tau-b is not covered in the text), or **Spearman's rho** (used for two ordinal level variables reported as rank order). After selecting the appropriate correlation coefficient you will need to choose a two-tail (the system default) or one-tail test of significance. You may also have SPSS report the actual level of significance (the default), or report the results at the .05 and .01 alpha levels. Options include some choice of summary statistics, and two alternative ways of handling missing values. Select the "OK" button to complete the analysis. The output consists of a correlation matrix where every variable is correlated with every other variable.

To conduct a **partial correlation analysis** click on "Analyze" and move the cursor down to "Correlate" then click on "Partial." A dialog box will open allowing you to specify the desired variables for the correlation analysis. At least two variables must be highlighted and moved to the selected list by using the directional arrow, but more than two variables may be included in the analysis. The partial correlation dialog box also allows you to select one or more variables whose effects will be controlled. Highlight the desired control variables and move them to the selected list by using the directional arrow. The partial correlation analysis is based on the Pearson r, so the procedure requires variables measured at the interval or ratio level. A two-tail or one-tail test of significance may be selected, and the actual level of significance may be displayed or you may opt for a .01 and .05 level of significance. The "Options" button allows you to specify summary statistics and choose one of two methods of handling missing values. A typical situation where a partial correlation coefficient might be used is to examine the correlation between education and income, controlling for the number of years of work experience. Click on the "OK" button to complete the procedure.

The next statistical procedure of interest is the "**Regression**" option. To conduct a regression analysis select "Analyze" and then the "Regression" option. SPSS can conduct several types of regression analysis including logistic regression mentioned briefly in the text. The primary regression technique is the "Linear" regression option, and it is the one that will be discussed in detail here. Note that SPSS uses the term "Linear" regression to refer both to what the text calls "**linear regression**," and "**multiple regression**." Click on "Linear" and a dialog box opens allowing you to specify one variable as the dependent variable. Highlight the desired variable and use the directional arrow to move it to the selected box. Then select one or more variables to serve as independent or predictor variables by highlighting them and moving them to the selected box. Selecting a single independent variable results in what the text refers to as "linear regression" (one dependent variable and one independent variable), and selecting two or more independent variables results in what the text refers to as "multiple regression" (one dependent variable and two or more independent variables). A number of options may be applied to regression analysis, and only some of them will be discussed here because many of them go far beyond the discussion in the text. One option is to specify the method of entering the independent variables into the regression equation (this option assumes two or more independent variables). The default for the "Method" parameter is "enter" which means that all of the independent variables you specify will be used in the analysis. Other options include only entering independent variables that have a significant predictive effect on the dependent variable. The "Statistics" option allows you to request a variety of summary statistics in addition to the regression coefficients and R squared value. The "Plots" option allows you to specify a number of different scatter plots that can help you visualize how good the regression model is. For example, you might want to see a scatter plot of the actual dependent variable value and the predicted dependent values. The "Save" option allows you to save a number of values associated with the regression analysis including the predicted values of the dependent variable, the value of the "residuals" (the difference between the predicted values and the actual values), and several summary statistics. The "Options" button provides you with: an opportunity to set the criteria for including a potential independent variable into the equation (if you choose that method of entry), two alternative methods for handling missing data, and the choice to include the "constant" (the Y axis intercept) in the regression equation.

The final option under the "Analyze" procedure that we will examine in this appendix is "**Nonparametric Tests**." Nonparametric procedures encompass several techniques that are discussed in Chapter 13 of the text, along with several other techniques that are not covered in the text. I will only discuss the nonparametric procedures here that are included in the text. Each of the tests discussed here are found by selecting "Nonparametric Tests" and then the "Legacy Dialogs" option (depending on which version of SPSS you are using). The **chi-square test** was mentioned briefly when the crosstabs procedure was discussed as part of the options under "Descriptive Statistics." The chi-square test in the "Nonparametric Tests" is based on a similar concept, but tests the distribution of cases in a single variable to determine if the observed results are significantly different from what you would expect by chance. This particular application of the chi-square is seldom used, and not discussed in the text. Chapter 13 includes two nonparametric tests of significance: the **Mann-Whitney U test** for two independent samples, and the **Wilcoxon T** test for two related samples. Each is available in SPSS. To conduct a Mann-Whitney U test click on "Analyze" and move the cursor to "Nonparametric Tests" and then "Legacy Dialogs," then select the "2 Independent Samples" option and a dialog box will open. Highlight one or more dependent variables and move it or them to the selected list by selecting the appropriate directional arrow. Then highlight the "grouping variable" and move it to the selected box by using the appropriate directional arrow. The grouping variable is the variable that defines the two groups that will be compared. For example, the dependent variable might be income, and "sex" might serve as the grouping variable. The Mann-Whitney U test would then compare the distributions of income for significant differences for males and females. After the grouping variable has been selected, it is necessary to select the "Define Groups" option to identify the coding for the two groups. For example, if "sex" is the grouping variable, you will need to tell SPSS that group 1 is coded as "1" for male, and group 2 is coded as "2" for female. It is permissible to use a grouping variable with more than two categories, but only two groups may be compared at any one time. For example, "religion" might be the grouping variable, and group 1 might be coded as "1" for protestant, and group 2 might be coded as "3" for Jewish. An "Options" button will allow you to specify desired statistics, and to choose between two methods of handling missing data. Select on the "OK" button to complete the procedure.

To conduct a Wilcoxon T test select on "Analyze" and move the cursor to "Nonparametric Tests." Then select on the "2 Related Samples" option and a dialog box will open. Highlight the two variables to be compared and move them to the selected list by using the directional arrow. The "Options" button will allow you to specify desired statistics, and to choose between two methods of handling missing data just as above. Select on the "OK" button to complete the procedure. The typical application of the Wilcoxon T test is a comparison of pretest and posttest values for the same sample, or a comparison of values for a matched pairs situation as described in the text.

Graphs

The "Graphs" procedure on the text-based menu bar provides a number of options to generate graphics from SPSS. Bar graphs, histograms, frequency polygons (line graphs), and scatter plots are all available from the "Graphs" procedure. SPSS also includes a number of other graphing techniques that are not mentioned in the text such as pie charts, boxplots, and others. In this Appendix we will only examine the techniques mentioned in the text. Depending on which version of SPSS you are using you may generate graphs by using the "Chart Builder" process, or by selecting the "Legacy Dialogs" option. The instructions in this appendix assume you are using the "Graphs" and then "Legacy Dialogs" option. To **generate a bar graph** select on the "Graphs" procedure and then select "Legacy Dialogs" and then the "Bar" option. A dialog box will open giving you the choice of several styles of bar graphs including: simple, clustered, or stacked graphs. Simple bar graphs depict a single variable, whereas clustered and stacked bar graphs present two or more variables on the same graph. You select the type of graph you want by selecting on its representation in the dialog box, and then select the type of data to be presented in the graph. Choices include: summaries for groups of cases, summaries of separate variables, or values of individual cases. Next select on the "Define" button to select the variable to be graphed. For a simple bar graph highlight the variable to be graphed by selecting it, and then move it to the "Category Axis" box by using the directional arrow. The dialog box also provides you with several choices for what the bars on the graph represent including: the number of cases, the percent of cases, and so on. The "Titles" button will allow you to create multiline titles and footnotes, and a single line subtitle. The "Options" button will provide you with a choice on handling missing values, and some other display options. Select the "OK" button to complete the procedure and the graph will be created and placed on a "Chart Carousel" allowing you to preview or edit the chart prior to printing, saving, or pasting to another application program.

To create a **frequency polygon or line graph** select on the "Graphs" procedure then "Legacy Dialogs" and then select the "Line" option. A dialog box will open giving you the choice of several styles of line graphs including: simple, multiple, or drop-line. You may select the style of line graph you want, and then select the "Define" button to select the variable to be displayed. The dialog box that opens when you select on the "Define" button is the same as that for the bar graph. The only difference is that the data will be represented by a line rather than bars. Select the "OK" button to complete the procedure. To create an **ogive chart** simply choose "cumulative number of cases," or "cumulative percent of cases" as the format for the data.

To **create a histogram** select on the "Graphs" procedure and then select on the "Histogram" option. A dialog box will open allowing you to select the variable to be displayed. The "Titles" button will allow you to supply the same type of titles, footnotes, and subtitles as before. One interesting feature of the "Histogram" option is the opportunity to have the normal curve super imposed on the data. This is accomplished by selecting on the appropriate box in the "Histogram" dialog box. Select on the "OK" button to complete the procedure.

To **create a scatterplot** of two variables select on the "Graphs" procedure and then "Legacy Dialogs" and then select the "Scatter" option. A dialog box will open giving you the choice of several styles of scatterplots including: simple, matrix, overlay, or 3-D. You may select the style of scatterplot you want, and then select the "Define" button to select the variables to be displayed. A dialog box opens and allows you to highlight the "Y Axis" variable and then select it using the appropriate directional arrow. Then highlight the "X Axis" variable and select it by using the appropriate directional arrow. Remember that we conventionally assign the dependent variable to the Y axis, and the independent variable to the X axis. The dialog box allows you the option of selecting variables to serve as "Set Markers," and "Label Cases." The "Set Markers by" parameter allows you to select a variable whose values will be used as the symbols on the scatterplot. The "Label Cases by" parameter allows you to select a variable whose values will be used to label individual points on the scatterplot. The "Titles" and "Options" buttons operate as before with the previous graphs. Select on the "OK" button to complete the procedure.

Miscellaneous Utilities

In this section several useful options are discussed that may be found in a variety of locations on the SPSS Menu Tool Bar. In some previous versions of SPSS all of these functions were found under the "Utilities" option, but in the current version most of them have been incorporated into other menu choices. One function still located under "Utilities" is the "**Define Sets**" option that allows you to assign a name to a set or group of variables. The defined set name can then be used in some dialog boxes associated with other applications, which will save you the time and effort of naming the same group of variables over and over. For example, if you are analyzing the same three or four variables with a number of different procedures, you can define them as a set and use the set name. To define a set select "Utilities" and then on "Define Variable Sets." A dialog box will open allowing you to enter a set name, and to select the variables assigned to the set. Previously created sets may also be changed or deleted from the dialog box. To use a set select "Utilities" and then on "**Use Variable Sets.**" A dialog box will open allowing you to specify the set to be used.

"**Fonts**" is an option that may be found under the "View" tab. "Fonts" will allow you to select a font style and size different from the system default. To enter the fonts option select "View" and then select "Fonts." A dialog box will open with a number of alternative fonts. The **"Variables" option** is also found under the "View" tab and allows you to change the appearance of the SPSS worksheet from "data view" where the variable names appear across the first row of the worksheet and the data appear for each variable appear in the appropriate column to the "variable view" in which each row presents information on each variable, but the actual data do not appear. The same effect can be realized by using Control T to toggle back and forth from data view to variable view, or even more simply by selecting the data view or variable view tab at the bottom left of the SPSS worksheet.

Two additional options under the "View" tab affect the appearance and function of the Data Editor Window. To add or remove the grid lines separating cases and variables in the Data Editor select "View" and then **"Grid Lines."** The grid lines will appear when a checkmark is next to the "Grid Lines" command. Selecting on "Grid Lines" toggles the feature on and off. The SPSS default is to use grid lines. Similarly, to display assigned value labels rather than their numeric values in the Data Editor Window select "View" and then **"Value Labels."** Selecting on "Value Labels" toggles the feature on and off. When "Value Labels" is selected the assigned label will appear in the SPSS worksheet rather than the associated numeric value that was originally entered. The SPSS default is to use the numeric values.

In addition to opening and saving files, the "File" tab has two functions that might be of use. The first is the **"Display File Information"** option that is accessed by selecting "File" and then "Display File Information." The option will display a list of the variables in the file, and some information on the file formatting. The option may be used for the existing open file, or any other file that you can access on your hard drive or portable storage media. The second option under the "File" tab is **"Set Viewer Output Options,"** which allows you to set a number of parameters including: an output page header or footer; portrait or landscape orientation; page margins; starting point for page numbering; and, inter-item spacing for output.

The final procedure on the text-based menu bar is **"Help."** The "Help" feature is common to almost all windows-based applications, and provides information through a table of contents or a separate search feature. "Help" is also the path to the SPSS tutorial if that feature has been loaded on your computer (some systems may not have the tutorial available due to space limitations).

Summary

Appendix D provides you with a brief introduction to the SPSS software package for Windows, but it is not intended to be a substitute for the User's manual, the online SPSS tutorial, or the help features available on the menu bar or within each procedure. Hopefully, it will provide you with the information necessary to conduct commonly used procedures in SPSS, and will be a useful reference to complete the computer application exercises that are found at the end of most of the chapters in the text.

Glossary

a See Y axis intercept

A Normal Distribution Any distribution wit some mean (X), and some standard deviation (s), and whose raw scores are distributed around their mean in the same proportion as the normal distribution.

Accidental Sampling A nonprobability sampling plan in which any available members of the population are selected for the sample. Accidental sampling is also referred to as "haphazard sampling."

Alpha Level See level of significance.

Analysis of Variance (ANOVA) A procedure used to examine a group of means to determine if any of the pairs of means are statistically significantly different.

Asymmetrical Measure of Association A statistic indicating the amount of variance one variable can explain in another variable, usually the amount of variance in the dependent variable Y that is explained by the independent variable X.

Attributes The actual level or state observed for a member of the sample for a particular variable.

b See Regression coefficient

Bar Graph A common graphing technique in which the height of bars displayed on an axis represent the number of frequencies in the category being represented by the bar.

Beta The name for the regression coefficient when standardized regression is used. Beta indicates the size of the change in standard deviation for the dependent variable for each unit change in standard deviation in the independent variable.

Bimodal Distribution A distribution of scores with two major concentrations of scores called modes.

Binomial Experiment A probability experiment involving a fixed number of repeated, independent trials which can result in only two outcomes: success or failure.

Binomial Probability Distribution The probability distribution based on binomial experiments involving a fixed number of independent trials, each of which will result in one of only two possible outcomes, success or failure.

Captive Sampling A nonprobability sampling plan in which members of the population who are already selected for some other purpose are selected for the sample.

Cartesian Coordinate System A set of two axes set at right angles to each other with the horizontal axis representing an independent variable, and the vertical axis representing the dependent variable.

Census A research study in which data are collected from all members of the population.

Central Limit Theorem A mathematical theorem describing the nature of the sampling distribution of the mean. As the sample size "n" increases, the sampling distribution of all possible samples of size "n" becomes normal in form, with a mean ($\mu_{\bar{x}}$) equal to the population mean (μ) from which the samples were drawn, and a standard error ($\sigma_{\bar{x}}$) equal to the population standard deviation (σ) divided by the square root of the sample size.

Chi Square Test A nonparametric statistical procedure used to determine if the pattern of observed results in a contingency table is statistically significantly different from what would be expected by chance.

Cluster Sampling A probability sampling plan often used for large populations. Areas or groups of individuals are selected, perhaps in several stages, until a small enough unit has been selected that will allow selection of individual units.

Coefficient of Contingency A measure of association indicating the strength of the relationship between two variables based on the observed value of the chi square statistic. The coefficient of contingency may be applied on contingency tables of any size.

Coefficient of Determination r^2 The coefficient of determination r^2 indicates the amount of variation in the dependent variable Y that is explained by the independent variable X. The value of r^2 is the chief way of assessing the quality of a linear regression model.

Compound Probability The probability based on satisfying two sets of outcomes with a single event. Compound events include: P(E or F); P(E and F); and P(E|F).

Concepts General level, abstract aspects subject to study in a research effort. Concepts serve as "mental handles," and help us focus on the major research questions which we will investigate.

Confidence Interval The estimation of a population parameter by creating a range of values on either side of a sample statistic. The size of the range is determined by our degree of confidence that it contains the population parameter being estimated.

Contingency Table A method of presenting the relationship between two categorical variables by cross tabulating one variable with the other. The logic behind the construction of the contingency table is similar to that of the Cartesian coordinate system for continuous variables.

Continuous Measurement Continuous measurement occurs on a scale with an infinite number of points. A measured variable may logically take on any value along the scale.

Control Group The group in an experimental design exposed to a placebo rather than the stimulus.

Control Variable A variable presumed to have an effect on the apparent relationship between two other variables. The control variable is conventionally designated as "Z."

Correlation Coefficient The general name for measures of association which indicate the strength of the relationship between two variables.

Critical Region The area in the distribution associated with a test statistic corresponding to the level of significance chosen for a statistical test.

Critical Value The value of the test statistic marking the beginning of the critical region. The observed value of the test statistic must equal or exceed the critical value in order to reject the null hypothesis.

Data Reduction A group of procedures used to put raw data into a form that is more easily interpreted. Common types of data reduction include frequency distributions, and various types of graphic displays.

Degrees of Freedom The number of values in a sample that are free to take on any value and still represent an unbiased estimate of a population parameter. In a contingency table, degrees of freedom are the number of cells that are free to take on any value before the remaining cells are determined given the marginal totals.

Dependent Variable A variable whose value is presumed to be determined by some other variable (the independent variable). The dependent variable is conventionally designated as "Y."

Descriptive Statistics A set of statistical procedures that describe a set of data with no attempt to generalize the results to any other group.

Deviation from the Mean A measure of each score's distance from the mean. The sum of the deviations from the mean for a set of scores is always zero.

Dichotomous Variable A variable that has only two possible values.

Discrete Measurement Discrete measurement occurs on a scale with separate categories. A measured variable may logically fit in any category along the scale, but not between categories.

Eta Squared A descriptive statistic indicating the amount of variance in the dependent variable explained by the independent variable in Analysis of Variance.

Event An outcome from a probability experiment to which we can assign a probability.

Experimental Group The group in an experimental design exposed to the stimulus or independent variable.

F Distribution The distribution for the F statistic which is used in Analysis of Variance. A statistically significant F value indicates that at least one of the possible pairs of means is statistically significant.

F Statistic The test statistic for Analysis of Variance. F is equal to the Mean Square Between divided by the Mean Square Within.

Factor A term for the independent variable in an Analysis of Variance Procedure

Fisher's Protected t-Test An ANOVA post hoc test appropriate for groups of differing or equal sample sizes.

Frequency Distribution A method of data reduction in which raw data are organized into ordered categories.

Frequency Polygon A common graphing technique in which a line is used to connect points representing the number of frequencies at each point in a distribution.

Goodman's and Kruskal's Gamma A measure of association for two ordinal level variables. Gamma is a symmetrical measure of association, and has a PRE interpretation.

Grand Mean A mean calculated from a set of means rather than from a set of individual scores.

Guttman's Coefficient of Predictability, Lambda A measure of association for two nominal level variables. Lambda has a PRE interpretation, and can be computed as a symmetrical measure of association (mutual predictability between the two variables), or as an asymmetrical measure of association (one variable used to predict the other).

Histogram A graphing technique similar to a bar graph except that the bars in a histogram are contiguous representing the continuous nature of the underlying variable. Histograms are appropriate for data of the ordinal, interval, or ratio level.

Homoscedasticity A characteristic of linear regression where the actual Y values are normally distributed around the predicted Y' values.

Hypothesis A hypothesis is a statement of relationship between two variables. A hypothesis may also be stated for a single variable.

Independent Events The outcome of which event is not affected by the outcome of a prior event. Independence of events is often achieved by sampling with replacement.

Independent Variable A variable whose value is presumed to determine or cause some response in another variable (the dependent variable). The independent variable is conventionally designated as "X."

Inferential Statistics Statistical procedures designed to allow us to make generalizations (inferences) regarding population parameters based on an analysis of sample statistics.

Interquartile Range The distance from the 75th percentile to the 25th percentile in a distribution.

Interval Estimation See Confidence Interval

Interval Level Data The data measured on a continuous scale with equal appearing intervals, but without an absolute zero point. The difference between two intervals at any point along the scale represents the same amount of change in the phenomenon being measured.

Interval Midpoint The center point of an interval in a frequency distribution.

Judgmental Sampling A nonprobability sampling plan in which members of the population are selected for the sample bases on special characteristics or expertise they might possess.

Level of Significance (Alpha Level) The amount of error one is prepared to accept in hypothesis testing when reaching a decision. The level of significance is the probability of rejecting the null hypothesis when the correct decision should be to fail to reject the null hypothesis.

Linear Regression A statistical procedure in which a single independent variable (Y) is used to predict a dependent variable (X).

Linear Regression Equation: $Y' = bX + a$ An equation for a straight line resulting from linear regression analysis used to describe the relationship between a single independent variable and the dependent variable.

Logistic Regression A type of nonlinear regression analysis. Logistic regression is often used for dependent variables measured at the ordinal level.

Lower Limit The actual beginning point of an interval in a frequency distribution as opposed to the apparent limit which is the point where an interval appears to begin.

Mann-Whitney U Test A nonparametric test of significance for two independent samples based on the difference of the sums of ranks associated with each sample.

Mean A measure of central tendency consisting of the arithmetic average of a set of scores.

Mean Square The value resulting from dividing the Sum of Squares by its associated degrees of freedom. Mean Square values are computed for Between Groups and Within Groups.

Mean Square Between Groups A measure of the amount of the total variation in a set of scores due to group membership.

Mean Square Within Groups A measure of amount of the total variation in a set of scores due to random variation.

Median A measure of central tendency indicating the midpoint in a distribution of scores. Half of the scores will be above the median and half of the scores will be below the median.

Minimizing the Sum of Squares A linear regression line is said to provide the best fit to the data if it minimizes the sum of squares (the sum of the squared deviations from the actual Y values to the predicted Y values along the regression line).

Mode The simplest measure of central tendency representing the most frequently observed value of a variable.

Multiple R Squared The amount of variation explained by all of the independent variables included in a multiple regression analysis.

Multiple Regression A type of regression analysis when several independent variables are used to predict a single dependent variable.

Negative Relationship A relationship between two variables in which the variables move in opposite directions.

Nominal Level Data The most basic type of measurement possible allowing us to simply categorized our observations. Nominal level data are measured at the discrete level with separate categories with no underlying order.

Nonprobability Sampling A group of sampling methods in which the sample is not selected on a random basis. Major methods of nonprobability sampling include: accidental sampling, captive sampling, quota sampling, spatial sampling, systematic sampling, judgmental sampling, and snowball sampling.

Nonparametric Statistics A group of inferential statistics requiring less rigorous assumptions. Nonparametric statistics usually involve measurement at the nominal or ordinal level and do not require normality.

Null Hypothesis (H_0) A statement of no relationship between variables crafted to state the opposite of what the researcher expects to find. The Null Hypothesis is subjected to statistical testing.

Observed Test Statistic The computed value of a statistical test which is then compared to a critical value to determine if a statistically significant result is present.

Ogive Chart A graphing technique that is used to display cumulative frequencies rather than simple frequencies.

Omega Squared A parametric statistic in Analysis of Variance indicating the amount of variance in the dependent variable explained by the independent variable for the population from which the sample under investigation was drawn.

One-Tail Test A test of the hypothesis used when the research hypothesis states a direction (a positive relationship, or a negative relationship) for the relationship between variables.

Open Interval An interval without an upper or lower limit found at the top or bottom of a frequency distribution.

Ordinal Level Data The data measured at the discrete level with separate categories and an underlying rank order among the categories.

Outliers Observed values of a measured variable that lie outside the typical range of most of the other scores. Outliers are sometimes removed from the data prior to analysis.

P(E and F) A type of compound probability satisfied when any event common to both E and F occurs.

P(E or F) A type of compound probability satisfied if any event in E occurs and event in F occurs or any event common to both E and F occurs.

P(E|F) A type of compound probability equal to the probability of E being satisfied given the fact that F has been satisfied.

Parameters Characteristics of a population.

Parametric Statistics A group of inferential statistics requiring rigorous assumptions regarding the measurement of the dependent variable. The two most common assumptions are measurement in a continuous fashion at the interval or ratio level, and a normal distribution of the variable around its mean.

Partial Correlation Coefficient A measure of association based on the Pearson r which provides evidence of an association between two variables while controlling for the possible effects of a third variable.

Pearson Product Moment Correlation Coefficient A measure of association for two variables measured at or above the interval level. The Pearson r is the most widely used measure of association in behavioral science research.

Pearson r See the Pearson Product Moment Correlation Coefficient.

Percentile A measure of the percentage of a distribution equal to or below a given point. The 75th percentile is the score in a distribution with 75 percent of all scores in the distribution at or below it.

Phi Coefficient A measure of association indicating the strength of the relationship between two variables based on the observed value of the chi square statistic. The phi coefficient may only be applied on 2X2 tables, and has a proportional reduction in error (PRE), or variance explained interpretation.

Placebo A treatment given to the control group in an experimental design intended to mimic the appearance of the stimulus or independent variable.

Point Estimate The estimation of a population parameter by using the observed value of a sample statistic.

Point-Biserial r_{pb} A measure of association appropriate for a dependent variable "Y" measured at the interval or ratio level, and an independent variable "X" which is expressed as a dichotomy (only two values are possible).

Population The total group to which a research project refers. (Also called the Universe)

Positive Relationship A relationship between two variables in which the variables move in the same direction.

Post Hoc Test One of several procedures used to uncover all of the statistically significantly different pairs of means in an ANOVA procedure.

Power of a Test The power of a statistical test is the probability of rejecting the null hypothesis when the correct decision is to reject the null hypothesis.

PRE See Proportional Reduction in Error

Probability Sampling A group of sampling methods in which all members of the population have a chance of being selected for the sample. Major methods of probability sampling include: simple random sampling, stratified random sampling, systematic sampling, and cluster sampling.

Proportional Reduction in Error A characteristic of some measures of association allowing an estimate of the amount of variance in the dependent variable explained by the independent variable.

Proposition A proposition is a statement of relationship between two concepts. The proposition expresses our expectation of how we think two concepts are related.

Quota Sampling A nonprobability sampling plan in which members of the population are selected based on certain characteristics, but are not considered representative of the population. Quota sampling is the nonprobability equivalent to stratified random probability sampling.

r^2 The square of the Pearson r indicates the amount of variation in the dependent variable "Y" that can be explained by the independent variable "X."

Random Number Table A table of random numbers used for some types of probability sampling plans, or to assign members of a sample to various treatment groups.

Range The range is the distance from the smallest score in a distribution to the largest score. It is one of the simplest measures of variation.

Ratio Level Data The data measured on a continuous scale with equal appearing intervals and an absolute zero point. Data measured at the ratio level allow us to make ratio type statements of the form: A is twice as much as B.

Reducing Variance See Variance Explained

Regression Equation See Linear Regression Equation

Related Events The outcome of which events is related to the probability of some other event.

Research Hypothesis (H_1) A hypothesis stating the researcher's expectation regarding the relationship between variables.

Robustness The ability of some parametric statistics to provide valid results when some of the assumptions associated with the procedure are violated.

Sample A subgroup of the population that is selected in some fashion for direct observation and measurement.

Sample Space The set of all possible outcomes for a probability experiment.

Sampling The process of selecting elements from a population.

Sampling Distribution The set of all possible samples of size "n" taken from a particular population.

Sampling Population The population from which a sample will actually be drawn.

Sampling Unit The actual element drawn from a population in order to obtain a sample.

Scales of Data The four levels at which data may be measured: nominal, ordinal, interval, and ratio.

Scatterplot The result of plotting variables on a Cartesian Coordinate system. In the case of two variables, the dependent variable (Y) is plotted on the vertical axis, and the independent variable (X) is plotted on the horizontal axis.

Semi-Interquartile Range The interquartile range divided by 2.

Simple Probability The probabilities based on a single set of outcomes.

Simple Random Sampling A probability sampling plan in which all members of the population have an equal chance of being selected for the sample.

Skewed Distribution An asymmetrical distribution in which a large concentration of scores is found at one end of the distribution.

Snowball Sampling A sampling plan in which earlier members selected for the sample lead to additional members.

Spatial Sampling A nonprobabilility sampling plan in which members of the research team select members for the sample as they move systematically through a large area.

Spearman r_s A measure of association for two variables measured at the ordinal level and expressed as rank values.

Spurious Relationship A situation where a relationship appears to exist between two variables when none actually exists. The relationship appears to exist due to the relationship each of the observed variables has to a third variable.

Standard Deviation A measure of variation computed by taking the square root of the variance.

Standard Error The standard deviation of a sampling distribution. The standard error ($\sigma_{\bar{x}}$) is equal to the population standard deviation (σ) divided by the square root of the sample size.

Standard Error of the Estimate A measure of the variation of the predicted values of Y' compared to the actual values of Y in a linear regression analysis. The size of the standard error helps assess the quality of the regression model.

Standardized Regression A type of linear regression performed by converting the raw scores to normal scores prior to the regression analysis. One result is that the best fit regression line will always pass through the origin.

Statistical Significance A result indicating that a difference observed in sample data is sufficiently large to suggest that a similar difference exits in the respective population parameters.

Statistics Characteristics of a particular sample taken from a population. Since many samples are possible, statistics represent estimates of a particular population parameter.

Stimulus The name often given to the independent variable in an experimental design. Exposure to the stimulus is expected to be the cause of some response in the level of the dependent variable.

Stratified Random Sampling A probability sampling plan in which a series of random samples are drawn from subgroups of the population.

Sum of Squares The sum of the squared deviations from the mean used in Analysis of Variance. Sum of Squares are computed for the total sample, and then partitioned into that due to group membership (Sum of Squares Between Groups), and that due to random error (Sum of Squares Within Groups).

Sum of Squares Between Groups The portion of the total sum of squares due to group membership. Sum of squares between groups represents explained variance.

Sum of Squares Within Groups The portion of the total sum of squares that is due to random variation. The sum of squares within groups represents unexplained variance.

Survey A research study in which data are collected from the members of a sample drawn from a population.

Symmetrical Measure of Association A statistic indicating the amount of mutual predictability between two variables; i.e. the amount of variance in the dependent variable Y that is explained by the independent variable X, and the amount of variance in the independent variable X that is explained by the dependent variable Y.

Systematic Sampling A sampling plan in which members of the sample are systematically selected from a list, or as they move by. Systematic sampling may be considered both a probability sampling plan, or a non-probability sampling plan depending on its execution.

t Distribution A set of sampling distributions based on small sample sizes ($n < 30$) when the population standard deviation (σ) is unknown and must be estimated by the sample standard deviation (s).

t-Test A parametric statistical test used when the population standard deviation is unknown and the sample size is small to evaluate a sample mean for a statistically significant difference from a population mean. Versions also exist to evaluate two independent sample means or two related sample means.

Test Statistic The value from a statistical test used to test the null hypothesis.

The Normal Distribution A theoretical distribution with a mean $\mu = 0.00$, and a standard deviation $\sigma = 1.00$ with the scores distributed symmetrically around a center point where the mean, mode, and median of the distribution are found. The normal distribution resembles the shape of a bell-shaped curve.

Treatment A given level associated with the independent variable in an Analysis of Variance procedure. The number of treatment levels is equal to the number of group means (k) being compared.

Tukey's HSD Test An ANOVA post-hoc test appropriate when all groups have the same sample size.

Two-Tail Test A test of the null hypothesis when the research hypothesis only states an expectation of a difference, but no statement of the direction of the difference.

Type I Error The error of rejecting the null hypothesis when the correct decision is to fail to reject the null hypothesis. Type I error is equal to the level of significance chosen for the hypothesis test. Type I error is often referred to as alpha error.

Type II Error The error of failing to reject the null hypothesis when the correct decision is to reject the null hypothesis. Type II error is often referred to as beta error.

Upper Limit The actual ending point of an interval in a frequency distribution as opposed to the apparent limit which is the point where an interval appears to end.

Variables Measurable indicators of concepts.

Variance A measure of variation equal to the average of the squared deviations from the mean for a set of scores. The smaller the variance the more closely the scores of a distribution are to their mean.

Variance Between Groups The variance in a sample due to group membership.

Variance Explained A measure of the reduction in error when predicting a dependent variable by making use of information obtained from an independent variable as opposed to predicting the dependent variable by guessing the mean. The amount of variance that can be explained in the dependent variable "Y" by the independent variable "X."

Variance Within Groups The variance in a sample due to random error.

Wilcoxon T Test A nonparametric test of significance for two related samples. The test is based on the sum of positive and negative ranks associated with the difference in pretest and posttest scores for a sample.

Yate's Correction for Continuity A method of adjusting the computation of a Chi square for 2 X 2 tables when the sample size is small (usually between 25 to 75).

Z Scores The distance of which scores from the mean is expressed in units of standard deviation.

Z Test A parametric statistical test used when the population standard deviation is known or when the sample size is large ($n > 30$) used to evaluate a sample mean for a statistically significant difference from a population mean. Versions also exist to evaluate two independent sample means, and two proportions for statistically significant differences.

Index

A

accidental sampling, 24, 38
accuracy of political polls, 159–160
Add Cases option, SPSS, 392
Add Variables option, SPSS, 392
alpha level, 168, 179–180, 182, 251, 256, 279
analysis of variance (ANOVA), 283–284, 300
  assumptions of, 286–287
  computational example (equal sample size), 288–291
  computational example (unequal sample size), 294–296
  one-way analysis of variance, 284–286
  summary table, 290–291
  two-way, 299
Analyze command, SPSS, 394–399
A normal distribution, 113, 114, 117, 122–128, 139
  comparing scores from one to another, 125–126
ANOVA. *See* analysis of variance
asymmetrical measure of association, 321–322, 333
attribute, 7–8, 18
Automatic Recode option, SPSS, 394

B

b (regression coefficient), 225–229
bar graph, 60–64, 67
behavioral science research, 97
beta (standardized regression coefficient), 236–237
bimodal, 73, 85
bimodal distribution, 65, 68
binomial probability
  applying, 153–160
  formula for, 154, 156
  properties of binomial experiment, 153
bivariate data plots, 189–193
  cutting points added, 306
  of education and income, 304
Bivariate option, SPSS, 397

C

captive sampling, 24, 38
Cartesian coordinate system, 169, 189, 216, 304–305, 307
"CAT" variable, 66–67
census, 6, 22
census block, 29
central limit theorem, 32, 38
central tendency, measures of, 71–85

Charts option, 395
chi square, statistical table, 359
chi-square test, 399
chi square test for independence (χ^2), 311, 333
  assumptions of, 311, 317
  computational example, 312
  computing expected results, 312–314
  critical value, 315–317
  formula, 314–315
  Yates' correction for continuity, 317–318
cluster sampling, 28–29
coefficient of contingency (C), 304, 318–319, 333
coefficient of determination (r^2), 229–232, 241, 243
Compare Means, 396–397
compound probability, 149–153, 161
  applying, 152–153
computational formula
  for standard deviation, 99–100
  for variance, 95–96
Compute option, SPSS, 393
concepts, 4–5, 7, 8, 18
confidence interval, 133–136, 140
  formula for, 135
contingency table, 191, 333
  construction and presentation of data, 304–308
  degrees of freedom, 309–311
  difficult to interpret table, 308
  measures of association for, 318–327
  percent down and compare across, 308–309
continuous measurement, 12, 19
continuous scale, 12, 19
controlling for a third variable, 213
Correlate option, SPSS, 397
correlation coefficients, 172, 187–188
  choosing proper, 188–189
  guidelines for interpretation, 200–201
count option, SPSS, 393
critical region, 168, 182, 250, 251, 279
critical value, 168, 176, 182, 249, 279
Crosstabs option, SPSS, 395
cumulative frequency, 45, 46
curvilinear relationships, 192–193

D

data
  graphic display of, 60
  graphic patterns in, 64–66
Data command, SPSS, 392–393
Data Editor window, SPSS, 389–390
data reduction, 44, 68
deciles, 52–54
  percentiles and quartiles relationship, 54
Define Dates option, SPSS, 392
Define Sets option, SPSS, 400
degrees of freedom, 97, 107, 257
  in contingency table, 309–311
  between groups, 286
  within groups, 286
  "n – 1," 97–98
  total, 286
dependent variables, 9, 19
descriptive statistics, 10, 19, 23
Descriptive Statistics option, SPSS, 395–396
deviation from the mean, 88, 99, 107
discrete measurement, 12, 19
discrete scale, 12, 19
dissemination of results, research process, 6–7

E

Edit command, SPSS, 391–392
education
  data on, 201
  scatterplot of, 190, 202
equation for a straight line, 222–223
error
  Type II, 178–180, 182
  Type l, 178–180, 182, 284
estimation, 134–135
  interval estimation, 134–135
  point estimation, 134–135, 140
Eta Squared (η^2), 293–294, 298–300
  computational formula, 293
event, 145, 161
  independent events, 148, 161
  related events, 148, 161
expected results, 311, 314
  computing expected results, 312–314
explained variance, 232–233, 244
Explore option, SPSS, 395

F

F
  computing, 290
  critical value, 290–291
  distribution, 290, 300
  ratio, 286
  statistic, 286, 346–354
factor, 284, 286, 300
File, SPSS, 390–391
Fisher's protected t-test, 287, 291, 296–300
  computational formula, 296–297
Fonts option, SPSS, 400–401
frequency, 45

frequency distribution, 44, 67, 68
  advantages of, 45–46
  construction of, 44–49
  simple, 45–46
  standard deviation for, 101–103
frequency polygon, 60–64, 67

G

Gamma. *See* Goodman's and Kruskal's Gamma (γ)
Gauss Karl Friedrich, 115
General Social Survey (GSS), 17–18, 332
Goodman's and Kruskal's Gamma (γ), 304, 318, 322, 333
  computing, 326–327
grade point average (GPA), 7–8, 83, 84
grand mean, 83–85
graphic display of data, 60
Graphs procedure, SPSS, 399–400
Grid Lines, SPSS, 401
GSS. *See* General Social Survey
Guttman's coefficient of predictability, lambda (λ), 304, 318, 320–322, 333
  computing asymmetrical, 326
  computing symmetrical, 325–326

H

Help menu, SPSS, 401
histogram, 60–64, 67
homoscedasticity, 238, 243
hypothesis, 5, 9, 18, 169–170, 172
  null hypothesis, 248, 251, 258, 279
  reject the null hypothesis, 252
  research hypothesis, 173–174, 182, 248, 249, 251, 258, 279
hypothesis testing, 160, 163
  error in, 178–180
  null hypothesis, 173–175, 182
  one-tail, 170–171, 182, 197–198, 217, 249–250, 258, 264
  review of logic, 248–250
  steps in, 174–175, 249–250
  two-tail, 170–171, 182, 197, 199, 217, 249–250, 258
hypothesis tests for means, 247–279. *See also* t-test; *Z* test
hypothetical data file, 8
hypothetical ogive charts, 64
hypothetical regression line, 224

I

income
  data on, 201
  scatterplot of, 190, 202
independent events, 148, 161
Independent Samples T Test, 396
independent variables, 9, 19
inferential statistics, 10, 19, 23, 163, 182
intelligence scores (IQ), 125
  test for, 7, 8
interpretation, 6
interquartile range (IQR), 88, 90–91, 107
interval level data, 13, 19, 188. *See also* levels of measurement
interval midpoints, 50, 67
IQ. *See* intelligence scores
IQR. *See* interquartile range

J

judgmental sampling, 25, 38

L

lambda. *See* Guttman's Coefficient of Predictability, Lambda (λ)
least squares, 224
level of education, 169–170, 172
level of income, 170, 172
level of significance, 168, 182, 198, 216, 251, 255, 256
levels of measurement, 12–14, 188
limits
  apparent limits, 51
  lower limits, 50, 52, 67
  real limits, 51
  upper limits, 50, 52, 67
linear regression, 221
  assumptions of, 238–239, 243
  point of, 228–229
  quality of regression model, 229–236
linear regression equation. *See* regression equation
linear regression model. *See* regression model, quality of
linear relationship, 223–225
line graph, 61
List Cases option, SPSS, 394–395
logical formula for standard deviation, 99
logistic regression, 239, 243
logit regression. *See* logistic regression
lottery odds, 36–37
lower limits, 50, 52, 67

M

Mann-Whitney *U* test, 327–328, 333, 399
  application, 329–330
  independent samples, 328–329
  statistical table, 360–367

MANOVA. *See* multivariate analysis of variance
marginal totals, 312
margin for error. *See* confidence interval
mean, 71, 79–82, 85
  compared with median and mode, 83
  frequency distribution computation, 80–82
  raw data computation, 79
mean square, 286, 300
  mean square between groups, 286
  mean square within groups, 286
measurement, 11–12
measures of association, 193, 216
measures of central tendency, 71–85
measures of variation, 87, 88
median, 71, 74–79, 85
  frequency distribution computation, 76–79
  raw data computation, 74–75
Merge Files option, SPSS, 392
midpoints, 50–52
mode, 65, 66, 71–74, 85
multiple regression, 239, 243
  conceptual example, 240–241
multiple *R* squared, 241, 243
multivariate analysis of variance (MANOVA), 299

N

negative, income, 236
negative relationship, 169–171, 189–193, 216, 307–308
nominal level data, 13–14, 19, 188
nonparametric statistics, 303–304
nonparametric tests, 398–399
  of significance, 327–332
nonprobability samples, 23
nonprobability sampling methods, 24, 37
  accidental sampling, 24, 38
  captive sampling, 24, 38
  judgmental sampling, 25, 38
  quota sampling, 24–25, 38
  snowball sampling, 25, 38
  spatial sampling, 25, 38
  systematic sampling, 25, 38
normal distribution, 65, 66, 68, 99, 113–140. *See also* a normal distribution; the normal distribution
normal in form, 99, 107
null hypothesis, 173–175, 182, 248, 251, 258, 279
  testing, 197–199

O

observed results, 311, 314, 321
observed test statistic, 168, 250, 255–256, 259
ogive chart, 60, 63, 67
Omega Squared (ω^2), 293–294, 298–300
  computational formula for, 293–294
One Sample T Test, 396
one-tail test, hypothesis testing, 170–171, 182, 197–198, 217, 249–250, 258, 264, 330, 331
One Way ANOVA, 397
open interval, 52, 68
ordinal level data, 13, 19, 188
outliers, 237, 243
Output Window, SPSS, 390

P

Paired Samples T Test, 397
parameters, 10, 19, 30, 180
parametric statistics, 303
partial correlation analysis, 398
partial correlation coefficient, 193, 213–216
  computer applications, 215–243
  formula, 214–215
Pascal, Blaise, 144
Pearson correlation coefficient *(r),* 187, 193–203, 216, 227, 230, 237
  computational example, 195–203
  computation and interpretation, r^2, 199–200
  example using education and income, 201–203
  formula, 194–195
  logic of, 194–195
  statistical tables, 344
  testing for statistical significance, 196–199
Pearson Product Moment Correlation Coefficient, 188
Pearson *r,* 188
  SPSS, 397
percentage, 52–54
"percent down, compare across" rule, 308–309, 324
percentiles, 52–54, 67
  equal to a given score, 57–59
  relationship between deciles, and quartiles, 54
  score equal to a given percentile, 54–56
  steps in finding, 56–57
phi coefficient (φ), 304, 318–320, 333
point-biserial correlation coefficient (r_{pb}), 187, 203–207, 216
  computational example, 205
  formula, 204–205
  logic of, 203–204
  sex and physical dexterity relationship measurement, 207
  testing for statistical significance, 206
population, 23, 37
population mean, 97
population of interest, 4, 11, 18, 21, 37
positive relationship, 169–171, 189–193, 216, 307

post hoc procedure, 286–287
post hoc test, 291–300
power of a statistical test, 179–180, 182
predicted annual sales, 242
prediction error, variance as, 103–107
probability, 143–182. *See also* binomial probability; compound probability; simple probability
  origins of probability theory, 144–145
probability samples, 23
probability sampling methods, 25, 37
  cluster sampling, 28–29
  simple random sampling, 25–26
  stratified random sampling, 27–28
  systematic sampling, 28–29, 38
proportions, 52–54
propositions, 4

Q

quartiles, 52–54
  relationship between percentiles, deciles and, 54
quota sampling, 24–25, 38

R

R, 241, 243
r^2, 230, 231, 241, 243
random number table, 26–27, 38
range, 88–90
ratio level data, 12–13, 19, 188
raw scores
  converting from one normal distribution to another, 126–127
  converting to a z score, 123–125
recode option, SPSS, 393
reducing variance, 107, 108. *See also* variance explained
regression coefficient (b), 225–229
  alternative formula for, 237–238
  formula for, 225–229
regression equation, 225
regression model, quality of
  coefficient of determination, 229–231
  explained and unexplained variance, 232–233
  predict income, 232–236
  standard error of estimate, conceptual formula, 232
  standard error of the estimate, 231
regression option, 398
reject the null hypothesis, 252
related events, 148, 161
replace missing values option, SPSS, 394
Reports procedure, SPSS, 394–395
research hypothesis, 173–174, 182, 248, 249, 251, 258, 279
research method, 6
research process, 4–7
  role of statistics in, 4
research strategies, 171–172
robustness, 14, 328
R Squared, 241, 243

S

sample, 5–6, 9, 11, 18, 21, 37
  good sample *vs.* bad sample, 130–132
sample size, large and small sample size, 256–257
sample space, 145, 161
sample statistics, 10
sampling, 21, 22, 37
  strategies, 23
sampling distribution, 21, 29–34, 37, 114, 128–134, 136–139, 160, 164–169
sampling population, 23, 37
sampling unit, 23–24, 38
scales of data, 12–14, 19. *See also* levels of measurement
scatterplot, 189–193
score's percentile rank, 53
semi-interquartile range (SIQR), 88, 91–92, 107
Set Viewer Output Options, SPSS, 401
"shotgun" pattern, 191, 192
simple frequency distribution, 44
simple probability, 146–149, 161
  rules for, 147–148
simple random sampling, 25–26
SIQR. *See* semi-interquartile range
skewed distribution, 65, 67, 68
slope, 223
snowball sampling, 25, 38
spatial sampling, 25, 38
Spearman rank order correlation coefficient (r_s), 187, 208–213, 216
  computational example, 209–210
  formula, 208–209
  logic of, 208
  statistical tables, 345
  testing for statistical significance, 210
  with tied ranks, 210–213
Spearman's rho, SPSS, 397
SPSS. *See* Statistical Package for the Social Sciences
spurious relationship, 193
squared deviation, 93
standard deviation, 92, 98–101, 107, 113, 127, 128, 134, 248, 258
  for frequency distribution, 101–103
standard error, 97, 99, 128, 129, 139, 164, 248

standard error of the difference, 262–263
for a proportion, 268–269
standard error of the estimate, 229, 231, 243
conceptual formula for, 232
standardized linear regression, 228, 236–237, 243
standardized regression coefficient, 236
Standard Normal Distribution, Pearson correlation coefficient *(r),* 194
statistical analysis, 6
basic terms in, 7–11
Statistical Package for the Social Sciences (SPSS), 17–18, 37
Analyze command, 394–399
basics, 389
Data command, 392–393
Data Editor window, 389–390
Edit command, 391–392
File, 390–391
Graphs procedure, 399–400
Output Window, 390
Tool Bar, 390
Transform command, 393–394
Utilities, 400–401
statistical significance, 114, 136–137, 140, 176, 180, 182, 217, 247, 279
finding of, 253–254
meaning of, 181
what statistical significance does not mean, 181, 254
statistical symbols, 15–17
statistical tables
chi square, 359
F statistic, 346–354
Mann-Whitney *U,* 360–367
of Pearson correlation coefficient, 344
proportional areas under standard normal curve, *Z* tables, 340–342
Spearman rank-order correlation coefficient, 345
studentized range statistic, q_k, 355–358
of *t* for two-tail and one-tail tests, 343
Wilcoxon *T,* 368
statistical testing, 181
need for, 180–181
statistical tests and procedures
chi square (χ^2), 304
coefficient of contingency (C), 304
Eta Squared (η^2), 293–294
Fisher's protected t-test, 287, 296–299
F test, 286
Goodman's and Kruskal's Gamma (γ), 304
Guttman's coefficient of predictability, lambda (λ), 304
Mann-whitney *U* test, 304
omega squared (ω^2), 293–294
partial correlation coefficient, 213–216
Pearson correlation coefficient *(r),* 193–203
phi coefficient (φ), 304
point-biserial correlation coefficient (r_{pb}), 203–207
Spearman rank order correlation coefficient (r_s), 208–213
t-test for related samples, 272–278
t-test for sample mean to known population mean, 258–260
t-test for two independent samples, 261–266
Tukey's HSD test, 291–292
Wilcoxon T test, 304
Z test for proportions, 267–272
Z test for sample mean and population mean, 250–256
Z test for two independent samples, 266–267
statistics, 3–4, 10, 19, 180
common symbols and mathematics used in, 14
descriptive statistics, 10, 19
inferential statistics, 10, 19, 143, 145, 160
nonparametric statistics, 303–304
parametric statistics, 303
stratified random sampling, 27–28
sum of squared deviations, 224
minimize, 223, 243
sum of squares, 233, 300
computing, 287–288
sum of squares error, 233
sum of squares explained, 233
sum of squares total, 233, 284
sum of squares between groups, 285
computing, 289
sum of squares error, 233
sum of squares explained, 233
sum of squares total, 233
sum of squares within groups, 285
computing, 289
symmetrical measure of association, 321–322, 333
systematic sampling, 25, 28–29, 38

T

t distribution, 137–138, 140, 256–257
test statistic, 168, 182, 249–253, 265, 279
The normal distribution, 113–140
Tool bar, SPSS, 390
total degrees of freedom, 286
Transform command, SPSS, 393–394
treatments, 286, 300
trimodal, 73, 85

t-test, 258
  assumptions of, 264
  for a sample mean to known population mean, 258–260
  for two independent samples, 261–266
  for two related samples, 272–278
Tukey's HSD test, 287, 291–292, 300
  formula, 291
two-tail test, hypothesis testing, 170–171, 182, 197, 199, 217, 249–250, 258, 330, 331
  critical regions for, 178
Type I error, 178–180, 182
Type II error, 178–180, 182

U

unexplained variance, 232–233, 244
unit of analysis, 22–24, 37
universe, 9, 23
upper limits, 50, 52, 67

V

value labels, SPSS, 401
variables, 5, 7–8, 18
  dependent variables, 9, 19, 169–170, 172, 304, 307
  independent variables, 9, 19, 169–170, 172, 304–305, 307
Variables option, SPSS, 401
variance, 92–98, 101–107, 230
  computational formula for, 95–97, 101
  for a frequency distribution, 101–103
  pooled estimate of, 261–262, 266
  as prediction error, 103–107
variance between groups, 285, 286, 300
variance explained, 108, 217, 230–233
variance within groups, 285, 286, 300

W

Wilcoxon T test, 327–328, 333, 399
  related samples, 330–332
  statistical tables, 368
writing of results, research process, 6–7

Y

Yates' correction for continuity, 317–318
Y axis intercept, 222
  formula, 226–228

Z

Z scores, 113–114, 128, 139, 164–165
  formula to convert raw score to *Z*, 123
  formula to convert sample mean to *Z*, 132–134, 165–169
  and percentile rank, 128
Z table (explained), 116–122
Z test, 168, 178
  examples, 175–178
  for a sample mean and population mean, 250–256
  for two independent samples, 266–267
  for two proportions, 267–272

Printed in the USA
CPSIA information can be obtained
at www.ICGtesting.com
LVHW060217080924
790282LV00001BA/3